PEST MANAGEMENT IN SOYBEAN

PEST MANAGEMENT IN SOYBEAN

Edited by

L. G. COPPING
Consultant, Saffron Walden, Essex, UK

M. B. GREEN
Consultant, Wallington, Surrey, UK

and

R. T. REES
Schering AG, Berlin, Germany

Published for SCI
by
SCI ELSEVIER APPLIED SCIENCE
LONDON and NEW YORK

ELSEVIER SCIENCE PUBLISHERS LTD
Crown House, Linton Road, Barking, Essex IG11 8JU, England

WITH 88 TABLES AND 45 ILLUSTRATIONS

© 1992 SCI
(except pp.137-146, 262-271)

British Library Cataloguing in Publication Data

Pest Management in Soybean
I. Copping, Leonard G.
632

ISBN 1-85166-874-8

Library of Congress CIP data applied for

Preface

This book is the third in a series of volumes on major tropical and sub-tropical crops. These books aim to review the current state of the art in management of the total spectrum of pests and diseases which affect these crops in each major growing area using a multi-disciplinary approach.

Soybean is economically the most important legume in the world. It is nutritious and easily digested, and is one of the richest and cheapest sources of protein. It is currently vital for the sustenance of many people and it will play an integral role in any future attempts to relieve world hunger. Soybean seed contains about 17% of oil and about 63% of meal, half of which is protein. Modern research has developed a variety of uses for soybean oil. It is processed into margarine, shortening, mayonnaise, salad creams and vegetarian cheeses. Industrially it is used in resins, plastics, paints, adhesives, fertilisers, sizing for cloth, linoleum backing, fire-extinguishing materials, printing inks and a variety of other products. Soybean meal is a high-protein meat substitute and is used in the developed countries in many processed foods, including baby foods, but mainly as a feed for livestock.

Soybean (*Glycine max*), which evolved from *Glycine ussuriensis*, a wild legume native to northern China, has been known and used in China since the eleventh century BC. It was introduced into Europe in the eighteenth century and into the United States in 1804 as an ornamental garden plant in Philadelphia. In 1905 the first commercial plant for extraction of oil from the seeds was developed, but it was not until around 1926 that it became an economically important crop. Now it is a major crop in the United States and in Brazil, Paraguay and Argentina, and its cultivation is being extended in southern Europe.

About 105 million tonnes of soybeans are currently produced world-wide from 55 million hectares, mostly between latitudes 25° and 45° and below 1000 metres. About one-third of this total area is in the United States, which produces about half the total tonnage, of which about 55% is used domestically and 45% exported. Soybean is an annual nitrifying legume which produces erect branching plants ranging in height from 30 to 200 cm. Planting density is normally 200 000 to 600 000 plants per hectare in rows 20 to 100 cm apart. Yields range from around 2000 kg/ha in America through about 1300 kg/ha in China and about 1000 kg/ha in India to as little as 300 kg/ha in other developing countries.

Soybean is attacked by a variety of arthropod pests which can cause extensive damage and crop losses. Widespread use of insecticides in the past has resulted in the emergence of resistant strains of pests, and the papers in this volume indicate that the emphasis nowadays is consequently on preventative rather than therapeutic tactics. A problem that is addressed in these papers is that the introduction of sophisticated pest management systems makes considerable demands on crop managers and growers, and that, although expert systems and simulation models are being developed to assist them in the decision process, there are formidable difficulties in getting these accepted and used. An interesting development which is described in two of the papers is the use of baculoviruses for pest control in soybean.

Both cyst and root-knot nematodes are an increasing and intractable problem in soybean in the United States, and three papers in this volume describe the current approaches to management of these pests. Soybean is also attacked by a large number of fungi, and the papers in this volume describe the management of soil-borne and foliar diseases by chemical, biological and cultivational methods in both temperate and tropical regions. The development and use of resistant varieties is crucial for the protection of soybeans against nematodes, viruses and plant pathogens, and molecular biology offers new opportunities in this area.

Because of the very narrow row spacing of soybean, mechanical weeding is generally not practicable, so herbicides are widely used. Weeds in South America and in Asia present different problems from those in the United States, and herbicides designed for the US market are often not appropriate for use in these other areas. Consequently, new approaches to major weed problems are being pursued. Mycoherbicides represent an interesting development which is described in two papers in this volume. In the mature US market, aspects of molecular biology which will lead to herbicide-resistant soybeans are under investigation. The papers in this volume show that the present-day trend is towards integrated management of weeds, and a number of strategies are described. The importance of determining economic thresholds for weed infestations is stressed.

The papers address all the current issues in pest management in soybean and give a good overview of recent developments in research and their application in the field and of the successes that have been achieved and the problems that remain in soybean pest management. It is a challenge to all those involved to ensure that cultivation of this valuable crop remains an economically sustainable activity and continues to make its contribution to farm incomes and the economies of the producing countries and to the task of supplying the nutritional needs of the ever-increasing world population.

L. G. COPPING
M. B. GREEN
R. T. REES

Acknowledgements

The Editors record their sincere thanks to the following companies for financial contributions to this venture:

Sandoz Agrochemicals Ltd
CIBA-Geigy AG
Duphar BV
Bayer AG
ICI Agrochemicals
Rhône-Poulenc Agriculture
American Cyanamid Company
Schering Agrochemicals Ltd
Monsanto Agricultural Company

L. G. C.
M. B. G.
R. T. R.

Contents

THE IMPORTANCE OF SOYBEAN AS A MARKET FOR AGRICULTURAL CHEMICALS

MALCOLM ROSIER
District Manager, West Europe
Uniroyal Chemical Limited
Kennet House, Langley Quay, Slough SL3 6EH

ABSTRACT

Soybean is one of the most important markets for agrochemicals in the world, worth over $US1.8 billion. It is dominated by the herbicide sector and the USA, which account, respectively, for 81% and for 63.5% of agrochemical usage by value. Soybean is expected to remain a significant market for herbicides, with many new compounds currently being developed and commercialised. Soybean is not a large market for fungicides or insecticides, although these sectors are growing rapidly as disease and insect problems become increasingly important.

INTRODUCTION

Soybeans are the world's most important source of food, oil and protein. They represent nearly fifty per cent of global oilseed production and fifty-eight per cent of global protein meal supply.

Soybeans have a long history as an important food crop. They originated in East Asia. In China they have been cultivated since before the time of written records. They have also been an important source of protein in Korea and Japan since the earliest times. Soybeans were first brought to the West in the eighteenth century, reaching the USA in 1804, but little interest was shown in them until the first shipments were made to Europe a century later.

There was little commercial production in the USA until the 1930s. The most rapid increase in production has been since 1942, initially in response to war-time demand for edible oils and fats and latterly due to the ease with which soybean production has been mechanized.

Soybeans have been introduced into most tropical countries

during this century, finding favour particularly in Latin America. Over three-quarters of the world crop is traded and this is proving a useful source of income to many Latin American countries.

As Table I shows, the USA is the world's largest producer of soybeans. Over 40% of the world's crop is grown in the USA and it produces nearly half of the world's soybean harvest. The Latin American producers, Brazil and Argentina, have almost another third of the planted area and contribute more than a quarter of the world's annual production. Other major producers are China and India. The USA, Brazil, Argentina and China, alone, account for 87% of the world's production of soybeans.

TABLE I
World Soybean Statistics - 1991 season [1]

Country	Area Grown (Ha million)	%	Production (MT million)	%
USA	22.9	41	51.8	49
Brazil	10.5	19	18.5	17
China	7.5	14	11.5	11
Argentina	5.0	9	10.5	10
India	2.0	4	2.0	2
USSR	0.8	1	0.9	1
Canada	0.5	1	1.3	1
Others	5.0	11	9.3	9
World	55.4		105.8	

U.S.D.A.

INSECT, WEED AND DISEASE PROBLEMS

The major problem encountered in soybean production is weeds. Young soybean seedlings are unable to compete with fast-growing weeds, and control of weeds during the first six to eight weeks following emergence is critical. Soybeans will effectively suppress weed growth, once the canopy has met over the inter-row area. Work completed in the 1960s at the Universities of Illinois and Nebraska [2] demonstrates the potential damage to yields of weed infestation, as well as the effect of shading on weed growth as the crop becomes established.

TABLE II
Effect of removing foxtail from soybeans

Height of Soybean crop when foxtail removed (inches)	Average yield (bushels/acre)
Weed free check	30
08	30
12	30
16	29
22	28
Left until beans were mature	12

University of Illinois

Table II demonstrates the effect on yield of weed populations. In this example, trial plots of soybeans had foxtail plants removed from them at different growth stages of the crop (expressed as crop height). If the weeds are removed early then there is no yield loss compared to a weed free crop, if the weeds are treated at a later stage or untreated then significant yield losses can occur. Once the stand is established, however, the soybean crop can keep itself free from weeds by shading.

TABLE III
Effect of shading on foxtail in soybeans

Time foxtail was seeded after crop was planted	Total dry matter produced (lbs/ac)	
	Foxtail	Soybean
Same day	2280	3970
3 weeks	30	5240
6 weeks	0	5390
12 weeks	0	5440
Weed free	0	5410

University of Illinois

Row spacing can also have a direct effect on weed control, as it affects the time it takes for the canopies to meet and shading to take effect.

TABLE IV
Days after planting before canopy covers inter-row area

Row width (inches)	Number of days
40	67
30	58
20	47
10	36

University of Nebraska

So farmers must compromise between ease of cultivation (wide-rows) and the shading effect.

The need to eradicate weeds at an early stage from soybean crops has made soybeans one of the largest herbicide segments. Broad-leaved and grass weeds both cause major problems. The most important weeds of soybeans worldwide are grasses and sedges. Sorghum halepense and Digitaria spp. are major problems in direct drilled crops. S. halepense is, arguably, the most important weed of soybeans in North and South America. Echinocloa crus-gali and Echinocloa colonum are now major problems in all soybean growing areas due to their ability to regenerate from small rhizome fragments and also because of the shift to minimum tillage systems. Eleusine indica and Cyperus rotudus are difficult to control by any means. C. rotundus is particularly problematic because of its ability to regenerate from rhizomes and underground tubers.

Broad-leaved weeds are less troublesome, although Rottboellia cochinchinesis is spreading throughout Asia, Africa and South America, as its control is hampered by the presence of sharphairs (which prevent hand-weeding) and widespread resistance to herbicides. Of lesser importance are Chenopodium album and Amaranthus spp. A number of locally important weeds are found in the USA and include Cassia obtusifolia, Abutilon theophrasti and Aeschyomene virginica. Euphorbia heterophylla is a major weed of soybeans in Brazil, infesting an estimated 200,000 ha and causing losses of approximately US$ 4 million annually.

Insect and mite control accounts for only 11% of agrochemical expenditure on soybeans. Of this, 60% is for the control of spider mites, principally the two spotted spider mite in the US. The most important insect pests are leaf-feeding lepidoptera and the pentatomid, pod-feeding stink bugs. In Europe, wireworms are also a problem.

Diseases do not represent a major problem for soybeans. Damping-off symptoms are produced by many pathogens and are evident in most soybean crops. A number of pathogens - Sclerotinia spp, Diaporthe phaseolorum var caulivora, Microsphaera diffusa, Phakospara pachyrhizini and Pyrenochaeta glycines - are becoming increasingly important in Asia and Africa as soybean production has increased [3].

AGROCHEMICAL MARKET

Soybeans form a significant market for agrochemicals. Total usage of agrochemicals on soybeans was worth, in the 1990 season, $US 1.9 billion, which represented 7% of the global market for agrochemicals in that year. It is the third largest herbicide market. However, despite its size and significance, the soybean agrochemical market is a single-sector market, with herbicides accounting for 81% of the total expenditure.

It is also a market dominated by a single outlet - the United States of America. The USA accounted for over 65% of all agrochemical expenditure on soybeans in 1990. Of the rest of the world, only Argentina and Brazil are truly significant markets, accounting for 5% and 9% expenditure respectively. China, although the third largest producer of soybeans, does not spend significant amounts on agrochemical. As a result, yields are low, 1.44 tonnes per hectare, compared to a world average of 1.9 tonnes per hectare.

Nearly 60% of all agrochemicals applied to soybeans are for the control of grass weeds, reflecting their importance as a yield damaging competitor.

TABLE V
Herbicide usage on soybeans by weed-timing segment [4]

Weed-timing segment	% total usage
Pre-emergence grass weeds	57
Pre-emergence broad-leaved weeds	19
Post-emergence grass weeds	10
Post-emergence broad-leaved weeds	14

The ten most widely used herbicides are, in order of world sales: triflulralin, imazaquin, metotachlor, metribuzin, bentazone, alachlor, imazethapyr, fluazifop- p-butyl, chlorimuron-ethyl and pendimethalin [5].

A detailed picture of the principal herbicide products used in soybeans is given in Table VI.

TABLE VI
Major herbicide active ingredients used in
soybeans

Group	Active ingredient	Brand example	Market share
Amide	alachlor	Lasso	6.5%
	metolachlor	Dual	8.5%
Diazine	bentazone	Basagran	8.0%
Diphenyl-ether	aciflurofen	Blazer/Tackle	1.5%
	diclofop-methyl	Illoxan	<1.0%
	formesafen	Flex	<1.0%
	lactofen	Cobra	<1.0%
Imadizolinone	imazaquin	Sceptre	14.5%
	imazethapyr	Persuit	6.5%
Phenoxy	2,4-DB	various	<1.0%
Sulfonyl-urea	chlorimuron-ethyl	Classic	4.0%
Toluidine	ethalfluralin	Sonalan	2.5%
	pendimethalin	Prowl	3.0%
	trifluralin	Treflan	20.5%
Triazine	metribuzin	Sencor	8.5%
Graminicides	fluazifop-butyl	Fusilade	5.5%
	sethoxydim	Poast	<1.0%
Others			<10.0%

Although the largest group of products used on soybeans remains the toluidines, it is the imadizolinones that are having the greatest effect on the market. From their launch in 1987, imazaquin and imazethapyr have achieved a 21% market share. Other new product introductions, such as members of the diphenyl-ethers (for example formesafen and lactofen), have not performed as well, probably as a result of early crop damage problems in the US.

The exceptions to the trend of unsuccessful introductions of new products have been the graminides. Led by sethoxydim and fluazifop-butyl, these products have achieved a healthy share of the post-emergence market. Many new products are being introduced in this sector and some fragmentation can be expected, although I believe it will be difficult to shake these products from their pre- eminent position.

The sulfonyl-urea chlorimuron-ethyl is performing well and has increased its market share during the last few seasons from 2% to 4%.

As soybean herbicides are such a large market segment, there has always been much development activity. Currently, there are a number of new compounds in trials and late-stage commercialisation, but all are analogues or members of current product groups and there is no novel chemistry apparent. A summary of development compounds for soybeans is given in Table VII.

TABLE VII
Herbicides under development for use in soybeans

Code	Company	Group	Usage
AKH-7088	Asahi	diphenyl-ether	post-em blw
F6285	FMC	triazine	ppi/pre-em blw/gw
HC 252	HCW	diphenyl-ether	post-em blw
HOK 1566	Hokko	graminicide	post-em gw
KIH 9201	Kumiai	not known	blw
MON-097	Monsanto	amide	pre-em blw/gw
MON-1320	Monsanto	not known	ppi/pre-em gw
Ro17-3664	Ciba	graminicide	post-em gw
S53482	Sumitomo	dicarboxamide	pre-em blw/gw
SAN 582H	Sandoz	chloroacetamide	pre-em blw/gw
UBI-C4874	Uniroyal	graminicide	post-em gw

At $US 125 million, the fungicides are the least important segment of the soybean agrochemical market. Agronomic methods and the climatic conditions of the major soybean growing areas means that diseases of soybeans are rarely significant from an economic point of view. However, studies completed in the late 1970s by the University of Arkansas [6] have demonstrated yield increases of up to 27% in crops created with foliar fungicides when compared to untreated controls.

Major methods of combatting disease tend to be cultural: the use of resistant varieties, rotation, good stubble hygiene, high seed-rates and seed with higher vigour.

Where fungicides are used they tend to be seed treatments such as PCNB, carboxin, captan and TMTD, although there is some use of benomyl and chlorothalonil as foliar treatments.

Insecticide sales in soybeans are mainly of commodities and this is reflected in the low value of the segment at $US 220 million. At the present time, with the possible exception of Agrostis spp and Heliothis spp, most soybean pests are effectively controlled with currently available insecticides such as carbaryl, endosulfan, methomyl, parathion and the synthetic pyrethroids, or cultural methods such as crop rotations, destruction of crop residues and the eradication of volunteers.

The single largest market for insecticides is for the control of spider mites in the Mid-West of America. Sales for this purpose are estimated to exceed $US 100 million. Insecticide usage on soybeans is growing at about 5% per annum (compared to herbicide usage at less than 1% per annum) and its use is predicted to continue to increase, particularly in South America, as recent studies show significant yield improvements following control of pests previously considered of little economic importance [3].

Biological control of pests is also being introduced to soybeans. Over 300,000 ha are treated in Brazil with a baculovirus for the control of <u>Anticarsia gemmatalis</u>, and there are significant opportunities for the control of this pest in the rest of Latin America and in South East Asia.

TABLE VIII
Insecticide usage on soybeans

Country	Market (US$ millions)
Argentina	18
Brazil	57
USA	132
Others	13
Total	220

CONCLUSIONS

Soybeans are going to remain the world's premier oilseed crop. Soybeans are widely traded and the trade is not distorted by tarrifs as much as other oilseed crops. Price levels are likely to remain where they are and the economics of maintaining and even increasing agrochemical usage are good. Thus, we can expect soybeans to continue to be a significant market for agrochemicals, particularly herbicides, but also of increasing significance for insecticides. We can also expect to see increased interest in the use of biological and IPM methods, particulary in tropical countries.

REFERENCES

1. United States Department of Agriculture, World Soybean Statistics, 1991.

2. Scott, W.O., Aldrich, S.R., Modern Soybean Production S & A Publications, Champagne, IL., pp111-124.

3. Biopesticides 2000, Landell Mills Market Research Ltd, Bath, 1988.

4. Rosier, M.J., Crop Profile: Soyabeans, Pesticide Outlook, Royal Society of Chemistry, pp31-45.

5. Unpublished information from Landell Mills Market Research Ltd, Bath.

6. Walters, H.J., Fungicides for Foliar, Pod and Stem Diseases of Soybeans, University of Arkansas Agricultural Experiment Station Report #250, 1980.

INTEGRATING PREVENTIVE AND THERAPEUTIC TACTICS IN SOYBEAN INSECT MANAGEMENT

LARRY P. PEDIGO
Department of Entomology
Iowa State University
Ames, Iowa 50011

ABSTRACT

Current production and environmental demands on agriculture are creating a need for improved IPM programs that integrate both preventive and therapeutic tactics. Research with the bean leaf beetle was used to exemplify development of such an integrated program in soybean. Key elements in development of therapeutics and prevention involve use of economic thresholds and late planting, respectively. A prototype management program for the bean leaf beetle is suggested that integrates these preventive and therapeutic tactics.

INTRODUCTION

The development of the pest management concept by Geier and Clark [1] is a signal event in the history of pest control. Indeed, the concept changed the course of pest control by providing both a framework in which to assemble effective pest-suppressive tactics and guidelines for environmentally safe manipulation of pest populations.

Although much has been accomplished since the inception of pest management, there is demand to speed development of new programs. This demand has arisen in a changing agricultural environment which challenges research with demands for high crop productivity, low inputs, high profits, environmental compatibility, and system sustainability.

To meet these and other demands, pest management research should place a new emphasis on developing programs that integrate multiple tactics. Although much has been written on integration and integrated pest management (IPM), too few multiple-tactic programs are being practiced. Most often, current IPM consists of pest scouting, use of economic thresholds, and proper application of pesticides when needed. Although this is an improvement over the identify-and-spray approaches of the past, IPM implementation needs to go further to achieve sustainable systems. The use of several well integrated tactics in addition to pesticides can cut production costs, prevent ecological backlash, and allow continued reductions of pesticide inputs.

There are several reasons for the slow development of programs integrating several tactics. Chief among these is the need for immediate alleviation of pest problems, often satisfied by a pesticide-based strategy. Pest managers have little incentive to change this strategy if it is profitable. Another reason may be the lack of an adequate conceptual framework for integration. For at least two decades, mathematical modeling and systems analysis have claimed almost exclusive domain over the theoretical development of integration. Yet, to date, few systems models have proven useful in addressing tactical management problems [2]. To advance practical integration, a conceptual basis is needed for combining potential tactics, followed by experimentation to test the outcome of the most promising combinations.

The purposes of this paper are to discuss an alternative conceptual basis for integration in IPM and to present a practical example of integrating management tactics. Thus, the paper's focus will be research with the bean leaf beetle, *Cerotoma trifurcata* (Forster), a pest of increasing importance in soybean.

CONCEPTS OF INTEGRATION

When a conceptual basis for integration in pest management is considered, it is useful to take into account other disciplines in which integration is a crucial part of problem solving. One such discipline, somewhat analogous to pest management, is human medicine. Medicine uses a broad range of tactics in managing diseases, and some of its most spectacular successes involve combining both preventive and curative procedures. In fact, the time-proven method of integrating tactics from preventive medicine and therapeutics is the cornerstone of modern medical practice.

Likewise, this powerful dualistic concept can be applied to IPM [3, 4, 5, 6]. With well defined goals relating to prevention on the one hand and cure on the other, pest management tactics can be investigated to produce an overall program, in which the integration of tactics is effective and practical.

IPM Prevention

IPM prevention (preventive pest management) involves preemptive actions before pest injury has occurred. Such actions are taken as a matter of course, often without specific knowledge of pest presence or status at a particular time. Preventive tactics are employed because the pest or pest complex has caused losses in the past and is likely to do so in the future. For example, seed of insect-resistant plants may be planted as a preventive to late season injury, even though potential for pest losses is unknown. To be effective, preventives must persist, at least for the growing season, or be applied regularly.

Ecologically, preventive tactics aim to prohibit establishment, limit growth, and/or reduce injuriousness of a given pest population. To prohibit establishment and population growth, tactics are aimed at making the pest's effective environment inhospitable, resulting in lower density, less total injury, and, consequently, a lower average damage level (Fig. 1). Such tactics rely on a detailed understanding of pest behavior, life cycle, seasonal cycle, habitat ecology, and population dynamics. Conversely, reducing pest population injuriousness involves lowering plant sensitivity to a given level of injury, i.e., reducing damage per unit of injury. Plant sensitivity can be lowered with cultural methods to improve crop vigor or by selecting cultivars more tolerant to injury. These approaches raise the level of economic damage [7] relative to the pest's average damage level, thereby preventing losses, at least to a degree. The specifics of pest injury and of resulting plant responses must be known to successfully use crop tolerance as a pest management tactic. Although difficult to develop, tolerance may be the ultimate sustainable strategy because pest resistance and other forms of ecological backlash will not likely result from application.

In total, there are several tactics that can be considered for use in preventive pest management. Many involve planning and modifications made early in a crop production cycle. The most common tactics include most biological controls; crop rotation; sanitation and tillage; planting date; trap cropping; plant spatial arrangements; nutrient inputs; and cultivar selection.

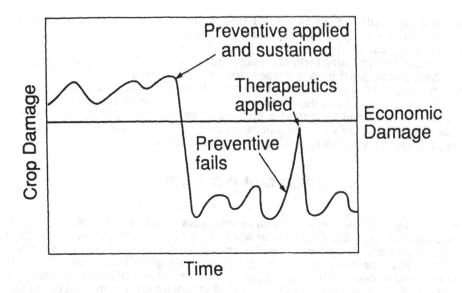

Figure 1. Effects of preventive and therapeutic tactics on pest damage.

Although pesticides have been used for prevention in the past, this strategy has resulted in pest resurgence and resistance to pesticides. Therefore, it is not a sustainable strategy and is not recommended for IPM prevention.

IPM Therapy

IPM therapy (therapeutics) seeks to cure an acute or chronic crop disorder. Unlike IPM prevention, therapy is applied only after pest assessment indicates that injury has occurred and/or economic damage is imminent; it is used any time in the crop production cycle when a favorable cost/benefit ratio exists. Therapeutics has been one of the most widely practiced forms of IPM and, because of its efficiency in averting crop losses, is usually the first phase of IPM developed.

This strategy relies on both well developed sampling programs and sound economic thresholds. For threshold development, an economic injury level should be calculated, requiring much information on pest injury rates and host reactions to injury.

From an ecological perspective, IPM therapy seeks to interrupt ongoing pest population growth and injury. It is applied when either natural control of the population or IPM prevention fails, and it serves to dampen damage peaks (Fig.1). The beneficial action of therapeutics is temporary and, if used infrequently, usually causes no significant ecological backlash.

Acceptable therapeutic tactics are those that are cost effective and produce few undesirable side effects on the pest population and the environment. By far, the most important therapeutic tactic is the selective use of conventional pesticides. Pesticidal action is rapid, cost effective, and environmentally safe when used judiciously. Other tactics used in IPM therapy include fast-acting, nonpersistent biological controls (e.g. microbial pesticides); early harvest; replanting; and mechanical removal of pests (e.g., cultivation, roguing, and pruning).

Integrating Tactics

IPM development should proceed by identifying potential tactics under both preventive and therapeutic categories, and then testing these individually. Data from such tests can be used to formulate either mathematical or conceptual models of systems integrating the tactics. From these models, field experiments can be conducted to measure cost, compatibility, and effectiveness.

A major goal in IPM program development is to discover preventive tactics that are persistent. Self-sustaining preventives, such as successful releases of natural enemies, are particularly desirable, because they have potential as a final solution to the pest problem. Even palliative tactics that are self-sustaining are of great value, in as much as they reduce frequency of nonsustaining tactics, e.g., pesticides.

The need for preventive tactics in the overall program is greatly determined by the pest-type involved. Occasional pests, which produce outbreaks sporadically, have a low general equilibrium position relative to their economic injury level owing to natural control factors [7]. These pests may not require preventive pest management and need only a dampening of damage peaks using therapeutics. Because corrective action is required only occasionally, ecological backlash is unlikely, and a multiple-tactic scheme is unnecessary. For severe and perennial pests, however, multiple tactics become increasingly important, and preventive tactics play a major role in a sustainable management system.

DEVELOPMENT OF AN INTEGRATED STRATEGY FOR THE BEAN LEAF BEETLE IN SOYBEAN

Pest control practice in soybean has followed the same trend recognizable in many other annual crops in the U.S. This has basically meant the use of pesticides as preventives for weeds and as therapeutics for arthropods and fungi [8]. Notable exceptions to this pesticide-based management are preventive strategies using resistant cultivars and cultural manipulations targeting diseases and nematodes.

Although there is less pesticide use in soybean than in many other U.S. crops (e.g., corn and cotton), use should still be reduced. This stems from concerns over real and/or perceived threats to the environment and public health, as well as undesirable side effects such as pesticide resistance [9]. Taken in total, these forces will undoubtedly result in stricter regulations, greater costs, and less availability of effective chemicals.

To reduce pesticide inputs and develop sustainable IPM systems, soybean pest research will need to emphasize practical nonpesticidal tactics and integrate them into existing programs. Certainly, pesticides will form the bulwark against unexpected problems, but preventive pest management, using nonpesticidal tactics, must assume a greater role in production practices. The development of such an integrated system can be exemplified by research done over the past six years with the bean leaf beetle, in Iowa soybean.

Bean Leaf Beetle Biology and Soybean Injury

The bean leaf beetle (BLB) is a widespread pest of soybean in the major production areas of the U.S. In Iowa, the BLB begins activity in April as adults leave their overwintering habitats, which can include woodlands, clumps of grass, leaf litter, and possibly soybean fields of the previous season [10]. At this time, they move into alfalfa or other suitable habitat, feed, and lay a few eggs. Upon soybean emergence and first cutting of alfalfa, beetles emigrate to the soybean, feed on leaves, and lay eggs in the soil. These eggs constitute the first generation, with larvae and pupae developing in the soil.

In Iowa, the BLB seasonal cycle involves two generations [11]. Most adults of the first generation emerge in mid July and lay eggs for the second generation, which produces adults from August to early September. These adults do not mate, but feed on soybean leaves and pods, moving to alternative hosts as soybeans mature. In October, these adults migrate to protected habitats, where they overwinter.

Both larvae and adults injure soybeans. Larvae consume roots, root hairs, and nodules, with a preference for nodules. Although nodule size and number are reduced, actual effect on yield from a natural field population has not been demonstrated. Currently, nodule feeding is an active area of research. The most obvious form of injury is defoliation by adults, who chew round symmetrical holes between major leaflet veins. The greatest potential for loss from defoliation is in soybean stages V1 to V3, when overwintering beetles colonize fields. Even then, loss is unlikely because of the tolerance of plants to defoliation during vegetative stages.

Based on current knowledge, pod feeding by second-generation adults is the most important type of BLB injury to soybean. Pod injury occurs as leaves mature and beetles begin feeding on younger tissues of the pod. They feed only on the pod surface, consuming tissue down to the endocarp, which directly encloses the seed. BLB pod lesions increase seed vulnerability to weather and secondary pathogens, particularly *Alternaria tenuissima* (Fries) Wiltsh. Seeds beneath the lesion become shrunken, discolored, and sometimes moldy. The result is loss of seed weight and quality. In many instances, beetles have also been observed feeding on the pod peduncle and surrounding tissue, an activity which can cause breakage and complete pod loss.

Bean Leaf Beetle Therapeutics

To alleviate immediate BLB problems, the first IPM strategy developed was therapeutics. The therapeutic program consists of regular sampling to estimate population densities, use of economic thresholds, and application of insecticides, when necessary.

As research priorities were developed, it was found that methods of sampling [12] and of insecticidal control [8] were well established. Existing economic injury levels (EIL) and economic thresholds (ET) for pod feeding, however, were subjective and not based on research data. Therefore, research was initiated that focused on understanding the impact of BLB pod feeding and proceeded to develop objective EILs and ETs.

The impact of pod feeding was studied in a preliminary experiment in 1986 to establish procedures; more thorough experiments were conducted in 1987 and 1988 [13]. Studies were conducted with 'Corsoy 79' soybean near Ames, Iowa, using field cages covering 2-m of row (50 plants). Cages (plots) were infested with feral BLB adults during mid-August, when soybean were in stages R5 to R6. Treatment densities consisted of 2, 6, and 12 beetles/plant in 1987 and 6, 12, and 18 beetles/plant in 1988. Both caged and uncaged check plots with no beetles were included each year. Plots were arranged according to a randomized complete block design, with four blocks. Plant samples were taken from all cages at harvest (5 to 8 plants) and number of BLB-injured pods per plant determined. A stationary thresher was used to collect seed from remaining plants in each cage, and this seed was combined with that from the plant samples to determine total seed weight per plot and weight per plant. Additionally, BLB adult abundance was monitored in designated cages to establish a BLB-density/time curve. The area under this curve was determined to estimate BLB-days in the plots.

Results of these experiments showed that in both years average seed weight per plant decreased linearly from the uninfested cage to the second highest infestation density, followed by an increase at the highest infestation density. This increasing trend was unexpected and is not easily explained. Possibly, the highest density triggered a plant compensatory response, or adult feeding was reduced because of density-related interference. In either instance, including the highest density in a regression analysis would substantially change the slope of the regression line and give an unrealistic coefficient for EIL calculations, i.e., densities causing yield increases were believed much higher than the expected EIL. Therefore, yield data from the highest density were considered no further in the impact analysis.

To determine loss per BLB, soybean yield was analyzed statistically by regressing seed weight against BLB days. Data from 1987 and 1988 were pooled (excluding the highest density treatment) and analyzed with a dummy-variable, which accounted for a y-intercept difference between years. Results of the analysis showed that BLB days and the year variable accounted for a significant proportion variation in the data ($r^2 = 0.765$, $P < 0.01$) (Fig. 2). Seed weights did not differ significantly ($P > 0.05$) between the uncaged check and the caged check for either year.

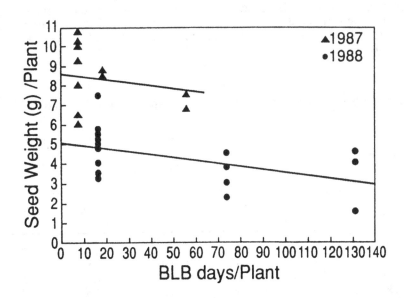

Figure 2. Relation of BLB days to loss in soybean seed weight in 1987 and 1988.

The seed yield-loss/BLB-day regression indicated that 0.0153 g of seed weight was lost for each BLB day per plant. This coefficient (B_1) was multiplied by the average BLB feeding period of 20 days [11], giving a 0.306 g yield loss per BLB per plant. This loss value was used to calculate EILs.

EILs for second generation BLB injury were calculated according to the equation [6]:

$$EIL = \frac{C}{V \times b \times K} \tag{1}$$

where EIL = economic injury level (no. BLB per m^2), C = cost of management ($/ha), V = soybean market value ($/kg), b = 3.06×10^{-4} (kg/BLB), and K = 1.00 (assumed reduction of population from the management tactic). EILs were calculated for a range of management costs ($17.30 to $24.70/ha) and soybean market values ($.019 to $0.38/kg), producing values from 14.9 to 42.5 BLB/m^2.

Subsequently, ETs were calculated to serve as practical decision guidelines. Because ETs must take into account future injury by a pest population as well as time delays in final suppression, they are usually set at a level lower than is the EIL. A conventional level used in soybean entomology has been 75% of the EIL [14];

accordingly, all EILs were multiplied by 0.75. Additionally, density per m^2 was converted to number per 30-cm of row for use with conventional ground-cloth sampling estimates.

The calculated ETs are presented in a decision table (Table 1). To use the table, a producer or manager first estimates the size of the BLB population by taking samples, at least weekly, from either late July or when plants reach stage R4, until pods yellow. A ground-cloth sample is usually taken by placing a heavy muslin or canvas cloth (1 m x 80 cm) between two adjacent rows. Plants within ca. 60-cm of row on both sides of the cloth are bent over the cloth and shaken vigorously to dislodge beetles, which drop onto the cloth and are counted. Eight such 120-cm samples are taken, and a 30-cm average count is calculated (total/32). The user compares this estimate with the ET given for the estimated-cost and market-value variables. If the estimated BLB density is equal to or greater than the ET, action is taken; if smaller, pest management activity is terminated.

TABLE 1

Bean leaf beetle economic thresholds on soybean in number/30-cm of row[a,b]

Market value	Management Costs $/ha			
$/kg	17.30	19.80	22.20	24.70
0.19	5.2	5.9	6.7	7.4
0.23	4.3	4.9	5.5	6.1
0.27	3.7	4.1	4.7	5.3
0.34	2.9	3.3	3.8	4.1
0.38	2.6	3.0	3.4	3.8

[a]Economic thresholds set at 75% of the EIL.
[b]Based on a soybean row spacing of 75 cm.

Preventive Pest Management of the Bean Leaf Beetle

After an IPM program based on therapeutics was established, attention was given to methods of preventing BLB injury. Of the potential tactics, one with proven effectiveness in the southern U.S. was trap cropping [8]. In this trapping program, a portion of a field (5 to 10%) is planted to an early maturing variety 10 days to 3 weeks earlier than the late-maturing main planting. Subsequently, overwintering BLBs enter the early planting and lay eggs. About 7 to 10 days after emergence of first-generation adults, insecticides are applied to the trap area, thereby preventing an economically damaging population from developing in the main planting. This program has been effective in preventing economic damage and in reducing amounts of insecticides applied.

Such a program was deemed impractical in the northern soybean regions because of larger soybean fields and labor demands at a time when corn, a major crop in the region, also requires planting. Corn usually is planted in mid-April to early May, soybean in late April to mid-May.

A variation of the trap cropping approach, i.e., late planting, seemed to have potential for northern soybean. It was reasoned that if soybean planting were delayed until late May, the active BLBs would either colonize wild hosts and not move into the soybean or suffer reduced fecundity by the time the late planted soybean emerged. In either instance, late-season injury potentially could be reduced or avoided completely.

To test this hypothesis, 0.23-ha soybean plots, with c.v. 'Corsoy 79' soybean planted in 75-cm-wide rows, were established in 1989 and 1990 near Ames, Iowa. Three blocks with two plots each were located next to alfalfa to facilitate the entry of BLB into the soybean. An early or late-planting treatment was randomly assigned to each plot area, creating a randomized-complete-block design. Soybean were planted to span the recommended time for maturity-group II cultivars, viz., early May and late May. Treatment planting dates were May 10 and May 30 in 1989 and May 3 and May 30 in 1990.

BLB populations were sampled twice weekly in both early and late plantings soon after plant emergence, as well as in adjacent alfalfa. Direct counts (four 5-m lengths of row) were taken when plants were too small to sweep with a net (stages VE to V2). Sweep-net samples were obtained after the plants reached stage V2 by taking four 50-sweep samples in each plot on each date until soybean maturity. Alfalfa was swept similarly from late April until frost.

At harvest maturity, plant samples were taken from all plots by removing 4 2-m lengths of row and returning these to the laboratory for analysis. Number of pods per plant and number of pods per plant injured by BLB were estimated from a 50-plant aliquot drawn from each harvest sample. After the count, the aliquot was returned to the sample, and the sample was threshed with a stationary thresher. After the threshing, harvested seeds were weighed, moisture readings taken, and seed weights adjusted to a standard 13% moisture.

Sampling in alfalfa showed that overwintered BLB were first active on April 27 in 1989 and on April 22 in 1990. In both years, peaks occurred the middle of May and then declined in alfalfa following soybean emergence and the first alfalfa harvest. BLB was absent in the alfalfa most of the season but returned on or about August 21 in 1989 and on August 31 in 1990, after maturation of the soybean.

Consistently greater BLB numbers occurred in the early soybean plantings than in the late plantings in both years. Three well defined peaks occurred in the early plantings, representing invading (overwintered) adults (June 15-21), F1 adults (July 23-24), and F2 adults (September 4-14). BLB adults in the late-planted plots were detected 3 to 4 weeks later than in the early planted plots and lacked a well defined peak of invading adults.

Effects of planting date on potential soybean pod injury are revealed in an analysis of BLB feeding days during the susceptible period (R5-R8). A BLB feeding day is the integral of one beetle count on a day. Total feeding days are determined by calculating the area under a density curve plotted through time. For calculation, sweep-sample counts were converted to number of beetles per meter of row according to the conversion of Rudd and Jensen [15], and the trapezoid method was used to calculate area under the curve. The results gave means of 38.6 (1989) and 101.6 (1990) BLB-feeding days/m of row in the early plantings and 15.0 (1989) and 60.4 (1990) BLB-feeding days/m of row in the late plantings. These means represent BLB-feeding reductions, from late plantings, of 61.1% and 40.6% in 1989 and 1990, respectively.

Treatment differences in BLB pod injury BLB-feeding days were consistent (Fig. 3). Pod-injury rates in the early planted treatments were 5% and 10.7% versus the late-planted rates of 1.3% and 5.3% in 1989 and 1990, respectively. This translates into 73% and 50% reductions of pod injury by planting late in the two respective years. An analysis of variance of percentage injured pods per plot over the two years indicated that differences were significant ($F = 10.88, P = 0.02$).

No significant differences between treatments were found in either number of pods per plant or yield. This finding indicates that even the highest BLB densities were below the EIL. Nevertheless, density and injury differences between planting dates would be expected at population levels above the EIL, just as they are below it.

The Integrated BLB Strategy

Although much more information is needed regarding soybean planting date and BLB pod injury, these early results show much promise for delayed planting as a preventive tactic. If this tactic proves economical and reliable, a prototype IPM system for BLB in the north-central U.S. can be synthesized. The prototype would direct producers to

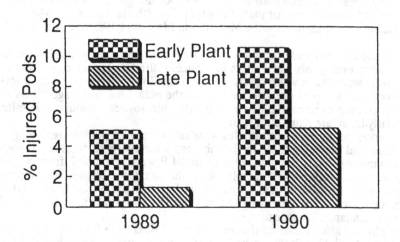

Figure 3. Differences in BLB pod injury between planting-time treatments in 1989 and 1990.

plant soybean as late as possible in the recommended planting period. Care must be taken not to plant later than dates recommended because yield losses from very late planting could be greater than those caused by BLB adults. When plants reach the R4 stage, sampling is begun using the prescribed ground-cloth techniques[10]. Weekly samples still are to be taken to detect any failures of the late-planting tactic. Subsequently, estimates of population or injury status are compared with ETs for selected management-cost and market-value parameters in the ET decision tables. If estimates are below the ET, no action is taken, but if they are above the ET, an insecticide is applied. Compounds allowing greatest personal and environmental safety are selected, and lowest effective rates used.

REFERENCES

1. Geier, P. W. and Clark, L. R., An ecological approach to pest control, *Proc. Tech. Meeting Intern. Union for Conserv. of Nature and Natural Res., 8th, 1960, Warsaw,* 1961, 10-8.

2. Worner, S. P., Use of models in applied entomology, *Environ. Entomol.*, 1991, **20**, 768-73.

3. Rabb, R. L., A sharp focus on insect populations and pest management from a wide-area view, *Bull. Entomol. Soc. Am.*, 1978, **24**, 55-61.

4. All, J. N., Importance of designating prevention and suppression control strategies for insect pest management programs in conservation tillage, *Proc. 1989 Southern Conservation Tillage Conf., Tallahassee, FL*, 1989, 1-3.

5. Nyrop, J. P. and Binns, M., Quantitative methods for designing and analyzing sampling programs for use in pest management, In *CRC Handbook of Pest Management in Agriculture*, 2nd ed., ed. D. Pimentel, CRC Press, Boca Raton, FL, 1991, pp. 67-132.

6. Pedigo, L. P., *Entomology and Pest Management*, Macmillan, New York, 1989, pp. 517-21.

7. Stern, V. M., Smith, R. F., Van Den Bosch, R. and Hagen, K. S., The integrated control concept, *Hilgardia*, 1959, **22**, 81-101.

8. Hammond, R. B., Higgins, R. A., Mack T. P., Pedigo, L. P. and Bechinski, E. J., Soybean pest management, In *CRC Handbook of Pest Management in Agriculture*, 2nd ed., ed. D. Pimentel, CRC Press, Boca Raton, FL, 1991, pp. 341-472.

9. Wintersteen, W. and Higley, L. G., Advancing IPM systems in corn and soybeans, In *IPM: Environmentally Sound Agriculture*, ed. A. Leslie and G. Cuperus, Am. Chem. Soc., Lewis Publishers, Boca Raton, FL, 1991 (in press).

10. Pedigo, L. P., Zeiss, M. R. and Rice, M. E., Biology and management of bean leaf beetle in soybean, *Proc. 1990 Crop Production and Protection Conf.*, Iowa State Univ. Coop. Ext. Serv., Ames, IA, 1990, 109-117.

11. Smelser, R. B. and Pedigo, L. P., Phenology of *Cerotoma trifurcata* on soybean and alfalfa in central Iowa, *Environ. Entomol.*, 1991, **20**, 514-9.

12. Kogan, M., Waldbauer, G. P., Boiteau, G. and Eastman, C. E., Sampling bean leaf beetles on soybean, In *Sampling Methods in Soybean Entomology*, 1980, Springer-Verlag, New York, pp. 201-236.

13. Smelser, R., B., *Phenology, Herbivory, and Bioeconomics of the Bean Leaf Beetle*, 1990, Ph.D. Dissertation, Iowa State Univ., Ames, IA, pp. 32-51.

14. Pedigo, L. P., Higley, L. G. and Davis, P. M., Concepts and advances in economic thresholds for soybean entomology, In *Proc. World Soybean Res. Conf. IV*, 1989, **3**, 148-93.

15. Rudd, W. G. and Jensen, R. L., Sweep net and ground cloth sampling for insects in soybeans, *J. Econ. Entomol.*, 1977, **70**, 301-4.

Acknowledgment

Thanks are extended to the Iowa Soybean Promotion Board and to the Leopold Center for Sustainable Agriculture for partial support of the research reported herein. I am also indebted to my graduate students, J. Browde, T. DeGooyer, T. Klubertanz, and M. Zeiss, for discussing many ideas and reviewing the manuscript. Additionally, I thank R. Smelser for his comments and suggestions on the manuscript. Journal Paper J-14658 of the Iowa Agriculture and Home Economics Experiment Station, Ames, Iowa; Project Nos. 2580 and 2903.

THE PRESENT SITUATION OF SOYBEAN PEST MANAGEMENT IN JAPAN

O. MOCHIDA AND A. KIKUCHI
National Agriculture Research Centre,
Ibaraki-ken, Tukuba-si, Kan'nondai, 305 Japan

ABSTRACT

Soybean was cultivated in 140,800 ha with the total production of 197,300 tonnes (t) and the average yield of 1.40 t/ha in Japan in 1991. As pests on soybean 271 invertebrates (245 insect, 12 nematode, 7 Stylommatophoran, 5 mite, and 2 Isopod species), 9 birds, and 13 mammals including 2 rodents were recorded. Since the 1st October, 1991, 55 insecticides, 4 nematicides, 3 acaricides, and a female-sex-pheromone product have been registered for controlling 22 groups of soybean pests (20 insect, a nematode, and a mite groups). Two to 4 frequency of insecticide applications are recommended, depending on pest situations in different regions and prefectures. Fulltime farm households follow to the recommendations but most parttime ones apply commonly less than in recommendations and lose timely applications. Key pests are soybean podborer, aphids or dwarf, and beanfly in the north, stinkbugs, Limabean podborer, soybean podgallmidge, and tobacco leafworm in the central, and stinkbugs and tobacco leafworm in the south regions. The yield was frequently lost totally without insecticide application in south. The gross income/plant protection cost ratio was roughly estimated at 8-11: 1 at the NARC in central Japan where pests and diseases usually are not very serious. Labour shortage was one of constraints for both fulltime and part-time farm households, especially in the south regions with low yields of 1.1-1.3 t/ha and average harvested areas of 0.07-0.18 ha/household.

INTRODUCTION

Soybean is traditionally one of the five important food crops in Japan. As consumption was/is bigger than domestic production, however, soybean was imported mainly from China before 1945. In 1990 crop year soybean of 220,400 t was produced in Japan but 4,681,382 t was imported; 74, 18, and 6% from USA, Brazil, and China, respectively. Japan consists of 47 prefectures. Soybean is cultivated widely in all of the 47 prefectures. However, soybean yield and cultivating conditions vary in regions/prefectures and year by year. Yields were usually less than 1.3 t/ha before synthetic insecticides started to be used for soybean crop in 1950s. Yield was usually higher in north than in south and in upland than in lowland or paddy fields. Heavy infestations of various kinds of insect pests were frequently recorded before mid 1970s. Stinkbugs and the tobacco leafworm (abbreviation, T-leafw, Spodoptera litura) have been recently remarked as key pests in most areas.

Soybean cultivated area was 0.8 ha/farm with yield of 2.6 t/ha in the north (Hokkaido) and 0.2 ha/farm with yield of 1.6 t/ha in the south (kyuusyuu) on an average in 1990. In general insect pests are more severe in the south than in the north. Timely applications are frequently losed because of shortage of labour. Thus, pest management has one of key roles for soybean growing farmers, especially in the south.

SOYBEAN CULTIVATION AND YIELD

Cultivated area, production, and average yield for 1878-1991 are illustrated in Figure 1. Cultivated area reached the bottom of 79,000 ha in 1977 and afterwards increased slightly due to the government policy for promoting area converted from rice to other crops including soybean because of overproduction of rice since 1970. Annual production was 197,300-187,200 t for last 5 years (1987-91). Yield fluctuated annually but increased about 3 times for last centuries; 0.51 t/ha in 1878 and 1.71-1.79 t/ha in 1984-89.

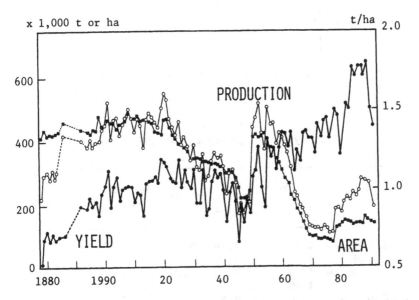

Figure 1. Soybean cultivated area, production, and average yield in Japan from 1878 to 1991.

The maximal records of cultivated area, production, and average yield were 491,700 ha in 1908, 550,900 t in 1920, and 1.79 t/ha in 1989. Those average values for recent 5 years (1987-91) were 152,680 ha, 250,700 t, and 1.64 t/ha.

Yields varied in regions (Figure 2 and Table 1). Okinawa, located in the most south or region X, produced the lowest yield of 1.07 t/ha, whereas Hokkaido, located in the most north or region I, produced the highest of 2.27 t/ha on an average for last 3 years. Relative value for the average yields was 100:227 in Okinawa:Hokkaido (Table 1). Average soybean harvested area in regions VIII and IX was 0.07-0.18 ha/soybean farm household.

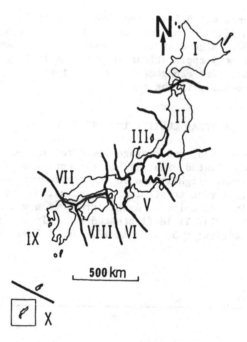

Figure 2. Ten regions of Japan. See Table 1 for region names.

TABLE 1
Soybean yield in 10 regions in 1989-91

Region*	Yield (t/ha)				Relative value
	1989	1990	1991	Avg	
I Hokkaido	2.52	2.60	2.19	2.44	227
II Toohoku	1.56	1.63	1.38	1.52	142
III Hokuriku	1.89	0.67	0.92	1.16	108
IV Kantoo	1.90	1.80	1.52	1.74	162
V Tookai	1.46	1.22	1.31	1.33	124
VI Kinki	1.63	1.18	1.45	1.42	132
VII Tyuugoku	1.66	1.14	1.30	1.37	127
VIII Sikoku	1.39	1.11	1.26	1.25	117
IX Kyuusyuu	1.95	1.58	1.09	1.54	143
X Okinawa	1.00	1.10	1.12	1.07	100
Avg	1.79	1.51	1.40	1.57	-

* See in Figure 2.

Soybean was traditionally cultivated in upland fields and also on le-
vees within paddy rice fields for under intensive farming in small-scaled

agriculture in Japan. Since 1970, however, soybean has been cropped in low-land/paddy fields widely because of the pressure of overproduction of rice. Annual cultivated area was 67,000 ha in upland and 16,000 ha in lowland for 1975-77 but 42,000 ha in upland and 104,000 ha for 1987-91. During last 14 years, namely, soybean cultivated area in lowland fields increased about 6 times but that in upland fields decreased to 2/3. Production decreased to 7/9 in upland but increased about 8 times in lowland. Yields increased from 1.39 to 1.64 t/ha in upland and from 1.37 to 1.53 t/ha in lowland (Table 2).

TABLE 2
Changes of soybean cultivated area, production, and average yield in upland and lowland fields for last 14 years

Year	Cultivated area (ha)		Production (t)		Yield (t/ha)	
	Upland	Lowland	Upland	Lowland	Upland	Lowland
1975	71,200	15,600	104,000	21,600	1.48	1.38
1976	66,800	16,000	88,400	21,000	1.32	1.31
1977	63,900	15,300	89,400	21,500	1.40	1.41
Avg	67,300	15,633	93,933	21,361	1.39	1.37
1989	41,800	109,900	71,700	200,100	1.72	1.82
1990	40,900	105,000	68,700	151,600	1.68	1.14
1991	43,200	97,600	65,500	131,700	1.52	1.35
Avg	41,967	104,167	68,633	161,133	1.64	1.53

PEST SITUATION

Pest species and their annual fluctuations in occurrence area
As soybean pests, 271 invertebrates (245 insect, 12 nematode, 7 Stylomato-phoran, 5 mite, and 2 Isopod species), 9 birds, and 13 mamals were recorded.
Heavy infestations and/or outbreaks were reported in some species; Sca-rabaeids [Anomala rufocuprea in 1938-39 (11), A. rufocuprea and Maladera ca-stanea in 1967-69 (7)], Lepidopterous defoliators [Heliothis maritima in 1908-09 and Pyrrhia umbra in 1908-09, 1920, 1932, and 1950 (8)], bean-web-worm (abbreviation, B-webw., Pleuroptya ruralis) (9), and tobacco-leafworm (T-leafw, Spodoptera litura) recently very often], and soybean stemmaggot (Melanagromyza soiae) in 1905, 1914, 1921, 1923, 1928, 1931, 1934-36, and 1941 (10). Except for the T-leafw and stinkbugs, any outbreak of these insect pests were not reported in these 5 years.
Major invertebrate pests are tentatively categorized into 14 groups in the viewpoint of occurrence areas (Table 3); 1) aphids, 2) stinkbugs, 3) T-leafw, 4) 2-striped soybean beetle (2-ssb), 5) Lepidopterous pests (Lepests), 6) soybean podborer (S-podborer), 7) spider mites, 8) Scarabaeids, 9) B-webw, 10) soybean podgallmidge (gallmidge), 11) soybean podworms (S-podws), 12) Limabean podborer (L-podborer), 13) beanfly, and 14) others.
The soybean cultivated area and occurrence of these pest groups for last 14 years are given in Table 4. The occurrence area of aphids varied from 25,758 ha in 1980 to 49,438 ha in 1991. Occurrence areas of some other pest groups fluctuated considerably year by year but others varied slightly in acreage. In 1991, the occurrence areas of pest groups of 1 to 13 were 49438,

TABLE 3
Major pests on soybean in Japan

Group [Abbreviation]	Order	Family	Major species
1. Aphids	Hom	Aphididae	Aulacorthum solani, Aphis glycines
2. Stinkbugs	Het	Alydidae	Riptortus clavatus
		Pentatomidae	Piezodorus hybneri, Dolycoris baccarus, Nezara antennata, N. viridula, Holyomorpha mista
3. Tobacco leafworm [T-leafw]	Lep	Noctuidae	Spodoptera litura
4. 2-striped soybean beetle [2-ssb]	Col	Chrysomelidae	Medythia nigrobilineata
5. Lepidopterous pests other than 3, 6, 9, 11, 12, & 14 [Lepests]	Lep	Noctuidae	Heliothis maritima, Plusiinae spp.
		Geometridae	Ascotis selenaria
6. Soybean podborer [S-podborer]	Lep	Tortricidae	Leguminivora glycinivorella
7. Spider mites	Aca	Tetranychidae	Tetranychus spp.
8. Scarabaeids	Col	Scarabaeidae	Anomala spp., Maladera spp. Popillia japonica
9. Bean webworm [B-webw]	Lep	Pyralidae	Pleuroptya ruralis
10. Soybean podgallmidge [gallmidge]	Dip	Cecidomyiidae	Asphondylia sp.
11. Soybean podworms [S-podws]	Lep	Tortricidae	Matsumuraeses spp.
12. Limabean podborer [L-podborer]	Lep	Pyralidae	Etiella zinckenella
13. Beanfly	Dip	Anthomyiidae	Delia platura
14. Cutworms & others	Lep	Noctuidae	Agrotis spp.

43650, 42724, 32844, 31696, 21532, 20413, 19481, 15417. 10301, 7167, 3388, and 3164 ha, respectively.

Relative occurrence areas (occurrence area/soybean cultivated area, %) of these pest groups for last 14 years are also given in Table 4. Pest groups of 1 to 3 occurred in more than 30% of cultivated area, whereas those of 4 to 9 occurred in 10-24% and those of 10 to 13 in less than 10%. Table 4 indicates that T-leafw, stinkbugs, and 2-ssb have been recently increasing their occurrence areas, whereas S-podws, L-podborer, and beanfly keep their occurrence areas less than 10% of cultivated area.

Pest occurrence in different regions
Various insect pests appear on soybean in different regions. Tables 5 and 6 indicate pest species/groups, which farmers targeted to apply insecticides to in 1991. Namely, regarding 4 pod/seed-feeding insect pests, relative occurrence area (actual occurrence area/soybean cultivating area, in %), treated area, and estimated number of frequency of insecticide treatments are shown (Table 5). S-podborer in regions II and I, gallmidge in IV, VI,

TABLE 4

Actual (ha) and relative occurrence area (%) of soybean pests in Japan in 1978-91

Year	Cultivated area ha	Relative value, %	Aphids	Stink-bugs	T-leafw	2-ssb	Le-pests	S-pod-borer	Spider mites	Scara-baeids	B-webw	Gall-midge	S-podws	L-pod-borer	Bean-fly
								Actual occurrence area (ha)							
1978	127,000	78.0	—	11,408	5,950	—	—	11,445	—	—	—	4,892	—	1,654	—
1979	130,300	80.0	—	18,171	7,189	—	—	13,623	—	—	—	9,429	—	7,460	—
1980	142,200	87.4	25,758	20,216	9,020	—	—	14,007	—	—	—	8,292	—	5,537	—
1981	148,800	91.4	35,281	56,555	24,264	—	—	16,276	—	—	—	11,264	—	11,468	—
1982	147,100	90.4	46,331	33,768	18,452	—	—	24,689	12,654	12,450	11,094	16,001	11,162	9,618	3,948
1983	143,400	88.1	40,609	39,152	41,105	—	—	22,814	20,011	20,030	11,926	16,589	12,984	13,458	8,441
1984	134,300	82.5	45,129	46,458	34,128	—	—	24,226	25,182	23,750	17,260	19,664	8,529	9,933	4,769
1985	133,500	82.0	43,331	41,736	32,768	—	—	24,607	22,889	18,211	9,526	15,366	10,592	11,019	4,452
1986	138,400	85.0	43,840	29,517	29,530	6,788	21,841	13,924	22,915	17,506	9,879	14,212	10,524	10,112	4,898
1987	162,700	100.0	36,758	30,631	23,374	11,109	21,776	15,968	28,889	17,066	10,213	12,416	9,715	10,457	4,420
1988	162,400	99.8	39,860	24,276	32,197	10,054	16,540	8,108	28,185	23,328	15,709	7,401	6,064	3,441	2,575
1989	151,600	93.1	44,839	34,536	24,601	13,327	12,709	8,401	19,599	18,726	13,913	8,327	9,428	4,022	2,019
1990	145,900	89.6	46,228	33,360	46,348	21,728	31,173	8,919	25,499	24,309	12,878	11,637	7,323	4,084	2,626
1991	140,800	86.5	49,438	43,560	42,724	32,844	31,696	21,532	20,413	19,481	15,417	10,301	7,167	3,388	3,164
								Occurrence area/cultivated area (%)							
1978			—	9.0	4.7	—	—	9.0	—	—	—	3.9	—	1.3	—
1979			—	14.0	5.5	—	—	10.5	—	—	—	7.2	—	5.7	—
1980			18.1	14.2	6.3	—	—	9.9	—	—	—	5.8	—	3.9	—
1981			23.7	38.0	16.3	—	—	10.9	—	—	—	7.6	—	7.7	—
1982			31.5	23.0	12.5	—	—	16.8	8.6	8.5	7.5	10.9	7.6	6.5	2.7
1983			28.3	27.3	28.7	—	—	15.9	14.0	14.0	8.3	11.6	9.1	9.4	5.9
1984			33.6	34.6	25.4	—	—	18.0	18.8	17.7	12.9	14.6	6.4	7.4	3.6
1985			32.5	31.3	24.6	—	—	18.4	17.2	13.6	7.1	11.5	7.9	8.3	3.3
1986			31.7	21.3	22.1	4.9	15.8	10.1	16.6	12.7	7.1	10.3	7.6	7.3	3.5
1987			22.5	18.8	14.3	6.8	13.3	9.8	17.7	10.4	6.2	7.6	5.9	6.4	2.7
1988			24.5	14.9	19.8	6.1	10.1	4.9	17.3	14.3	9.6	4.5	3.7	2.1	1.5
1989			29.6	22.7	16.2	8.8	14.3	5.5	12.9	12.3	9.1	5.4	6.2	2.6	1.3
1990			31.2	22.8	31.7	14.8	21.3	6.1	17.4	16.6	8.8	7.9	5.0	2.7	1.7
1991			35.1	30.9	30.3	23.3	22.5	15.2	14.4	13.8	10.9	7.3	5.0	2.4	2.2

TABLE 5

Occurrence of 4 soybean pod/seed-feeding insect pests and insecticide-treated areas in different regions in 1991

Region no.	Occurrence area/cultivated area (%)				Maximum value among 4 pests (%) 1/			Estimated frequency of treatments in farmers' practice
	S-pod-borer *	Gall-midge **	L-pod-borer ***	Stink-bugs ****	1st treated area /cultivated area	2nd/more treated area/cultivated area	2nd/more treated area/1st treated area	
I	12.0	0.0	0.0	0.0	66*	31*	46*	0 - 2
II	44.1	3.5	0.0	45.6	23*	4****	18****	0 - 2
III	4.5	2.0	0.3	13.8	79*, ****	61*, ****	96***	0 - 2
IV	4.3	16.8	2.2	11.2	52****	47****	88*****	0 - 2
V	5.8	4.3	3.3	26.2	29*****	13*****	45*****	0 - 2
VI	3.5	12.3	8.1	44.2	89***, ****	149***, ****	168***, ****	0 - 3
VII	5.5	7.5	9.0	58.6	35*****	18**, ****	94**	0 - 2
VIII	1.7	7.5	4.2	18.0	28*****	13****	113****	0 - 3
IX	5.0	12.1	0.6	46.9	55*****	48*****	88*****	0 - 2
X	?	?	?	?	?	?	?	0 - ?

1/ The maximum values were picked out to avoid duplicated counting of treated areas.

TABLE 6

Occurrence of dwarf virus disease, the beanfly, and T-leafw on soybean and insecticide-treated areas in different regions in 1991

Region no.	Occurrence area/culti-vated area (%)	1st treated area/culti-vated area (%)	2nd//more treated area/culti-vated area (%)	2nd//more treated area/1st treated area (%)	Estimated fre-quency of treat-ments in farmers' practice
Dwarf					
I	59.9	81.7*	27.8*	34.0*	0 - 2
II	12.0	11.6*	3.1*	27.4*	0 - 1
III	0.0	0.0	-	-	0
Beanfly					
I	7.5	40.8**	-	-	0 - 1**
II	0.4	6.9**	-	-	0 - 1**
III	10.0	0.7**	-	-	0 - 1**
IV	0.1	0.7**	-	-	0 - 1**
V	0.0	0.0	-	-	0
VI	0.0	2.5**	-	-	0 - 1**
VII	0.6	2.1**	-	-	0 - 1**
VIII	0.0	0.0	-	-	0
IX	0.0	0.0	-	-	0
T-leafw					
I & II	0.0	0.0	0.0	0.0	0
III	21.1	48.3	61.3	126.8	0 - 3
IV	20.6	43.9	34.6	78.7	0 - 2
V	8.2	28.3	17.1	70.6	0 - 2
VI	57.9	51.8	61.9	119.6	0 - 3
VII	56.5	25.0	3.6	14.4	0 - 2
VIII	65.1	29.6	16.5	55.7	0 - 2
IX	85.0	63.1	56.9	90.1	0 - 2

* Insecticide application to aphids, especially to the glasshouse-potato aphid (Aulacorthum solani), vector of dwarf virus disease.
** Seedcoating is common.

and IX, L-podborer in VII, VI, and VIII, and stinkbugs in III-IX were major. In order to control one or some of these pod/seed-feeding pests, farmers apply insecticide(s) with wide spectrum once to 3 times from late flowering to mid seed-growth stage. Though relative insecticide-treated area/ cultivated area varied region by region, the frequency of insecticide applications was esti-mated to be 0-3 times (Table 5).

Soybean dwarf is transmitted by the glasshouse-potato aphid and occurs widely in I and partially in II. Seed-furrow treatment was done at seeding in 81% of soybean cultivated area in I. Supplementary sprays were done in 28% of cultivated area with ethiofencarb (Allylmate*). Beanfly was major in I and treated once with seedcoating. T-leafw occurred widely through III-IX. Treatments were done in 25-63% of all the soybean cultivated area in III-IX. Three or more applications might be done in some areas (Table 6).

Judging from the data shown in Tables 5 and 6 and occurrence areas of pest species for last 3 years, currently important insect pests are listed up in Table 7. Some Scarabaeids were serious in some areas in regions III, IV, V, and IX. Late maturity cultivars, however, have been recently cultivated widely to avoid severe defoliation by the adult feeding.

TABLE 7
Important insect pests in different regions in 1989-91

Region no.	Pest species and/or group name
I	1) S-podborer* 2) beanfly* 3) dwarf/glasshouse-potato aphid*
II	1) S-podborer* 2) Stinkbugs*
III	1) L-podborer* in hot summer, or S-podborer* in cool summer, 2) T-leafw* 3) Stinkbugs* 4) Scarabaeids 5) B-webw 6) 2-ssb
IV	1) Stinkbugs* 2) Gallmidge* 3) T-leafw* 4) Scarabaeids
V	1) Stinkbugs* 2) T-leafw* 3) S-podborer* 4) Scarabaeids
VI	1) Stinkbugs* 2) T-leafw* 3) S-podws 4) Gallmidge 5) 2-ssb
VII	1) Stinkbugs* 2) T-leafw* 3) L-podborer
VIII	1) Stinkbugs* 2) T-leafw*
IX	1) Stinkbugs* 2) T-leafw*

* Pest(s) are targeted for insecticide application in normal pest occurrence.

PEST MANAGEMENT

Insecticides and pest management

Currently 55 insecticides, 4 nematicides, 3 acaricides, and one female-sex-pheromone product are registered against soybean invertebrate pests. Out of 55 insecticides, single formulations are 46 in number, 6 are mixtures of 2 kinds of insecticides, and 4 are mixtures of insecticides with fungicides (Table 8). Except for the pheromone product, these 62 products in total consist of 33 kinds of chemicals and 5 fungicides.

Against pod/seed-feeding insect pests and T-leafw, 35 insecticides and a pheromone product are registered (Table 9). Out of 35, 28 insecticides are commonly effective on 2 or more soybean pod/seed-feeding pest species. The pheromone product is used not for killing but monitoring of T-leafw adult males.

To control pod/seed-feeding pests (frequently together with T-leafw larvae), one to 3 applications of insecticides with broad spectrum are popularly recommended, depending on pest situations, starting at the beginning pod growth stage at an interval of 7-10 days (1, 2, 5, 12).

Regarding T-leafw, older larvae are not killed by any insecticides except for mesomyl (Lannate*). Aggregated young larvae feed on leaves and make them white or "White leaves". The largest peak of "White leaves" appears frequently in south (Kagosima and Miyazaki prefectures) of region IX in late August to mid September and in most parts other than the 2 prefectures in region IX in late September to mid October. The 2nd largest peak of "White leaves" appears about one month or one generation ahead from the largest peak. One to 2 applications of effective insecticide(s) are recommended during the 2nd largest peak of "White leaves" at an inverval of 7-10 days.

When T-leafw is abundant, 3-4 applications are recommended (12). In Hukuoka prefecture of region IX for 3 years of 1984-86, however, 29, 38, 26, 6,
and 0.3% of the soybean cultivated area were treated 0, 1, 2, 3, and 4 times
on an average by insecticides, respectively (12). It means that about 1/3
of soybean cultivated area was not applied by insecticides.

The peaks of "White leaves" fluctuate less or more in time, height or
population, and shapes frequently. Under outbreak conditions like in 1989,
3 applications sometimes were not enough to prevent soybean crop from T-
leafw's feeding. If farmers lost the good timing for insecticide application, soybean plants were destroyed totally (Figure 3). Most parttime farmers applied insecticides once or twice and afterwards abandoned their crops,
because of shortage of labour.

TABLE 8
Pesticides registered for soybean pest control in Japan since 1st October,
1991

Chemical group	No. pesticides registered	No. of kinds of chemicals
OP	24	15
Botanical	8	3
Carbamate	7	3
Synthetic pyrethroids	4	3
OC	1	1
IGR	1	1
Female-sex-pheromone product	1	1
Acaricides	3	2
Nematocides	4	4
Synthetic pyrethroid + OP	1	1
Batanical + OP	1	0
OP + OP	3	0
Insecticide + fungicide(s)	4	(5)*
Insecticide + insecticide	1	0
Total	63	34 + (5)*

* Chemicals as fungicides.

Against the beanfly, 8 insecticides are registered; 1) chlorfenvinphos
[CVP, another common name](Vinylphate*, trade name) D (dust), 2) diazinon
(Diazinon*) D, G (granule), 3) dichlofenthion [ECP](VC*) D, 4) dichlofen-
thion + kasugamycin + tillam [rebulate], (Pairkasumin*) D, 5) dichlofenthi-
on + tillam, (Nomart*, VCT*) D, 6) ethylthiodemeton [disulfoton] + diazinon,
(Ethimeton*) G, 7) isoxathion (Karphos*) D, D-G mixture, and pyridaphenthion
(Ofunack*) D.

Against the glasshouse-potato aphid as the vactor of dwarf, 4 are recommended throughout region I; seed-furrow treatment with either ethylthiodemeton (Disyston*, Ekatin TD*) G, ethiofencarb (Allylmate*) G, or dimethoate (Dimethoate*, Dimethoate S*) G at seeding and foliar spray with ethofen-
carb (Allylmate*) EC (emulsifiable concentrate) in late June to mid July.

TABLE 9

Registered insecticides and a pheromone product against soybean pod/seed-attacking insect pests and the T-leafw since 1st October, 1991

Chemical	S-pod-borer	Gall-midge	L-pod-borer	Stink bugs	T-leafw
1 carbaryl(Sevin*, Denapon*) WP	•				
2 cyanophos[CYAP](Cyanox*) EC	•				
3 cycroprothrin(Cyclosal*) EC				•	
4 cyfluthorin(Baythroid*) EC	•			•	
5 diazinon(Diazinon*) G	•	•	•	•	
6 endosulfan[benzoepin](Malix*) EC	•				
7 EPN(EPN*) D	•	•	•		
8 " EC	•				
9 " WP	•	•			
10 etophenprox(Trebon*) D	•		•	•	•
11 " EC	•	•	•	•	•
12 fenitrothion(Sumithion*) D, EC	•	•	•	•	
13 fenitrothion + thiabendazole(fungicide), (Sumitect*) D	•	•	•	•	
14 fenitrothion + thiophanate-methyl (fungicide), (Sumitop M*) D	•	•	•	•	
15 fenthion[MPP](Baycid*) D	•	•			
16 " EC	•	•		•	
17 fenvalerate(Sumicidin*) + fenitrothion, (Parmathion*) WP	•	•	•	•	•
18 isoxathion(Karphos*) D	•	•	•	•	•
19 litlure[female sex pheromones] (Pherodin SL*)					•
20 malathion(Malathon*) D	•		•		
21 " EC	•				
22 malathion + cyanophos, (Cyathon*)	•		•	•	
23 malathion + fenitrothion, (Sumithon*, Tracide A*) D	•		•	•	
24 methomyl(Lannate*) D-G mixture			•	•	•
25 " WP	•		•	•	•
26 nicotine sulfate	•		•	•	
27 phenthoate[PAP](Elsan*, Paption*) D, EC	•		•	•	•
28 phosmet[PMP](Appa*, PMP*) D	•			•	
29 prothiophos(Tokuthion*) D	•	•			
30 " EC	•	•			•
31 pyrethrins EC	•		•		
32 pyrethrins + rotenone EC	•		•		
33 rotenone Powder	•		•		
34 " D	•		•	•	
35 " EC			•	•	
36 teflubenzuron(Nomolt*) EC					•

Note: Name bracketed indicates another common name. Name(s) parenthesized indicate trade name(s) with an asterisk.

Some recommendations for pesticide applications in 4 regions are shown in Table 10.

TABLE 10

Recommendations of pesticide application for soybean pest control in usual occurrence

Re-gion	Application	Target main pests
I	1st: Seedcoating with insecticide/fungicide mixture like Pairkasumin*, Nomart*, or VCT* seed-treatment formulation.	Beanfly & purple seed stain.
	2nd: Seed-furrow treatment of insecticide G like Disyston*, Ekatin TD* or others.	Glasshouse-potato aphid, the vector of soybean dwarf.
	3rd: Treatment with insecticide/fungicide mixture like Sumitop M*.	S-podborer & purple seed st.
	4th: "	"
II	1st: Seedcoating with fungicide like Benlate T*, Homai*, or others.	Purple seed stain.
	2nd: Treatment with insecticide/fungicide mixture like Sumitop M*.	S-podborer & purple seed st.
	3rd: "	"
IV	1st: Seedcoating with fungicide like Benlate T*, Homai*, or Topzin M*.	Purple seed stain.
	2nd: Treatment with insecticide like Elsan*, Karphos*, or Parmathion* or with insecticide/fungicide mixture like Sumitop M* or Sumitect*.	Stinkbugs, gallmidge, T-leafw, & purple seed stain.
	3rd/4th: "	"
IX	1st: Treatment with insecticide like Elsan*, Karphos* Parmathion* or Trebon* in mid Aug, when T-leafw-population is high.	T-leafw
	2nd: Treatment with insecticide like Elsan*, Karphos*, or etc. in late Aug - early Sep when pest populations are high.	T-leafw & stinkbugs
	3rd: Treatment with an insecticide mixed with a fungicide like Elsan* EC + Topzin M* WP in mid Sep.	T-leafw, stinkbugs, & purple seed stain.
	4th: Treatment with an insecticide mixed with a fungicide in late Sep when pest populations are high.	Stinkbugs, T-leafw, & purple seed stain.

Note: Name with an asterisk parenthesized is trade name. Prefectural governments provide their own recommendations. Thus, no common recommendation(s) are throughout each region, except for region I.

Figure 3. Soybean field damaged seriously in Hukuoka prefecture (region IX) in October 1989 by T-leafw (Photo by O. Mochida)

Economics of insect pest management with pesticides

Kikuchi and Kobayashi (1) showed that normal seeds were 88% in plots treated 3 times with Parmathion* and 58% in untreated plots at our NARC on 2 years' average. Referring to this information, gross income/plant protection cost ratios in soybean were roughly estimated by the data from field experiments at our NARC in 1991 (Table 11). The cultivar "Tatinagaha" was seeded on 17 June and harvested on 24 October. Gross income/plant protection cost ratios were 11:1 and 8:1, depending on which pesticides were chosen and frequency of applications. However, these values might vary also, depending on other factors like pest situations, whether conditions affecting yields, and social and economic factors like shortage of labour, pesticide price, etc. Anyhow, the values of 11:1 and 8:1 shown in Table 11 suggest that applications of insecticides promise the great benefit if appropriate pest management is implemented. During the session of the Symposium on Soybean pests held in Kumamoto (region IX) in October 1989, however, they pointed out that most of farmers or most of parttime farmers applied insecticides at most 2 times because of shortage of labour.

In region IX, technical recommendations indicate to apply insecticides 2 times in normal years but 4 times in years with higher pest populations (see Table 10). Under such conditions, Yamanaka (12) showed that normal seeds/total seeds in soybean were 33-89% with an average of 69% for 11 years, when no insecticide was applied. An average of 22% of the total seeds was damaged by pests, whereas about 9% was immature and/or sterile seeds. Out of 22% seeds damaged by pests, 11, 4, 3, 2, and 2% seeds were damaged by stinkbugs, S-podws, gallmidge, T-leafw, and others (purple seed stain, 2-ssb, L-podborer, etc.). When 33-89% of seeds were infested by pests, all of seeds (normal + infested seeds) had no market value. It means that soybean production in untreated plots with insecticide(s) may have no meaning in the commercial viewpoint but lost all the capital invested.

Market price of soybean produced in Japan was/is about 7 times higher

than the international one (6). This fact, the current situation of shortage of labour, small scaled farmers in soybean production, etc., may make
pest management on soybean with pesticides hard, even though farmers know
pest management has very important roles in soybean production in Japan.

TABLE 11

Estimated gross income/plant protection cost ratios in soybean production
at NARC in Tukuba, Japan in 1991

Application		Choice of pesticides			
1st: Seedcoating					
Pesticide cost/ha	¥	–		–	Pairkasumin* 1610
Labour cost/ha	¥	–		–	1125
2nd: Foliar spray					
Pesticide cost/ha	¥	–	Elsan*	1428	Parmathion* 4630
Labour cost/ha	¥	–		2000	2000
3rd: Foliar spray					
Pesticide cost/ha	¥	–	Baycid*	1490	Parmathion* 4630
Labour cost/ha	¥	–		4500	4500
4th: Foliar spray					
Pesticide cost/ha	¥	–	Elsan*	1428	Parmathion* 4630
Labour cost/ha	¥	–		4500	4500
Total:Pesticide cost/ha	¥	0		4346	13890
Labour cost/ha	¥	0		11000	12125
total (A)	¥	0		15346	26015
	(US$)			(118)	(200)
Yield	t/ha	1.55		2.25	2.36
Yield increase more than in untreated plots	t/ha	±0		+0.70	+0.81
Increased gross income in treated plots/ha, (B)	¥	±0		175000	202500
Gross income/pl prot cost ratio (B/A)		–		11 : 1	8 : 1

Note: Labour cost was ¥10,000/8 hrs. Manpower/ha needed 54 minutes for
seedcoating, 96 min. for 2nd, and 216 min. for 3rd/4th application each.
US$1.oo = JPN ¥130. Price of soybean was the standard market price plus
subsidy = ¥15,000 (=US$115.34)/60 kg.

Pheromone traps or traps with the pheromone product (Pherodin SL*) are
known to catch male moths of the T-leafw efficiently. The interactions
among seasonal occurrence of T-leafw male moths by pheromone traps, eggmass
appearance, appearance of "White leaves", and timing of insecticide application were examined. Thus, pheromone traps are available for determining
the timing of insecticide application to T-leafw on various crops. Regarding T-leafw on soybean, however, the observation of the appearance of
"White leaves" is simple and enough to determine the timing in farmers'
practice.

Insect pest management with methods other than pesticide application
Varietal difference in soybean is known regarding some insect pest infestations. Kobayashi (4) mentioned of some cultivars, less damaged by the gallmidge, soybean beetle (<u>Anomala rufocuprea</u>), L-podborer, and stinkbugs. As
far as we checked 156 main soybean cultivars cropped in Japan, a few are
resistant and moderatly resistant to insect pests in literature (Table 12).

TABLE 12
Soybean cultivars* resistant (R) and moderately resistant (MR) to insect
pests as of March 1992

Insect pests	Names of cultivars	
	R	MR
S-podborer	Nattoosyooryuu, Tamamusume, Hokkaihadaka, Kosuzu	-
Stinkbugs	Nattoosyooryuu, Tachinagaha, Kosuzu	-
T-leafw	Kin'nari No. 1, Akiakane	-
Soybean beetle	Norin No. 2	
Dwarf	-	Tsurumusume

* Out of 96 cultivars registered at the Ministry of Agriculture, Forestry,
  and Fisheries, Japan, in 1939-91, plus 60 other main cultivars cropped
  in Japan (13, 14).

Kobayashi and Oku (5) indicated that the S-podborer is the only one
key pest on soybean in the area north to the January isotherm of -1.5^{o}C in
region II. In 4 (Aomori, Akita, Iwate, and Miyagi) prefectures with abundant S-podborer populations in region II, Kosuzu, resistant to S-podborer,
was cropped only in 2.2% of soybean cultivated area in 1990 (14). Any other
cultivars resistant to insect pests were grown in relatively small areas.
It suggests that farmers may choose cultivars for the reason other than
resistant to insect pests, in spite of importance of insect pests in soybean cropping.

Kobayashi (4) also mentioned of some cultural methods to expect less
insect infestations; intercropping, mixed cropping, rotation, plant density, escape of the crop from heavy infestation based on the interactions of
insect occurrence and the developmental stages of soybean, and avoidance
of application of immature organic fertilizers. However, most of such methods look not to be practical in farmers' level.

REFERENCES

References involved in statistical data in agriculture are not listed up
except for (14).

1. Kikuchi, A. and Kobayashi, T., [Simultaneous control of some pod/seed-
   feeding insect pests and purple seed stain disease on soybean]. Converted upland field Res. Inf. Ser., MAFF AFF Res. Council, Tokyo, 1984.
   No. 61, 2 pp. In Japanese.

2. Kikuchi, A. [Control of soybean stinkbugs. In Abstracts of main res-

earches on upland crops converted from paddy fields — Fundamental technology in agriculture —]. ed. H. Ikeda, NARC, Tukuba, 1988, pp. 66-68. In Japanese.

3. Kobayashi, T. and Oku, T. Studies on the distribution and abundance of the invertebrate soybean pests in Tohoku district, with special reference to the insect pests infesting the seeds. Bull. Tohoku Natl. Agric. Exp. Stn. 1976, 52:49-106. In Japanese with English summary.

4. Kobayashi, T. [Insect pests. In Handbook of soybean pests and diseases]. ed. T. Kurita, JPPA, Tokyo, 1979, pp. 1-14. In Japanese.

5. Kobayashi, T. and Naito, A. [Simultaneous control of soybean pod/seed-feeding insect pests. In Abstracts of main researches on upland crops converted from paddy fields — Fundamental technology in agriculture —]. ed. H. Ikeda, NARC, Tukuba, 1988, pp. 74-75. In Japanese.

6. Mochida, O. Recent trends of soybean insect pest occurrence in Japan. A textbook for Plant Protection Training course. MAFF, Tokyo, 1988, 51 pp. In Japanese.

7. Okuhara, K. [Current situation of soybean cultivation and its pests and diseases in Kumamoto prefecture. In Symposium on soybean pests and diseases]. JPPA Symp., Kumanoto Oct 1989, pp. 1-10. In Japanese.

8. Sakurai, K. and Nishijima, Y. [Soybean pests and their infestation in Hokkaido]. In Survey on the fauna of soybean insect-pests in Japan. ed. S. Kuwayama, Yokendo, Tokyo, 1953, pp. 29-49. In Japanese.

9. Sekiya, I. [Soybean pests and their infestation in Tozan district]. Ditto, 1953, pp.81-94. In Japanese.

10. Sugawara, H. [Soybean pests and their infestation in Tohoku district]. Ditto, 1953, pp. 51-68 plus a plate. In Japanese.

11. Tamura, I. Ecological studies on the insect injury of soy bean plant. Kanto-Tozan Agric. Exp. Stn. 1952, 287 pp. In Japanese with English summary.

12. Yamanaka, M. [Key insect pests and diseases on soybean and their management in Kyuusyuu. In Symposium on soybean pests and diseases]. JPPA Symp., Kumamoto Oct 1989, pp. 11-26. In Japanese.

13. Anonym [Varietal characteristics of luguminous crops in 1988 FY]. MAFF, Tokyo, 1989, 99 pp. In Japanese.

14. Anonym [Data on soybean in 1990 FY]. MAFF, Tokyo, 1991, 195 pp. In Japanese.

IMPLEMENTING INNOVATIVE INSECT MANAGEMENT SYSTEMS IN SOYBEAN IN THE SOUTHEASTERN U.S.

T. P. MACK
Department of Entomology, 301 Funchess Hall, Auburn University
Alabama USA 36849-5413

ABSTRACT

A computerized management aid, called AUSIMM was developed for insects, diseases, and nematodes attacking soybeans in Alabama. AUSIMM was validated in two year's of validation studies with more than 11 experiments. Version 2.1 of AUSIMM was released to interested growers in 1989. Very few growers have requested a copy of the program. Why was this innovative technology not embraced by soybean growers in Alabama? Implementation was impeded by a severe lack of funds, technical complexity of the product being implemented, inadequate collaboration with extension specialists, institutional inertia towards research /extension cooperation, inability to demonstrate great increases in net profits, and the requirement of a computer for growers to use the software. Future efforts at implementing innovative IPM systems should involve extension specialists and social scientists at the outset, and should be directed towards crops with net profit margins that are not shrinking over time.

INTRODUCTION

Soybeans are attacked by a variety of insects, weeds, diseases, and nematodes in the Southeastern U.S. [5, 14], so management of these pests must emphasize complexes rather than individual species. Much research has focussed on multi-tactic approaches and on pest interactions on soybeans, primarily through the support of a regional project for soybean insect IPM. A systems approach to soybean pest management was initiated in the national CIPM project in the 1970's [9]. Research models for soybean plant growth [25] and for pesticide application [12], and for specific insect pest of soybeans [18, 22] were developed as a result of this and other systems research projects. The development of models to aid in decision-making for soybeans was stimulated in the 1980s by declining world market prices and decreasing profits to growers. Advances in computer technology and the development of expert system software for microcomputers also spurred interest in decision aids [16]. In 1985, a project was initiated for the development of a computerized management system for insects, diseases, and nematodes attacking soybeans in the southeastern U.S. The model developed from this project was AUSIMM, for the Auburn University Integrated Management Model (Table 1).

AUSIMM is an insect, disease, and nematode decision aid that assists growers in selecting varieties and in managing pests [3]. The relative productivity of more than 120 cultivars representing four maturity groups is used for selecting appropriate varieties for planting. The program will automatically select varieties that are resistant to cyst and rootknot nematodes, stem canker, and powdery mildew if requested. A midseason disease module in AUSIMM aids growers in managing Septoria brown spot and Anthracnose. It accomplishes this by calculating the economic benefits of fungicide applications based on predicted weather for the next five days. An insect module in the program calculates the profitability of insecticide applications for the guild of defoliators. It includes a soybean plant growth model, temperature-dependent growth and development and feeding for the defoliating insects, and stage-specific insecticide-induced mortality data.

Table 1. History of the development of AUSIMM.

YEAR	ACTIVITY	RESULT
1985	1. AUSIMM Project funded by the Ala. Research Institute. 2. M. Schwartz, modeler, hired. 3. Extension specialists told of project.	P. Backman (Plant Pathologist), R. Rodriguez-Kabana (Nematologist) work together.
1986	1. USDA co-funds AUSIMM Project. 2. Ala. Research Institute funds second year.	Postdoctoral Fellow hired to oversee field studies.
1987	1. Ala. Research Institute Funds third year. 2. Field trials at 11 locations in Ala. conducted.	AUSIMM version 1.0 developed.
1988	1. Field trials continue. 2. Extension specialists formally involved. 3. Proposal submitted twice to Southern Region LISA program. Both rejected-major criticism not 'whole farm' system. 4. Ala. Coop. Extn. Serv. provides seed monies for implementation	AUSIMM revised based on field data, version 2.0 developed. Extension specialists interested in project, but need time and money to fully participate.
1989	1. AUSIMM brochure sent to ca. 5,600 growers. Less than 20 requested the program. 2. Postdoctoral Fellow obtains Assist. Professor position.	Growers aware of AUSIMM. Implementation studies loses its coordinator due to loss of Postdoctoral Fellow.
1990	1. No funding opportunities for implementation.	Interest in AUSIMM wanes.

The insect module in AUSIMM was tested in 11 field experiments done in 1987 [8]. AUSIMM accurately predicted leaf consumption over all experiments. This finding was representative of all locations, varieties, and growing conditions used in the validations conditions. The nematode and midseason disease modules have been validated in a similar manner.

The code for AUSIMM was rewritten in structured PASCAL by C. McGraw of Optimization technologies, and version 1.0 of AUSIMM was released to extension specialists in 1988 for their evaluation. Versions 2.0 and 2.1 of AUSIMM were produced in spring 1989, and a brochure describing the program [11] was sent to ca. 5,600 growers in Alabama at this time. AUSIMM and its user's guide were offered to growers free of charge. As of May 1991, fewer than 20 growers have requested the program.

WHY WAS AUSIMM NOT READILY ACCEPTED BY GROWERS?

Why was this innovative technology not embraced by soybean growers in Alabama? The reasons for this lie in three areas: development, implementation, and acceptance of the program.

DEVELOPMENT OF THE PROGRAM: AUSIMM was developed in response to declines in grower net profits in the early 1980s. Three research scientists developed the program, with the assistance of one research associate, two postdoctoral fellows, and one computer consultant. No extension specialists or social scientists were *directly* involved in the development of AUSIMM. In retrospect, this was a mistake. Wearing [23] noted that close communication and coordination of all groups involved in IPM implementation has been cited by many authors as critical for successful implementation of an IPM system. AUSIMM represented a radical change in the way IPM decisions can be made for insects, diseases, and nematodes attacking soybeans. It was as much of a change for extension specialists as it was for growers. Wearing [23] described IPM as ". . . a complex innovation requiring intensive education of users to ensure effective implementation". We did not do this. Direct involvement of extension specialists in the development of the program would have helped, but even this would probably have not increased the acceptance rate by growers. Research and extension positions are traditionally separate at many land-grant institutions in the U.S., with limited contact. For example, a regional project on IPM of insects attacking soybeans in the southern U.S. has been meeting annually since at least 1980, but extension specialists rarely attend. This is not their fault, because most of them are not even notified about the meeting. Models such as AUSIMM and the development of new technology for pest management are discussed in these meetings for several years before they are ready for implementation, so increased participation by extension specialists is needed. However, out-of-state travel for extension specialists is often severely limited due to budgetary constraints. The net result is a lack of awareness on both research and extension scientists on what each other is doing.

This lack of awareness extends to the administration as well. The southern association of experiment station directors and the southern cooperative extension service director met together *for the first time* in 1990. This is a step in the right direction, and it should promote an improved level of cooperation between these two groups. This scenario is common; Wearing [23] said that the need to improve collaboration within IPM programs was cited as a major impediment to progress in a his survey of 104 IPM specialists in 14 countries.

No social scientists were involved in the development of AUSIMM. Many IPM systems have been developed without the aid of social scientists. What are the consequences of not involving social scientists in the development process? A slow rate of adoption of IPM programs was noted as early as 1965 [21], and this concern has been voiced by numerous other scientists [e.g.; 10, 13, 24]. The value of social scientists in the development of IPM programs has been documented [7], as well as the very negative effects of not using them [4]. We did not involve social scientists until we attempted to implement the system and wanted to gauge grower acceptance. Surveying grower/extension specialist acceptance is important, but we should have determined what the growers/extension specialists wanted and were willing to accept. Had we done so, we would have developed a simpler IPM system that did not require routine use of computers.

We did some things right in the development process. The problems of software incompatibility and poorly written software were avoided. The software was developed for IBM PC and compatible microcomputers, which were (and still are) the most common machines available in the 1980s. We hired a professional programmer to improve the look and operation of the program so that it would operate in a similar manner to word processing and spreadsheet software available to growers. Compiled computer code for AUSIMM was distributed instead of interpreted code so that it would be very difficult to alter the program. Finally, we demonstrated AUSIMM to growers at field days to increase interest in the program.

IMPLEMENTATION OF THE PROGRAM: Funds from the Alabama Research Institute and the USDA enabled us to develop and validate AUSIMM. The funds did not provide for implementation. The process of implementation must necessarily involve both research and extension scientists, where extension specialists gradually take over responsibility for the new technology. It is analogous to one runner handing off a baton to the next in a relay race. Both runners must hold the baton for a brief time to ensure that the hand-off is successful. Our hand-off was not. Why? We were not able to overcome the financial and technical obstacles to successful implementation.

1. Financial Obstacles: Frisbie [6] cited the funding shortage for technology transfer as a one of the most critical issues facing IPM in the 1990s; this was true in our case. We proposed to the southern region LISA program to implement AUSIMM. Our proposal involved three research scientists, one social scientist, and three extension specialists. We proposed to work with ca. 20 growers, to survey their acceptance of AUSIMM, and to improve AUSIMM based on their suggestions. The funding of this grant would have gone a long way towards implementing our IPM system because it would have established a core of users who could have spread the system. However, our proposal did not meet the funding criteria for the LISA project. We found no other agency or grants program where we could apply for money. I conclude that the USDA pays for the development of an IPM system, but not for implementation. The Cooperative Extension Service in the U.S. receives funds for implementation, but these are diffuse and are sometimes used for salaries of current extension specialists. These monies are not available to fund the supplies, travel, and equipment needed to implement an innovative IPM system. A chasm exists between research and extension, with many IPM systems falling in the chasm. A new grants program is needed in the U.S., where implementation of IPM systems is a primary goal. Lack of funds has been cited by many authors as a major obstacle to implementation [e.g.; 7, 17, 23].

2. Technical Obstacles: AUSIMM is technically complex, because it calculates thresholds based on crop value and expected yield. This is the major strength and the major weakness of AUSIMM. Thresholds should incorporate changing crop value and yield [15], but their incorporation makes calculation of a threshold a complex process. The increased complexity is unfortunately coupled with a lack of readily observable advantages. For example, our validation studies showed that AUSIMM makes decisions as well as the best soybean growers, but it is not noticeably better than them. The lack of obvious benefits and increased complexity has slowed the acceptance of many IPM programs [19].

ACCEPTANCE OF THE PROGRAM: Few growers even tried to use AUSIMM. There are more than 20 growers in Alabama with microcomputers, so one should ask 'Why did not more growers with microcomputers at least try AUSIMM?' The answers lie in an understanding of a grower's incentives to use new technology. Innovative IPM systems have several competitive disadvantages over traditional, chemically-based methods of managing pests. Wearing [23] noted that there is a high ratio of chemical advisors, distributors, and salesman to IPM personnel available to growers, that farmers have confidence and much experience with chemicals, that chemical companies have good marketing skills, and that chemicals are simple and easy to use. AUSIMM requires regular scouting and the collection of some data on percent defoliation. These data are simple to collect, but it does require additional work in the field. This substitution of labor (scouting) for capital investments (chemicals) is counter to modern trends in the business world [23].

A major problem with AUSIMM was that it did not greatly increase grower net profits. Our validation studies with AUSIMM showed us that some insecticide applications are applied unnecessarily, but that most are economically justified. AUSIMM was best at determining if a mixed population of defoliating insects could cause economic damage. The dominant factor in the lack of greatly improved profits from use of AUSIMM is the small profit margin associated with soybeans in the southeastern U.S. Soybeans are a more costly crop to grow in the southeastern U.S. than in the midwestern U.S. because there is more nematode, disease, and insect pressure, and often more drought stress on soybeans in the South. Low worldwide prices for soybeans in the 1980 coupled with these higher input costs have decreased net profit for soybean growers. Yields of soybeans declined over time from 1970 to 1981 in several southern states [1]. The low market prices, decreasing profit margin, and increasing production costs have greatly affected the number of hectares of soybeans grown in Alabama (Fig. 1).

Reduction of Hectarage in Alabama, USA
1981 to 1991

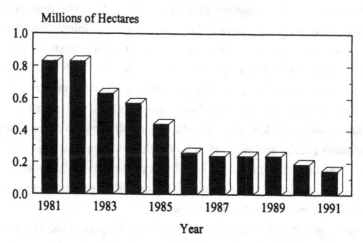

Figure 1. Decrease in soybean hectarage in Alabama from 1981 to 1991 [2, and personal communication]

Soybeans have changed in Alabama from being the dominant crop in the early 1980s to a minor crop in the early 1990s [20]. It is difficult to implement innovative technologies in a crop that growers are abandoning.

A second factor influencing acceptance of AUSIMM is that AUSIMM required the use a computer to make decisions. Use of computers for modeling and for making management decisions in entomology has greatly increased in the past decade, but grower acceptance of computers in IPM systems is low [23]. The current interest in expert systems should accelerate the trend towards computerized decision-making. The use of a computer program enables growers to make decisions using economic algorithms for calculating thresholds, with these algorithms being transparent to the user. The negative effects of using a computer program are that programs require updating after implementation as new chemicals are registered and older ones are lost; new, resistant varieties are released, and more information is made available on alternative methods of managing pests. Who updates the program, the researcher or the extension specialist? More importantly, is repackaging current information into computer software research or extension work?

RECOMMENDATIONS FOR THE FUTURE

There is an inherent risk in the development of innovative IPM systems- do we try and fail or do we fail to try? The latter path is the easiest, because it results in few problems in the short term. Extension specialists do not have to learn innovative technologies because none are being developed. Administrators do not have to concern themselves with communication and collaboration because the programs underway require no changes in the areas. Researchers will not become frustrated by spending time and money on a product the growers do not use.

The long term problems of this approach are obvious. Extension and research evolve into separate and non-collaborative units, administrators interact little with each other, and research scientists do low risk, low reward research. We must not fail to at least try.

Several recommendations are in order for improving the implementation process. First, IPM practitioners should work with social scientists to query growers as to their wants and needs, and involve extension specialists in the research from it's initial phases. This will enable them to develop IPM systems that growers will use and that extension specialists will trust. Second, a competitive grants program should be initiated for the implementation of innovative technologies. This will provide badly need funds and release time for extension specialists to work with researchers and implement the new technology. Finally, IPM practitioners should begin the process of educating growers about computers and computerized decision-making. Several Cooperative Extension Services in the U.S. have listed regaining agricultural profitability and developing human resources in their states as a priority for the future. This can be accomplished by either a massive effort to educate farmers or to provide them with tools that aid in making sound management decisions. Zalom [26] concluded that 'pest management professionals will need computers and appropriate applications to a greater extent in the future as American agriculture moves away from reliance on pesticides as its primary control tactic.' We must prepare growers for the new technology, the innovative IPM systems, that are on the horizon. We should not fail to try, again.

REFERENCES

1. Anonymous, Acreage yield production, 1983, American Soybean Association, St. Louis, MO, 90 pp.

2. Anonymous, Alabama agricultural statistics, Alabama agricultural statistics service, Montgomery, Ala., 1990, 60 pp.

3. Backman, P. A., Mack, T. P., Rodriguez-Kabana, R., and Herbert, D. A., A computerized integrated pest management model (AUSIMM) for soybean grown in the southeastern United States, Proceedings of the World Soybean Research Conference IV, 5 to 9 March, 1989, Buenos Aires, Argentina, Pp. 1494-1499.

4. Bentley, J. W. and Andrews, K. L., Pests, peasants, and publications: anthropological and entomological views of an integrated pest management program for small-scale Honduran farmers, Human Organization 1991, 50, 113-124.

5. Coble, H. D., Development and implementation of economic thresholds for soybeans. In Integrated Pest Management on Major Agricultural Systems, eds. R. E. Frisbie and P. L. Adkisson, Texas Agric. Expt. Stn. MP-1616, 1986, Pp. 295-307.

6. Frisbie, R. E., Critical issues facing IPM technology transfer, Proceedings of the National IPM Workshop, April, 1989, Las Vegas, Nev., Pp. 157-162.

7. Goodell, G., Challenges to international pest management research and extension in the third world: Do we really want IPM to work? Bulletin of the Entomol. Soc. Am., 1984, 30(3), 18-26.

8. Herbert, D. A., Mack, T. P., Backman, P. A., and Rodriguez-Kabana, R., Validation of a model for estimating leaf-feeding by insects in soybean, 1992, Crop Protection, in press.

9. Huffaker, C. B., New Technology of Pest Control, Wiley-Interscience, N. Y., N.Y., 1980, 500 pp.

10. Klein, K. K., Economic principles in entomology, Can. Entomol. 1985, 117, 885-891.

11. Mack, T. P., Backman, P. A., and Rodriguez-Kabana, R., AUSIMM, the Auburn University Soybean Integrated Management Model, Ala. Agric. Expt. Stn, Auburn, Ala, 1989, pp.

12. Marsolan, N. F., and Rudd, W. G., Modeling and optimal control of insect pest populations, Math. Biosci. 1976, 30, 231-244.

13. Newsom, L. D., The next rung up the integrated pest management ladder, Bull. Entomol. Soc. Am. 1980, 26, 369-374.

14. Newsom, L. D., Kogan, M., Miner, F. D., Rabb, R. L., Turnipseed, S. G., and Whitcomb, W. H., General accomplishments toward better pest control in soybean, 1980, In New Technology of Pest Control, ed. C. B. Huffaker, Wiley-Interscience, N. Y., N.Y., Pp. 51-98.

15. Pedigo, L. P., Hutchins, S. H., and Higley, L. G., Economic injury levels in theory and practise, Ann. Rev. Entomol. 1986, 31, 341-368.

16. Plant, R. E., and Stone, N. D., Knowledge-based systems in agriculture, McGraw-Hill, Inc., N.Y., N. Y., 1991, 364 pp.

17. Poe, S. L., An overview of integrated pest management, HortScience 1981, 16, 501-506.

18. Stinner, R. E., Bradley, J. R., Jr., and Van Duyn, J. W., An algorithm for temperature-dependent growth rate simulation, Can. Entomol. 1974, 106, 519-524.

19. Stoner, K. A., Sawyer, A. J., and Shelton, A. M., Constraints to the implementation of IPM programs in the U.S.A., a course outline, Agric. Ecosystems Environ., 1986,17, 253-268.

20. Traxler, G. J. and Molnar, J. J., Agriculture sector resilient during 1980's, Highlights of Agric. Res. 1991,38(4), 15.

21. Van den Bosch, R., Practical application of the integrated control concept in California, Proceedings of the 12th International Congress of Entomology, London, England, 1965,Pp. 595-597.

22. Waddill, V. H., Shepard, B. M., Lambert, J. R., Carner, J. R., and Baker, D. N., A computer simulation model for populations of Mexican been beetles on soybeans. South Carolina Agric. Expt. Stn. Bull. 1976, 590, Clemson, S. C.

23. Wearing, C. H., Evaluating the IPM process, Ann. Rev. Entomol. 1988, 33, 17-38.

24. Whalon, M. E., and Croft, B. A., Apple IPM implementation in North America, Ann. Rev. Entomol. 1984, 29, 435-470.

25. Wilkerson, G. G., Jones, J. W., Boote, K. J., Ingram, I. I., and Mishoe, J. W., Modeling soybean growth for crop management, Trans. ASAE, 1983, 26, 63-73.

26. Zalom, F., Expectations for computer decision aids in IPM by the turn of the century, 1989, Proceedings of the National IPM Workshop, April, 1989, Las Vegas, Nev., Pp. 189-192.

CONSERVATION TILLAGE, RELAY INTERCROPPING AND ALTERNATIVE CROPPING SYSTEMS: THEIR POTENTIAL FOR PREVENTIVE ARTHROPOD MANAGEMENT

RONALD B. HAMMOND
Department of Entomology
Ohio Agricultural Research & Development Center
The Ohio State University
Wooster, OH 44691 USA

ABSTRACT

Soybean growers employ various management practices which serve as preventive tactics against a variety of soybean arthropod pests. Conservation tillage practices and the use of cover crops will most often be useful as preventive practices against soil pests due to their direct impact on arthropods. Relay intercropping can have dramatic impacts on certain foliar pest species and should be further examined for its effect on other pests. As analyses which focus on the overall farming system lead to the adoption of alternative cropping practices, such practices might prove useful in preventing losses due to other pests, including weeds and plant pathogens. They may also have beneficial impacts on other parameters related to the cropping system, such as nutrient availability and grower profitability.

INTRODUCTION

Because of numerous concerns related to using insecticides, researchers and growers alike are examining how various non-insecticidal management practices can be used as preventive tactics to lower the carrying capacity of a pest's environment or to increase the tolerance of crops to pest injury. Grower practices with potential use include change in planting dates, use of different tillage procedures, utilizing crop rotations, better sanitation, etc. (see the chapter by Pedigo on "Integrating Preventive and Therapeutic Tactics in Soybean Insect Management" for a general discussion of preventive pest management concepts).

Relevant to the use of such practices for preventive purposes is a more complete understanding of the complex interactions between pests, cropping systems, and their environment. Earlier studies on how various practices would effect soybean pests were usually carried out as component research, examining the effect that an individual practice had on pest insects and crop damage [1,2], rather than how it fits into the overall cropping system.

These studies, although not using an interdisciplinary approach, did provide information pertinent to the grower. They often resulted in recognition that a cropping practice increased the potential for specific pests to reach economically damaging levels. In a talk to no-tillage growers for example, it would then be stated that the potential for damaging levels of slugs was high if other environmental conditions (moisture and temperature) were favourable for slugs. However, seldom was the use of a cropping practice suggested specifically for the prevention of an insect problem on soybean; insect problems were but one management concern for the grower. More often the likelihood of a pest problem was anticipated, scouting recommended, and if a problem developed, therapeutic control measures were taken (i.e., use of an insecticide).

Having further studied the effects of various management practices on soybean arthropods, and having determined the life cycles, behavior and ecology of the arthropods, researchers are beginning to recommend some of these cropping practices as preventive measures against specific pest species. The overall goal will be to understand how these practices fit into the whole farming process, rather than examining them as individual components. Although there are several cropping practices that could be considered preventive management tactics, this paper will focus on conservation tillage, cover crops, and relay intercropping, with a general discussion of alternative cropping practices where the focus is on the integration of all farming operations into a whole-farm system.

Conservation Tillage

Conservation tillage practices have been gaining grower acceptance over the past 10-15 years. In general, these practices are those that allow for greater than 30% ground cover and include both reduced tillage and no-tillage practices. Most often the cover is provided by the previous crop residue that is allowed to remain on the soil surface between growing seasons. Benefits of conservation tillage practices include, but are not limited to, reduced soil erosion and greater soil moisture conservation. The various relationships between these practices and both foliar and soil pests have been examined in the past, with more emphasis on the soil arthropod community. In many instances, growers using conservation tillage practices continue to routinely employ therapeutic tactics for pest control with or without prior knowledge of pest occurrence. However, there is the possibility that tillage practices can be used preventively if growers become better educated about the practices' impact on arthropod pests.

Foliar Arthropods: The influence of preplant tillage practices on foliar arthropods on soybean is often thought to be an indirect effect since many above-ground pests immigrate into a field following plant emergence. In one of the earliest reported studies, foliar arthropod guilds from no-tillage soybean plots showed a higher diversity compared with those from conventional tillage plots. This was thought to be due to the greater structural diversity provided by weeds and litter [3]. The presence of weeds in conservation tillage fields is often a result of inadequate weed control and can have a dramatic impact on populations of certain insect pests. Indeed, this was the reason suggested for the higher numbers of Japanese beetle, *Popillia japonica*, in reduced and no-tillage plots, but the lower numbers of potato leafhopper, *Empoasca fabae*, in Ohio [4]. Although other studies have reported on foliar insects in tillage comparisons, reasons for the often conflicting results were not completely known. Contrary results have been reported for the green cloverworm, *Plathypena scabra*, with larger populations in conventional than no-tillage in Kentucky [5], but higher levels in no-tillage soybean in a study in Louisiana [6]. The findings in Kentucky were thought due to greater levels of egg and larval mortality in no-tillage soybean, although no data or evidence were presented indicating higher levels of predators or parasites in those systems. This contrasts to a study in Florida that reported higher populations in disk tillage systems (compared with no-tillage systems) of bigeyed bugs, *Geocoris* spp., and damsel bugs, *Nabis* and *Reduviolus* spp., common predators found in soybean [7].

The presence of crop residue and weeds increases the diversity of the plant community, providing alternative shelter, food and oviposition sites for both pest and beneficial organisms. Depending upon the interactions among the various trophic levels, there can be an increase in the population of a pest species, a reduction of the pest population due to greater activity of natural enemies and/or presence of a non-host, or no effect at all.

Because most soybean foliar arthropods immigrate into fields following plant emergence, many of them relatively late in the growing season, the potential for conservation tillage practices to have a <u>direct</u> impact on them is slight. Because there has not been evidence of consistent changes in pest status with foliar arthropods due to changes in tillage practices in most cases, it would be unwarranted to recommend the use of a such cropping practices for prevention of foliar pests. However, as further studies delineate the mechanisms involved in these complex interactions, avenues in which to use tillage against foliar insects in an indirect manner might evolve.

<u>Soil Arthropods</u>: The situation is different for soil arthropod pests because tillage practices can have direct and often immediate effects on the pest status of numerous species. In midwestern states of the U.S., both the gray garden slug, *Derocerus reticulatum*, and seedcorn maggot, *Delia platura*, can reach economic levels due to the use or non-use of certain tillage practices.

We have seen an increase in damaging populations of slugs because of the greater residue cover in conservation tillage systems which provides for a more favorable habitat for the slugs [8]. Slug problems were virtually nonexistent prior to the adoption of conservation tillage practices for soybean production. In early work, slug densities were found to be highest in no-tillage plots, followed by reduced tillage, with few in conventional (plowed) areas [9] (Table 1). Slug populations commonly reach economic levels in fields with heavy residue cover during May and June when weather conditions are favourable (viz., wet and cool conditions). The obvious suggestion to no-tillage growers who have experienced losses from slugs would be to forgo no-tillage practices and employ farming operations that reduce the amount of residue cover. Indeed, tillage has been the only means of control, serving as a preventive tactic, since the removal of labels of all molluscicides for use on midwestern soybean.

While using tillage reduces the potential for slug damage, it increases the

TABLE 1
Numbers of slugs per 0.1 m^2 trap for various tillage systems in 1984

	May		June		
Tillage System	31	7	14	21	28
No-tillage	5.2	3.8	7.4	4.4	1.8
Reduced Tillage	2.2	1.1	1.8	1.5	1.4
Conventional Tillage	0.2	0.1	0.2	0.3	0.9

Taken in part from Hammond and Stinner [9]

likelihood of larger populations of seedcorn maggot under the proper conditions. In studies to determine the effect of tillage on seedcorn maggot populations, greater numbers of adults emerged from reduced and conventional tilled soil than from no-tillage fields [9,10]. However, this increase in maggot numbers will not usually develop into an economic loss unless a green cover crop is incorporated into the soil during the tillage operations.

Cover Crops

A cover crop is grown to fill gaps in either time or space when the lack of such would leave the ground bare. Most cover crops are grown during the winter months in northern latitudes. Not only do cover crops provide a residue that aids in the conservation of water and reduction of soil erosion, but they also aid in the amelioration of the soil structure, improvement of soil fertility, and in the suppression of weeds (10). Cover crops can be killed prior to planting the cash crop or used as "living mulches" that remain wholly or partly alive during the growing season (this discussion will focus only on the former).

The method in which a cover crop is eliminated can determine whether certain pest populations increase. As introduced in the previous section, of concern in the midwest are potential problems with seedcorn maggots. Greatest numbers of maggots are found when a green, living cover crop has been tilled into the soil [10] (Table 2). Adult flies are believed to be attracted to the decaying organic matter provided by the cover crop for both feeding and oviposition; they are not attracted to the soybean. If the cover is killed with a herbicide application and left on the soil surface, maggot problems are unlikely. However, if the cover is soil incorporated prior to planting, tactics should be used to prevent seedcorn maggot problems. One way to safeguard the soybean is to use a seed protectant or apply a soil insecticide. Although "preventive" in nature, these methods do not fit our general definition of preventive, non-pesticidal tactics.

Another possible tactic which uses the knowledge of the life cycle of the maggot to lower the damage potential would be to delay planting following tillage until the majority of the seedcorn maggot population are pupae, a time when feeding would be reduced [12]. Adults emerge after approximately 400 heat units have accumulated

TABLE 2

Numbers of adult seedcorn maggots per trap (1.0 by 0.3 m) averaged over a 3 year period for the interaction between tillage and type of cover crop

Tillage	Cover Crop - Residue Type				
	Bare	Soybean	Corn	Alfalfa	Rye
No-Tillage	1.2	1.9	1.2	1.7	1.7
Plowed	2.5	8.8	4.1	24.2	14.3

Taken in part from Hammond [10]; soybean and corn refer to the type of residue

following oviposition [13], with the insect entering the pupal stage after roughly 234 heat units. Waiting to plant soybean until 250 heat units have accumulated following the incorporation of a cover crop would decrease the potential for soybean damage since most of the insects would be in the pupal stage (Fig. 1). By the time adults emerge, mate, and oviposit, the soybean should have emerged. The tactic of delaying planting is another common approach to insect pest management which, by itself, is considered a preventive tactic for numerous other insect pests (see Pedigo's chapter for its use as a tactic for control of the bean leaf beetle, *Cerotoma trifurcata*).

Because of our better understanding of when tillage practices increase the likelihood of damage from soil pests compared with those occurring on the foliage, we can make recommendations that encourage or discourage use of tillage as a preventive tactic. This will often be dictated by field history (in the case with slugs) or presence and subsequent soil incorporation of cover crops (in the case of seedcorn maggot). Indeed, we are currently recommending the use of tillage as a preventive tactic against slugs in Ohio during years when conditions are favourable for slugs [14].

Relay Intercropping

Relay intercropping, planting soybean in the spring into winter wheat, is an alternative to double cropping in northern U.S. states. Rather than planting soybean following wheat removal in early July, soybean is planted in gaps in the wheat planting (skipped wheat rows) during May. At wheat harvest, the soybean is actively growing, allowing for a longer season compared with double-cropped soybean. Benefits of relay intercropping include better moisture available for seed germination, a longer growing season for soybean allowing for maturity before an early frost, and increased soil conservation provided by the wheat cover during the winter.

Because the two crops coexist during May and June, the possibility exists that their proximity might effect foliar arthropod pests that exploit only one of them. This is suggested by work in alfalfa that indicated that the various grassy weeds [15,16] and oats [17] had population-reducing effects on the potato leafhopper, a serious alfalfa pest and an infrequent pest of soybean in the U.S. Therefore, studies were initiated to study potato leafhopper populations on soybean relay intercropped with winter wheat.

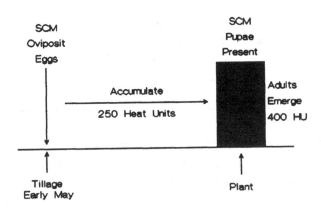

Figure 1. Using heat unit accumulations for seedcorn maggot management.

Pubescent and glabrous soybean were planted into winter wheat in early May during the time of normal soybean planting (i.e., non-double-cropped monoculture). Different pubescent types were used to provide a strong test of the hypothesis that relay intercropping could negatively effect the leafhopper. Glabrous soybean is extremely susceptible to potato leafhopper damage and is not grown commercially in the Midwest for that reason. Potato leafhoppers were sampled throughout the summer with a D-Vac sampling apparatus. Whereas populations were moderate on pubescent soybean and extremely heavy on glabrous soybean when the soybean lines were grown in monoculture, the insect populations were virtually undetectable on either line in relay intercropped systems prior to wheat harvest (Table 3) [18]. Following the removal of the wheat, potato leafhoppers reached relatively moderate levels on glabrous soybean in both cropping systems. The presence of wheat in a relay intercropping system protected the soybean from potato leafhopper injury. The lack of soybean injury to glabrous soybean in relay intercropped plots compared with the near destruction of monoculture glabrous soybean supports this conclusion [17].

Having determined that a full complement of wheat in these intercropped systems dramatically reduces leafhopper populations, we are now examining whether reduced levels of wheat can accomplish this decrease in insect populations. We are using wheat at a full, 1/2, and approximately 1/4 rate (14, 7, and 3 wheat rows per 3 soybean rows, respectively). Similar procedures are being used, except that only a pubescent soybean is being grown. Sampling techniques are comparable, although additional sampling is being done to study emigration and immigration of the potato leafhopper among the plots.

In results to date, leafhopper densities appear to be inversely proportional to the ratio of wheat rows to soybean rows (Miklasiewicz, unpublished data). These

TABLE 3
Numbers of adult potato leafhoppers per sample
from four cropping systems

Cropping System	June			July			August
	7	18	24	15	22	29	5
Glab, RI	366.3	185.5	300.3	451.3	85.8	66.9	67.9
Pub, RI	68.6	36.3	40.1	43.6	24.7	2.5	3.0
Glab, Mono	0.0	3.0	0.2	0.3	55.0	60.8	67.3
Pub, Mono	0.0	0.4	0.0	0.0	6.1	15.6	8.6

Glab = glabrous; Pub = pubescent; RI = relay intercropping; Mono = monoculture

differences are believed due to differential emigration. Indications are that even relatively small amounts of wheat (1 wheat row per soybean row, 3:3 ratio) have a substantial inhibitory effect on potato leafhopper density. A possible explanation for this reduction of leafhoppers is altered dispersal behavior of the insect caused by the lower light intensity within intercropped systems. Research on alfalfa has indicated that shading increased dispersal from alfalfa plots as does the presence of oats in a grass/alfalfa mixture (W.O. Lamp, personal communication). Similar mechanisms might be occurring in both systems.

Under prevailing conditions, only pubescent soybean is grown in the midwest U.S. because glabrous lines are too susceptible to the potato leafhopper and severe injury to plants occurs when they are grown. Nonetheless, the use of relay intercropping can be considered a preventive tactic if used in a situation where potato leafhopper might cause serious damage (albeit not that common an occurrence). However, the possibility exists that if relay intercropping becomes more widely utilized by growers, breeders might initiate programs to select for lines specific to relay intercropping. If so, their selection criteria might be altered because pubescent type would not be as important. Theoretically, a soybean line with less than normal pubescence could be used in conjunction with wheat in a relay intercropping system or when wheat is planted in lesser amounts. In either situation, the planting of wheat would function as a preventive tactic against the potato leafhopper.

There is a need to study the effect of relay intercropping on other soybean foliar arthropods that immigrate into the field when wheat is present. Because serious soybean pests in the Midwest (viz. the bean leaf beetle, *Cerotoma trifurcata*, and Mexican bean beetle, *Epilachna varivestis*) enter newly emerged soybean fields in late May - early June, the possibility exists that the presence of wheat might inhibit their immigration into the field. Because both are bivoltine generation pests, the reduction of the early season population might inhibit development of a large second generation (which causes most of the economic damage). Because relay intercropping holds promise as a preventive tactic in the soybean system, its' value as a preventive tactic in other cropping systems such as field or snap beans should be examined.

Alternative Cropping Systems

A recent development in agriculture has been an increased interest in alternative practices for the production of crops which will take advantage of natural processes and beneficial on-farm biological interactions, reduce the use of off-farm inputs, and improve the efficiency of operations. [19]. Another name for this concept, although not totally synonymous, is sustainable agriculture. Numerous practices are compatible with this concept of alternative agriculture, including conservation tillage for erosion control, relay intercropping and companion cropping to increase plant diversity within a field, crop rotation to provide alternative nutrient sources, use of organic residues and cover crops for erosion control, and leguminous green manure crops as a nitrogen source.

Much of the previous research in agriculture has been focused on individual farming practices, rather than on the development of an agricultural system, perhaps the most unique aspect of the alternative agricultural approach. Indeed, this individual focus is exemplified by the earlier examples in which the various cropping practices were studied specifically for their impact on arthropod pests. This is contrasted with sustainable agriculture where the focus is on the interaction and integration of all farm operations, and the whole farm system is evaluated within a matrix of resource management, productivity, environmental quality, and profitability criteria [11, 19]. This approach, therefore, will require an interdisciplinary team-approach much more extensively than previous work.

As part of an overall farm management system, the discussions on conservation tillage and relay intercropping are germane to a discussion on alternative agriculture. Under appropriate conditions, these practices can serve as preventive tactics in an overall agricultural system, and could reduce the reliance on therapeutic tactics. However, to better fit an alternative agriculture concept we need to determine their overall impact and consequences on all cropping and farming parameters, e.g., impact on weeds, pathogens, nutrient availability, profitability, etc. This will require a team-approach to examine the interactive impacts rather than only individual examinations of each parameter. However, past research that has taken the more individualistic approach will provide a basis for making decisions regarding grower practice's that can be employed as preventive tactics until research done via the systems approach becomes the norm.

REFERENCES

1. Hammond, R.B., and Funderburk, J.E., Influence of tillage practices on soil-insect population dynamics in soybeans, World Soy. Res. Conf. - III Proceedings, R. Shibles, Ed., Westview Press, Boulder, CO., 1985, pp. 659-666.

2. Hammond, R.B., Pest management in reduce tillage soybean cropping systems, In Arthropods in Conservation Tillage Systems, G.J. House and B. R. Stinner, Eds., Misc. Pub. Entomol Soc. Amer., 1987, 65, 19-28.

3. House, G.J. and Stinner, B.R., Arthropods in agroecosystems: community composition, seasonal dynamics, and ecosystem interactions, Environ. Management, 1983, 7, 23-28.

4. Hammond, R.B. and Stinner, B.R., Soybean foliage insects in conservation tillage: effects of tillage, previous cropping history, and soil insecticide application, Environ. Entomol., 1987, 16, 524-531.

5. Sloderbeck, P.E. and Yeargan, K.V., Green cloverworm (Lepidoptera: Noctuidae) populations in conventional and double-crop, no-till soybeans, J. Econ. Entomol., 1983, 76, 785-791.

6. Troxclair, N.N., Jr., and Boethel, D.J., Influence of tillage practices and row spacing on soybean insect populations in Louisiana, J. Econ. Entomol., 1984, 77, 1571-1579.

7. Funderburk, J.E., Wright, D.L. and Teare, I.D., Preplant tillage effects on population dynamics of soybean insect predators, Crop Sci., 1988, 28, 973-977.

8. Hammond, R.B., Slugs as a new pest of soybeans, J. Kansas Entomol Soc. 1985, 58, 364-366.

9. Hammond, R.B. and Stinner, B.R., Seedcorn maggot and slugs in conservation tillage systems in Ohio, J. Econ. Entomol., 1987, 80, 680-684.

10. Hammond, R.B., Influence of cover crops and tillage on seedcorn maggot (Diptera: Anthomyiidae) population in soybeans, Environ. Entomol., 1990, 19, 510-514.

11. Lal, R., Regneir, E., Eckert, D., Edwards, B., and Hammond, R.B., Expectations of cover crops for sustainable agriculture, In Cover Crops for Clean Water, Hargrove, ed., Soil Conservation Soc., 1991, pp. 1-10.

12. Hammond, R.B., Soybean insect pest management and conservation tillage: options for the grower, In Proceedings Southern Region No-till Conference, July 13-14, I.D. Teare, ed., Spec. Bull. 89-1, Inst. Food Agric. Sci., Univ. FL, 1989, pp. 3-5.

13. Hammond, R.B., Effects of rye cover crop management on seedcorn maggot (Diptera: Anthomyiidae) populations in soybeans, Environ. Entomol., 1984, 13, 1302-1305.

14. Wilson, H.R., Hammond, R., Flessel, J., and Stinner, B., Pest Control, In Crop Production Alternatives, D. Eckert, Coordinator, OCES Bull. 821, 1990, The Ohio State University.

15. Lamp, W.O., Barney, R.J., Armbrust E.J., and Kapusta, G., Selective weed control in spring-planted alfalfa: effect on leafhoppers and planthoppers (Homoptera: Auchenorrhyncha), with emphasis on potato leafhopper, Environ. Entomol., 1984, 13, 207-213.

16. Oloumi-Sadeghi, H., Zavaleta, L.R., Lamp, W.O., Armbrust, E.J., and Kapusta, G., Interactions of the potato leafhopper (Homoptera: Cicadellidae) with weeds in an alfalfa ecosystem, Environ. Entomol., 1984, **16**, 1175-1180.

17. Lamp, W.O., Reduced *Empoasca fabae* (Homoptera: Cicadellidae) density in oat-alfalfa intercrop systems, Environ. Entomol., 1991, **20**, 118-126.

18. Hammond, R.B. and Jeffers, D.L., Potato leafhopper (Homoptera: Cicadellidae) populations on soybean relay intercropped in winter wheat, Environ. Entomol., 1990, **19**, 1810-1819.

19. Alternative Agriculture, Committee on the Role of Alternative Farming Methods in Modern Production Agriculture, Board Agriculture, National Research Council, National Academy Sciences, 1989, 448 pp.

NEW UNDERSTANDINGS OF SOYBEAN DEFOLIATION AND THEIR IMPLICATION FOR PEST MANAGEMENT

LEON G. HIGLEY
Department of Entomology
University of Nebraska-Lincoln
Lincoln, Nebraska 68583-0816, USA

ABSTRACT

Uncertainties regarding the relationships between insect defoliation and yield loss limit existing management programs. Consequently, experiments were conducted over four years in multiple locations to better characterize soybean responses to defoliating insects. Results indicate that soybean yield reductions from insect defoliation primarily result from reduced light interception of defoliated plant canopies. Changes in leaf senescence were associated with defoliation and were identified as a compensatory mechanism. These results offer the potential for more accurate pest management guidelines and new approaches to managing defoliation through the development of defoliation-tolerant soybean varieties.

INTRODUCTION

Defoliation by insect pests is a major stress of soybean. Indeed, defoliators comprise the most abundant and diverse guild of insects that attack soybeans in the U. S. (1). Understanding the relationship between defoliation and plant responses, such as yield, is essential for better pest management. In particular, characterizing the relationship between defoliation and soybean yield loss is essential in determining economic injury levels and thresholds for defoliating insects.

Researchers have long recognized the importance of defoliating pests to soybeans, and over the past 40 years more than 50 research articles have addressed soybean defoliation (2). However, despite this volume of research, long standing questions regarding defoliation on soybeans persist. Often results from different studies do not agree, possibly because of differences in environment or methodology. For example, hail simulation studies, with injury imposed on a single day, clearly do not simulate injury by defoliating insects, which feed over many days (2). But even where actual insects or appropriate simulation techniques are used, reported relationships between defoliation by insects and yield vary greatly (2). Nor are soybean compensatory responses to defoliation well documented. Fundamentally, no clear understanding of the physiological effects of defoliation on soybeans has yet emerged.

The practical consequences of this situation are that calculated economic injury levels (EILs) and economic thresholds (ETs) for soybean pests differ widely among insects and across locations. It seems unlikely that defoliation/soybean response relationships are really as variable as existing thresholds suggest. Instead, these thresholds probably reflect our present uncertainty in how defoliation physiologically impacts soybeans. Consequently, to some degree existing thresholds are suspect. Moreover, the lack of more definitive understandings of how soybeans respond to defoliation impedes the implementation of more advanced decision tools, such as multiple species EILs (3,4). This latter limitation is especially important in many situations where soybean defoliators occur as a complex of species.

Two important points suggest that improved understandings of soybean defoliation are possible. First, in much of the previous research, injury was not precisely quantified and potential confounding factors were not eliminated (usually, because adequate methodologies were not available). Second, most experiments did not address soybean physiology in any quantitative sense; instead, final seed yield was the only soybean response measured. Consequently, by employing more quantitative experimental procedures and focusing on soybean physiological responses, it may be possible to greatly improve our understanding of the impact of insect defoliation on soybean.

In the late 1980's, advances in instrumentation and methodologies for imposing injury provided a solution to many experimental problems associated with defoliation studies. Research on simulation techniques (2) established requirements for simulating insect defoliation to soybean, which avoids problems of confounded treatment effects associated with the use of cages or insecticides to establish insect populations. In particular, this research demonstrated that the temporal duration of injury was a crucial factor in providing realistic plant responses to simulated injury. New instrumentation, such as leaf area meters, line quantum sensors, and portable photosynthesis meters, allowed precise, rapid measurements of important physiological parameters. Thus, by simulating defoliation and quantifying leaf tissue loss, simulated injury could be easily related to insect numbers for a variety of defoliating species. Further, soybean response to this injury could be better quantified through the use of new instrumentation.

Consequently, in 1988 studies on insect defoliation of soybean were initiated by using simulation techniques and focusing on physiological responses of soybean to injury. In 1988 and 1989 researchers in four states (Florida, Iowa, Nebraska, and Ohio) participated in these studies. Subsequently, the project was expanded and workers in an additional four states (Alabama, Arkansas, Illinois, and Indiana) participated in 1990 and 1991. Common experiments were conducted at these various sites to provide variable environmental conditions for characterizing responses.

Often, defoliation studies are designed to establish relationships between insect numbers and yield or percent defoliation and yield. However, 50% defoliation of a large soybean canopy is unlikely to produce the same response as 50% defoliation of a small canopy. Similarly, feeding by a given number of insects is unlikely to have the same impact on a small canopy as on a larger canopy. Consequently, neither insect numbers nor percentage defoliation provide physiologically meaningful measures of injury.

In conducting this study, we were anxious to provide physiological explanations for yield losses arising from defoliation. Crop productivity is a function of the interception of photosynthetically active radiation and the efficiency of use of that radiation (5). Research by Poston et al. (6) and in my laboratory (unpublished) demonstrate that leaf injury by most soybean defoliators does not reduce photosynthetic rates; therefore, the principal physiological impact of defoliation should be on light interception. Indeed, it seems reasonable to hypothesize that the general mechanism of yield reduction by insect defoliation is to reduce light interception of defoliated plant canopies. Aspects of this hypothesis have been proposed relative to plant diseases (7,8), but it has not been tested. One previous study (9) did include light interception effects of defoliation, but this was peripheral to the major direction of the study and cannot be used to test the defoliation-light interception hypothesis.

We tested the defoliation-light interception hypothesis by establishing treatments based on different levels of light interception after defoliation. If correct, we would expect to have consistent relationships between light interception after defoliation and soybean yields across years and locations. Because defoliation was imposed to simulate insect injury and all tissue removal was quantified, it is possible to relate treatments to levels of insect feeding (as is necessary for determining economic injury levels).

Even if the defoliation-light interception hypothesis is correct, it would be naive to think that defoliation does not influence other plant parameters. Therefore, we examined additional responses, such as growth and yield parameters, nitrogen relationships, and compensatory mechanisms. The following pages present some details on our methods, some results relating to the defoliation-light interception hypothesis and compensation, and some implications of these findings.

MATERIALS AND METHODS

Although experiments were conducted at multiple locations in all years, only selected results from Iowa in 1988 and 1989 and from Nebraska in 1990 and 1991 are presented here. Experimental design was a randomized complete block with 4 replications and 5 treatments (1988 and 1989) or 14 treatments (1990 and 1991). Row width was 0.76 m, and each plot was 4 rows x 7 m, with north-south orientation. The soybean cultivar used was 'Elgin 87'.

In 1988 and 1989, treatments consisted of simulated insect defoliation at stage R3 with four levels (defoliation to produce a leaf area index [LAI] of approximately 4.0, 3.0, 2.0, and 1.0) and an undefoliated check. In 1990 and 1991, treatments consisted of 4 defoliation patterns (simulated insect defoliation at stage R2, R4, R2+4 and one day defoliation at R2), three levels within each pattern (defoliation to produce an LAI of 3.4, 2.6. and 1.8 at late R4), and two undefoliated checks. Defoliation levels were chosen based on the measured LAI at late R1/early R2 and the projected LAI at R4 based on the R2 measurement. Defoliation levels were chosen to provide final LAIs after defoliation (R4) that include values above, at, and below the critical LAI (the LAI value at which 95% of incident light is intercepted by a plant canopy, estimated to be 3.5 for soybean). Soybeans were defoliated by leaflet, and leaf areas of all leaflets removed were measured.

To simulate insect defoliation we picked leaflets following an insect defoliation model after the method of Ostlie (2). Appropriate use of a simulation technique for insect defoliation requires proper location of defoliation (based on the location of actual insect defoliation in a plant stand), proper duration of defoliation (based on how long insects produce injury), and proper daily defoliation rates (based on insect development and consumption rates). Defoliation was limited to the upper two thirds of the canopy, but extended into the lower canopy if necessary for high defoliation levels. The center 4 row-m of the two middle rows were defoliated. Undefoliated plots received comparable handling (walking in plots and handling plants) as defoliated plots (to allow for compaction during defoliation and effects of touching plants). Areas adjacent to the defoliated region (border rows and ends of plots) were defoliated (stripping leaflets without quantifying area removed) to approximately the same level as the defoliated area.

Defoliation was imposed to simulate the appropriate duration of insect injury. Ideally, daily injury rates and duration should be based on temperature-driven consumption and development models. However, because temperature-driven models would result in substantial differences in injury rates among locations, standard durations and injury rates were employed at all sites. Many soybean defoliators have larval development times for the latter developmental stages (when >90% of consumption occurs) of approximately two weeks, at temperatures commonly occurring in mid to late summer. Consequently, defoliation was imposed over 12 days. Defoliation occurred at soybean stages R2 (full flower), R3 (beginning pod), and R4 (full pod) (depending on year and treatment) to correspond to the injury phenology for many soybean defoliators.

To properly simulate insect defoliation appropriate daily defoliation rates must be used. Daily defoliation rates depend on stage-specific consumption rates. In brief, to simulate insect feeding we needed to estimate what proportion of the total defoliation required should occur on each day. The rationale behind the values chosen is as follows. For this study, two aspects of development and consumption are pertinent. First, the proportion of total larval consumption in a stage, and second, the duration of developmental time in a stage. To determine the proportion of the total defoliation that should occur in each larval stage, an estimate of proportion of total consumption by stage is needed. Literature data on green cloverworm (GCW) (*Plathypena scabra*), corn earworm (CEW) (*Helicoverpa zea*), soybean looper (SBL) (*Pseudoplusia includens*), and velvetbean caterpillar (VBC) (*Anticarsia gemmatalis*) were used to determine appropriate values for this study. The proportion of total consumption by instars are: GCW 1-2=2%, 3-4=8%, 5-6=90% (10); SBL 1-2=1%, 3-4=9%, 5-6=90% (11); and VBC 1-2=3%, 3-4=5%, 5-6=92% (11). Because so little defoliation occurs in the first two larval stages (<3%), we considered defoliation only during the latter stages. Specifically, the proportion of defoliation by stage was 3-4=10%, and 5-6=90%.

The second issue of importance for simulating insect defoliation is duration of development time in a stage. The proportion of time spent in various instars are: GCW 1-2=29%, 3-4=26%, 5-6=45% (12); SBL 1-2=26%, 3-4=31%, 5-6=43% (11); and CEW 1-2=23%, 3-4=25%, 5-6=52% (11). Based on these values, an appropriate estimate of time spent in each stage is 1-2=25%, 3-4=25%, and 5-6=50%. We estimated development through stages 3-6 as requiring 12 days; therefore, the ratio of development times (25%:50% or 1:2) gives the number of days spent in each stage; specifically, stages 3-4=4 days and stages 5-6=8 days. Consequently, to provide an appropriate simulation of a lepidopteran defoliator of soybean (combining consumption and development data), we imposed injury over 12 days with 2.5% of the total defoliation occurring on each of the first 4 days and 11.25% occurring on each of the last 8 days. Although designed to simulate lepidopteran defoliation patterns, this simulation also is suitable for other species (such as bean leaf beetle [*Cerotoma trifurcata*] and grasshoppers). For adult defoliating insects (like beetles) the increasing defoliation rates simulate an increasing pest population rather than increasing consumption per individual pest.

A computer program, DEFOL, was written to calculate total leaf area to be removed, percentage of total leaf area to be removed per day per plot, conversion of leaf area to leaflets to be removed, and defoliation summaries. To adjust for possible discrepancies between projected and actual leaf area removed, all leaf area removed per plot per day was quantified and entered into the program to allow for daily adjustments. The program output leaf areas and numbers of leaflets to be removed for each plot on each day. Leaf area to be removed was based on target defoliation levels, appropriate injury rate, and previously removed leaf area. Leaflets to be removed were calculated from leaf area to be removed and a user-supplied estimate of average leaflet size on the first day. Subsequently, the program calculated the average leaflet size based on number of leaflets removed and measured leaf areas.

Light interception was measured in the plant canopy weekly from R2 to R6 to include measures at each reproductive stage. A line quantum sensor was centered across the row and a measure of photosynthetically active radiation (PAR) was obtained. Measurements were made for each of the two center rows of each plot. Additionally, a measurement was made outside each plot, in full sun, to indicate PAR with 0% light interception. All measurements were taken within one hour of solar noon.

Individual plant samples (for growth analysis) were taken at approximately weekly intervals at R2, R3, R4, R5, and R6. For each sample date 3 or 5 plants per plot were removed, using a stratified random sampling procedure. Leaf photosynthesis measurements were taken to examine the soybean compensation to defoliation through altered leaf photosynthesis. Leaflets at node 6 on two plants per plot were measured before, during, and after defoliation. Measurements were made in full sunlight at comparable times for each measurement using an LI-6200 Portable Photosynthesis System (LI-COR Inc.). Individual yield components and plot yields were determined, as were various other parameters.

RESULTS AND DISCUSSION

Defoliation-Light Interception Hypothesis

Significant linear relationships between light interception after defoliation and soybean yield losses were observed. Figure 1 presents data from Nebraska experiments in 1990 and 1991 on soybean responses to defoliation at stage R2. Although y-intercepts were different (reflecting differences in yield potential between the two years), the slopes of these lines were not significantly different (indicating a common relationship across years).

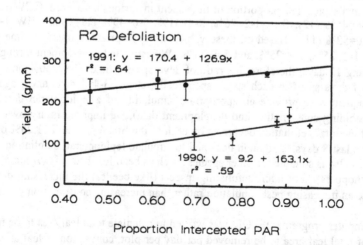

Figure 1. Relationships between soybean yields and different levels of interception of photosynthetically active radiation (PAR) after simulated insect defoliation at stage R2.

Comparable results were observed from Iowa experiments in 1988 and 1989 on responses to defoliation at R3 (Figure 2). Again, significant linear relationships were observed between soybean yields and light interception after defoliation, and the slopes of the lines for the two years were not significantly different. In 1988, a serious drought occurred, but in 1989 plots received near normal rainfall. Consequently, the consistency in the relationship between yield and light interception across years is particularly noteworthy, given the variability in environmental conditions.

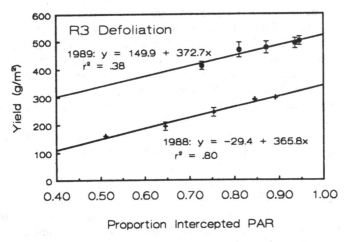

Figure 2. Relationships between soybean yields and different levels of interception of photosynthetically active radiation (PAR) after simulated insect defoliation at stage R3.

Similar results were observed for defoliation at R4. Additionally, preliminary analysis of data from other locations (not shown) indicates that relationships were consistent between locations, as well as across years. These results strongly support the defoliation-light interception hypothesis. Consequently, changes in soybean canopy light interception by insect defoliation do seem to have a major influence on subsequent soybean yields.

Given the consistency of the relationships observed, if insect injury can be directly related to changes in light interception, then reliable predictions of yield losses are possible. Because light interception is strongly correlated with leaf area indices for crop species, including soybean (13), relating the impact of insect injury on leaf areas (as is possible by considering insect consumption rates) provides one mechanism for relating injury to light interception. A remaining problem in the practical use of this information for managing insect pests is the need to assess the leaf area of a soybean canopy to make an appropriate management decision. However, instruments for direct measurements of canopy leaf areas are available and simple indices or sampling procedures may be feasible for making such estimates.

This observation on the importance of assessing soybean canopy size agrees with work by Herbert et al. (14). Based on tests of a computer model of insect leaf feeding on soybean, these workers observed that conventional measures of injury, such as percent defoliation, did not reliably predict yield losses. In contrast, estimates of remaining leaf area after defoliation did provide good yield estimates. They concluded that practical techniques for assessing canopy leaf areas by soybean producers were needed to provide reliable estimates on the impact of soybean defoliators. In addition to supporting the need for measures of soybean canopy size in managing defoliators, these results also indirectly support the defoliation-light interception hypothesis.

Compensation to Defoliation

Although soybean is known to tolerate substantial levels of defoliation, the mechanisms associated with this tolerance are not known. If the major impact of defoliation is to reduce light interception, then compensation might occur by improving the interception of light (such as by increasing leaf growth rates or altering leaf angles) or by improving the efficiency of light utilization. Results from our research indicate that changes in light utilization do occur in response to defoliation. However, these changes do not represent increases in leaf photosynthetic rates, as might be expected from theoretical arguments. Instead, soybean compensation to defoliation readily occurs through delayed leaf senescence, including delayed leaf abscission and altered leaf photosynthetic rates.

Normal leaf senescence in soybean is associated with declines in leaf photosynthetic rates and abscission of older leaves. We observed delays in both of these senescence phenomena in response to defoliation. Figure 3 presents data on the lowest leaf-bearing node of soybean plants before and after defoliation at stage R3 in 1988. Because the oldest leaves are on the lowest nodes, differences in lowest leaf-bearing node indicate differences in leaf abscission between treatments. Prior to defoliation (50 days post emergence) no differences were observed. After defoliation was completed (60 and 88 days post emergence) significant differences in leaf abscission were observed. Additionally these data indicate that delays in leaf abscission increased with increasing levels of defoliation. Similar results were observed from other years of the study.

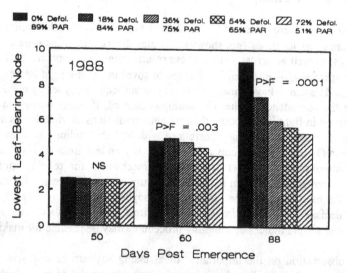

Figure 3. Differences in leaf abscission (indicated by lowest leaf-bearing node) before (day 50) and after defoliation (day 60 and 88) at R3 in 1988.

In addition to delayed leaf abscission, leaves on defoliated plants maintained higher photosynthetic rates than leaves on undefoliated plants (Figure 4). Although photosynthetic rates on remaining leaves of heavily defoliated plants may seem to have increased after defoliation (days 61 and 74), comparison with pretreatment values (day 47) indicates the actual response is a delay in the ordinary decline in photosynthetic rates associated with aging. Comparable responses were observed in other years of the study. Indeed, in some years no decline in leaf photosynthetic rates were observed in remaining leaves on plants in the most heavily defoliated treatments.

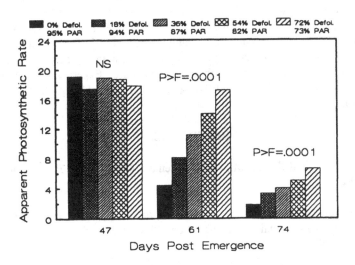

Figure 4.   Differences in leaf apparent photosynthetic rates ($\mu mol/m^2/sec$) before (day 47) and after (day 61 and day 74) defoliation at R3 in 1989.

Results for both leaf abscission and leaf photosynthesis indicate that the level of compensation depends of the amount of defoliation, and not on a qualitative or "wound-signal" response.   Given that compensation to defoliation occurred through a change in plant senescence, it seems likely that differences in hormone levels or responses to hormones were involved. Compensatory mechanisms to defoliation are largely unknown. Some studies have demonstrated an increase in photosynthetic rates of remaining leaf tissue after defoliation (15,16,17), although leaf injury in many of these studies probably does not correspond to arthropod leaf injury.

If genetic variability exists among soybean lines for compensation, then developing defoliation-tolerant soybean varieties might be possible. Although tolerance of high levels of defoliation may not be feasible, even moderate levels of tolerance would reduce the need for insecticides in many situations. Additionally, tolerance has the theoretical advantage of not imposing a selection pressure on pest populations.

CONCLUSIONS

Improving our understanding of the physiological responses of soybean to insect injury offers the prospect of greatly improved insect management guidelines and new alternatives to insect management. As results from this research indicate, using new understandings of relationships between insect injury and soybean physiology may require more consideration of soybean canopy size and similar parameters, as well as of pest populations. Consequently, the challenge in developing practical pest management programs based on new information is to weigh the need for sufficient data to make accurate predictions against considerations of time and cost necessary to obtain those data.

ACKNOWLEDGEMENTS

I thank P.M. Higley for her review of this manuscript. Research reported in this paper is that which I personally conducted; however, many cooperators have been involved in this project including: L. Bledsoe, J. Browde, R. Edwards, J. Funderburk, R. Hammond, C. Helm, M. Kogan, T. Mack, A. Mueller, L. Pedigo, R. Shibles, I. Teare, and J. Witkowski. This research was supported in part by the Iowa Agricultural and Home Economics Experiment Station, Project 2580, by the Nebraska Agricultural Experiment Station Projects NEB-17-052, NEB-17-053, and NEB-17-055, and by USDA-CSRS grant no. 90-34103-5040.

REFERENCES

1. Turnipseed, S.G., and Kogan, M., Soybean entomology. *Annu. Rev. Entomol.*, 1976, **21**, 247-282.

2. Ostlie, K.R., Soybean transpiration, vegetative morphology, and yield requirements following simulated and actual insect defoliation. Ph.D. dissertation, Iowa State University, Ames, IA, 1984.

3. Hutchins, S.H., Higley, L.G., and Pedigo, L.P., Injury-equivalency as a basis for developing multiple-species economic injury levels. *J. Econ. Entomol., 1988*, **81**, 1-8.

4. Pedigo, L. P., Higley, L.G., and Davis, P.M., Concepts and advances in economic thresholds for soybean entomology. In *Proc. World Soybean Res. Conf. IV.* , ed., A. J. Pascale , 1989, **3**, 1487-1493.

5. Monteith, J.L., Climate and the efficiency of crop production in Britain. *Philos. Trans. R. Soc.,* London, 1977, **281**, 277-294.

6. Poston, F.L., Pedigo, L.P., Pearce, R.B., and Hammond, R.B., Effects of artificial and insect defoliation on soybean net photosynthesis. *J. Econ. Entomol.*, 1976, **69**, 109-112.

7. Johnson, K.B., Defoliation, disease, and growth: a reply. *Phytopathol.*, 1987, **77**, 1495-1597.

8. Waggoner, P.E., and Berger, R.D., Defoliation, disease, and growth. *Phytopathol.*, 1987, **77**, 93-398.

9. Ingram, K. T., Herzog, D.C., Boote, K.T., Jones, W.J., and Barfield, C.S., Effects of defoliating pests on soybean canopy CO_2 exchange and reproductive growth. *Crop Sci.* , 1981, **21**, 961-968.

10. Hammond, R.B., Poston, F.L., and Pedigo, L.P., Growth of the green cloverworm and a thermal-unit system for development. *Environ. Entomol.*, 1979, **8**, 639-642.

11. Boldt, P.E., Biever, K.D., and Ignoffo, C. M., Lepidopteran pests of soybeans: consumption of soybean foliage and pods and development time. *J. Econ. Entomol.*, 1975, **68**, 480-482.

12. Hammond, R.B., Pedigo, L.P., and Poston, F.L., Green cloverworm leaf consumption on greenhouse and field soybean leaves and development of a leaf-consumption model. *J. Econ. Entomol.*, 1979, **72**, 714-717.

13. Shibles, R.M., and Weber, C.R., Interception of radiation and dry matter production by various soybean planting patterns. *Crop Sci.*, 1966, **6**, 55-59.

14. Herbert, D. A., Mack, T.P., Backman, P.A., and Rodriguez-Kabana, R., Validation of a model for estimating leaf-feeding by insects in soybean. *Crop Protect.*, in press.

15. Detling, J. K., Dyer, M.I., and Winn, D.T., Net photosynthesis, root respiration, and regrowth of *Bouteloua gracilis* following simulated grazing. *Oecologia*, 1979, **41**, 127-134.

16. Waring, P.F., Khalifa, M.M., and Treharne, K.J., Rate-limiting processes in photosynthesis at saturating light intensities. *Nature*, 1968, **220**, 453-457.

17. Welter, S.C. Arthropod impact on plant gas exchange. In *Insect-Plant Interactions*, ed., E.A. Bernays, CRC Press, Boca Raton, Florida, 1989, pp. 135-150.

MANAGEMENT OF INSECTICIDE RESISTANT SOYBEAN LOOPERS (PSEUDOPLUSIA INCLUDENS) IN THE SOUTHERN UNITED STATES

DAVID J. BOETHEL, JEFFREY S. MINK,
ALAN T. WIER, JAMES D. THOMAS, B. ROGER LEONARD,
Louisiana State University Agricultural Center
Louisiana Agricultural Experiment Station
Baton Rouge, Louisiana, USA 70803

FERNANDO GALLARDO
Puerto Rico Agricultural Experiment Station
Mayaguez, Puerto Rico 00709

ABSTRACT

The soybean looper, Pseudoplusia includens (Walker), ranks among the most important of the multitude of pests that threaten soybean annually in the southern United States. With the documentation of resistance to permethrin in 1987, this insect pest has developed resistance to virtually all classes of insecticides to include DDT, cyclodienes, organophosphates, carbamates, and most recently, the pyrethroids. This development is alarming in view of the infrequent use of insecticides on soybean and the rapidity with which resistance has appeared after adoption of an insecticide. The factors responsible relate to the insect's biology and ecology, which challenge assumptions frequently made when attempting to model development of insecticide resistance in insects. Research progress has been made toward formulation of an insecticide resistance management plan. However, it is evident that expertise and cooperation beyond the soybean community and quite possibly beyond the boundaries of the United States will be necessary to develop and implement a viable strategy.

INTRODUCTION

The soybean looper, Pseudoplusia includens (Walker), and the velvetbean caterpillar, Anticarsia gemmatalis Hubner, are important defoliators of soybean in the southern soybean growing region of the United States. Because both are annual migrants into the area, outbreaks of the pests are sporadic and are characterized by rapid population surges that require management to prevent losses to the crop. Insecticides have been the management tool most frequently used, and over the 25+ years that soybean has been intensively grown in the region, this approach has been effective and economical for the velvetbean caterpillar. Such has not been the case for the soybean looper because of this species' ability to develop resistance to insecticides.

HISTORY OF INSECTICIDE RESISTANCE

Organophosphate and Carbamate Resistance

Insecticide resistance in the soybean looper is a rather common phenomenon that has occurred several times in the southeastern United States since the early 1960's. Actually, resistance to DDT and cyclodienes may have preceded this time period. Georghiou and Mellon [1] indicated that the pest had developed resistance to these classes of insecticides but documentation was not provided to determine when resistance was first detected. In 1963, the Louisiana Cooperative Extension Service began publishing an annual "Insect Control Guide" that listed the recommended insecticides for insect pests on the important commodities in the state. For the first six years the guide was published, the recommendations for soybean looper control on soybean were methyl parathion or carbaryl with no mention of DDT and the cyclodienes. In 1970, and continuing until 1972, the following statement appeared in the section for soybean looper control: "None of the insecticides currently registered on soybean will give effective control. Recommendations for looper control will be issued later in the season." Thus, resistance to organophosphate and carbamate insecticides had developed before 1970. Later, Palazzo [2] confirmed resistance to methyl parathion.

During the 1970-1972 period, some of the microbial pesticides containing *Bacillus thuringiensis* (B.t.) were being investigated, and methomyl was emerging as the insecticide of choice for soybean looper management. In 1973, methomyl appeared in the Louisiana recommendations for looper control but was recommended under a "state label only." From 1974 until the present, B.t. products and methomyl have been recommended for control of the pest on soybean.

Methomyl Resistance

Field control problems with methomyl occurred shortly after its adoption, but these problems were more pronounced in the states of Georgia, South Carolina, and Florida than in the mid-South states of Louisiana and Mississippi. An account of the relative differences in susceptibility to methomyl among soybean looper populations from some of these states can be found in Newsom et al. [3]. These data revealed that Louisiana and Florida populations collected in 1970 exhibited 5- and 7-fold levels of resistance (LD_{50} level), respectively, when compared to a population of soybean loopers collected in 1976 in Louisiana. The 1970 Florida and Louisiana populations exhibited less than 2-fold resistance at the LD_{90} level but did demonstrate that differential response to methomyl existed in soybean looper populations prior to widespread adoption of the chemical for looper control on soybean. Chiu and Bass [4] reported that grower problems controlling soybean loopers with acephate and methomyl were widespread in Alabama in 1976. Likewise, in 1976, a year in which field control problems also became frequent in late season in Georgia, a soybean looper colony from that state exhibited 23- and 55-fold increases in response to methomyl at the LD_{50} and LD_{90} levels, respectively, when compared to the susceptible colony collected in Louisiana. These findings and reports prompted Newsom et al. [3] to state, "Insecticide pressure from treatment of soybean had been too little, affected too small a percentage of the total population of this highly mobile species, and had been applied over too short a time span to select resistant populations."

The question of how resistance in the Georgia population came about was answered with circumstantial evidence related to insecticide usage on crops in areas of southern Florida where the soybean looper overwintered before migrating into soybean fields in states further north during the summer. In these overwintering areas, some of the hosts of the soybean looper such as tomato received as many as 40 applications of insecticide per year while chrysanthemums in the Fort Myers area received as many as 100 applications of methomyl per year. Although this scenario was not thoroughly documented, the evidence strongly suggested that pest control practices in one state could result in a resistance problem that threatened soybean integrated pest management programs in another.

Permethrin Resistance

This scenario has been revisited recently with documentation of permethrin resistance in the late 1980's in Georgia [5], Louisiana [6], and Mississippi [7]. During the 1987 growing season, reports surfaced of failures to control soybean loopers with the standard insecticide, permethrin. It is highly unlikely that the insecticide selection pressure with permethrin on soybean in the south would have had a significant impact on resistance development. Generally, three [8] to five generations [9] occur on soybean annually with only one receiving an insecticide application. More importantly, permethrin had only been used on the crop since 1982.

Parallel to the situation that occurred with methomyl resistance in the 1970's, permethrin resistance has been most severe, in terms of field control, in the states of Georgia and South Carolina relative to the control in the mid-South. In Louisiana, the reports of problems encountered by growers were not associated with total control failures, but reflected noted reductions in permethrin efficacy. Characteristic of the reported field failures was the fact that the problem soybean fields were located adjacent to or in the vicinity of cotton fields [6]. Felland et al. [7] found higher levels of tolerance in populations from Mississippi than those found in Louisiana; however, field control problems were more pronounced in cotton growing regions of that state, also. McPherson and Herzog [10] also reported widespread control problems with permethrin in 1987 and to a greater extent in 1988 and 1989, throughout much of central and southeastern Georgia where large acreages of cotton and soybean were planted. Dosage-mortality data from Georgia [10] indicated 10-and 22-fold larger LD_{50} values for permethrin in soybean looper populations collected in 1987 and 1988, respectively. Resistance to permethrin resulted in only 8% control of soybean looper populations in South Carolina in 1988 [11]. Dosage-mortality data on a South Carolina colony collected following a field control failure in 1988 revealed significantly greater LD_{50} values compared with soybean looper populations collected from Louisiana in the same year and a Texas population that exhibited dosage-mortality response data no different than that of susceptible laboratory colonies (Boethel and Mink, unpublished). Thus, it appears that the intensity of pyrethroid resistance in the soybean looper populations increases for states closer to the east coast in the southeastern United States similar to the situation that occurred with methomyl a decade earlier. What is different in the case of permethrin resistance is the associations between cotton growing areas, presence of populations with increased tolerance, and field control problems.

The history of the development of insecticide resistance in the soybean looper reveals several scenarios not normally encountered when pest species develop resistance. Because the insect does not overwinter in the region [8] except in extreme southern regions of Florida and Texas (outside the traditional soybean producing areas) and selection pressure on the soybean crop per se has not been extensive [3], several hypotheses have been proposed to explain the factors contributing to the infestations of permethrin-resistant soybean looper in soybean. These hypotheses relate to 1) selection pressure on the soybean looper at sites where the source populations that eventually migrate into the southeast originate, 2) the role that cotton plays in the population dynamics of the pest, 3) the impact of cotton insecticide programs on soybean loopers in cotton-soybean agroecosystems, and 4) a combination of these factors.

If DDT resistance is present in this insect, as reported by Georghiou and Mellon [1], then cross-resistance to the pyrethroids should have been anticipated because both pyrethroids and DDT act at the same target site. Thus, with widespread use of the pyrethroids, both on cotton and soybean, a very rapid buildup of resistance could occur due to the presence of the DDT resistant gene in the population. Sparks [12] discusses this scenario for the development of pyrethroid resistance in Heliothis in the cotton growing region of the United States.

Although possible, the same set of circumstances being responsible for pyrethroid resistance in soybean looper is less likely. If resistance did result in the same manner, then selection occurred on hosts other than soybean. The era of intensive DDT use preceded the rapid expansion of soybean in the southeastern states in the mid-to-late 1960's, and the insecticide was not used on the crop to any great extent, at least in Louisiana (J. L. Bagent, Louisiana Cooperative Extension Service). Perhaps, as more toxicological data on mechanisms of resistance emerge, this aspect will be clarified.

Research to understand these factors that contribute to resistance development is essential and among the initial steps to formulate an insecticide resistance management strategy for this pest. However, other elements are necessary also, and the uniqueness of the situations described heretofore present ample challenge to entomologists involved in soybean IPM.

MANAGEMENT OF PERMETHRIN RESISTANCE

Immediately after it became apparent that permethrin control problems with soybean looper were becoming widespread, soybean entomologists throughout the southeastern states met to discuss prospects for the development and implementation of a resistance management plan for the soybean looper. At the initial meeting, Graves [13], one of the architects of the mid-South pyrethroid-resistance management plan for Heliothis virescens on cotton indicated that much of the information necessary for development of a plan to manage soybean looper resistance was not available. Among the principal data gaps he listed were 1) the source(s) of soybean loopers in the mid-South was unknown; 2) there was a lack of information on gene flow; 3) the mechanisms of resistance were poorly understood; 4) there was a lack of effective alternative management tactics, such as resistant varieties; and 5) there was a lack of effective alternative insecticides. Equally important was the fact that soybean producers were not the main source of the resistance problem. Some of these impediments and others not mentioned remain, but considerable progress has been made over the last four years to develop the research base necessary to cope with the resistance problem.

Mechanisms of Resistance

One study has been conducted on the insecticide resistance mechanisms in the soybean looper. Rose et al. [14] examined a soybean looper population from a field in northeast Louisiana wherein permethrin failed to provide adequate control and demonstrated an approximate 3-fold reduction in susceptibility to permethrin [6] when compared to a susceptible laboratory colony. The results indicated that a combination of target site insensitivity and increased activity of several enzymes [glutathione transferase (2.7-fold for the substrate 1-chloro 2,4-dinitrobenzene); monoxygenases (1.8-fold for p-nitroanisole O-demethylase); and hydrolases (1.5-fold for alpha-naphthyl acetate, 1.5-fold for p-nitrophenyl acetate, and 1.5-fold for permethrin)] involved in insecticide metabolism including a trans-permethrin hydrolase may have been contributing to the reduced susceptibility of the field population relative to the laboratory colony. Significant differences between populations were not observed for NADPH cytochrome c reductase or acephate hydrolysis. These data revealed multiple mechanisms were involved in resistance development. Graves [13] suggested that these mechanisms would be expected to cause resistance to pyrethroids, organophosphates, and carbamates. In view of this, he urged increased research on biologicals, insect growth regulators, and avermectin compounds and indicated synergists might provide temporary relief.

Synergists, Mixtures, and IGR's

Earlier, Dowd and Sparks [15] examined inhibition of trans-permethrin hydrolysis in the soybean looper and the use of inhibitors as pyrethroid synergists. Although a wide array of chemicals were investigated, it appeared that the most effective class of inhibitors of trans-permethrin hydrolysis in soybean looper midguts were the nonpolar (aryl or long chain alkyl) organophosphates. Stirophos, profenofos, crufomate, phosmet, dibenzoylmethane, piperonyl butoxide, phenyl TFP, and 4-chloro-PTEP, when mixed with the pyrethroid, were more toxic than the pyrethroid alone.

Of these potential synergists, only piperonyl butoxide has been tested to any extent in laboratory or field settings since permethrin resistance was documented. From 1987 through 1989, six insecticide screening trials were conducted in Louisiana using applications of permethrin at the standard dosage for looper control (0.11 kg ai/ha) plus piperonyl butoxide, generally at the rate of 0.28 kg ai/ha [16, 17, 18]. Only on two of the 15 evaluation dates did permethrin plus the synergist demonstrate a significant reduction in soybean looper populations compared with permethrin alone. Similar results were found in Georgia [19, 20, 21] and Alabama [22] when piperonyl butoxide was evaluated as a synergist. In 1990, Leonard [23] reported enhanced activity of permethrin when piperonyl butoxide was used at a rate (1.12 kg ái/ha) much higher than previously tested. These data suggest the proper ratio of the synergist to insecticide may be important.

In laboratory bioassays, the addition of piperonyl butoxide to permethrin (1:1 ratio) did not significantly lower LD_{50} values when compared to permethrin alone when the Paradise, Louisiana strain described by Leonard et al. [6] was tested (Leonard and Boethel, unpublished). Although this strain demonstrated that metabolic resistance mechanisms were involved [14], it did not exhibit extremely large resistance ratios (3-fold). When a population of soybean loopers from Puerto Rico that had an LD_{50} approximately 46.2 times larger than the Paradise, Louisiana strain was challenged with permethrin plus piperonyl butoxide (1:1 ratio), the LD_{50} was dramatically lowered but still remained 15.4 times larger than the Louisiana strain (Mink, Boethel, and Gallardo, unpublished). Thus, it appears that the Puerto Rico population may have had mixed function oxidases playing a greater role than that found in populations in the southeastern states. These findings further emphasize the need to study populations from suspected migratory source areas (covered in detail later in chapter), and the need to conduct expanded studies on the mechanisms of resistance in populations from various areas.

Since permethrin resistance has surfaced, several field control experiments have included mixtures of pesticides with permethrin, some of which were among the chemicals Dowd and Sparks [15] evaluated as synergists. Layton and Boethel [16] did not detect any synergism when acephate was applied with permethrin in Louisiana. Mink et al. [17] conducted two insecticide trials during 1988 wherein methyl parathion was added to permethrin treatments. Although one trial was conducted in Tensas Parish, Louisiana where resistant soybean looper were found, no synergism resulted from the mixture. In Mississippi, Felland and Pitre [24] tested numerous combinations of pesticides including several that contained mixtures of pyrethroids (permethrin and tralomethrin) with carbamates and organophosphates. No synergistic effects were observed for any of the mixtures. It appears that even though some organophosphates demonstrated synergistic activity in laboratory studies, field trials have not supported these findings. However, inappropriate ratios of the mixtures and the paucity of data on the level of metabolic resistance present in field populations may explain these results.

Diflubenzuron is the insect growth regulator that has received the most attention for control of soybean looper. Turnipseed et al. [25] examined it for control of soybean insects in Georgia, South Carolina, and Brazil during 1973 and 1974. It appeared that multiple applications and higher rates were necessary to provide control of the soybean looper, relative to the velvetbean caterpillar. Subsequent to this report, research in Louisiana demonstrated variable results ranging from suppression [26] to no activity [27]. Even though some of the early experiments with diflubenzuron revealed some activity against the soybean looper, the insect is not included on the product's label. With the onset of permethrin resistance, interest in the product again emerged, and several insecticide trials have included diflubenzuron alone or in combination with other pesticides. The recent results are characteristic of that seen in earlier trials wherein control is variable, and the best results were found on the latest evaluation dates (i.e., Insecticide and Acaricide Tests Vol. 14-16).

The use of insecticide mixtures has not been common on soybean. The need for mixtures to control individual pests has not existed. Another factor is economics. Newsom et al. [3] addressed this aspect when stating "...the per acre returns from soybean production are too low to allow for extensive use of conventional insecticides." The fact that insecticide alternatives for soybean looper control are more costly than permethrin has maintained interest in the product despite its current problems.

Monitoring and Detection of Permethrin Resistance

Field control with permethrin has steadily decreased in Louisiana since 1987 (Fig. 1). Similar and even greater reductions have occurred in other states with some, for example South Carolina, completely removing the chemical from its state's recommendations for soybean looper control. However, in the mid-South, the insecticide has provided adequate, though reduced, control. For example, in Louisiana [6] and Mississippi [7, 28], few control problems occurred in soybean fields that were

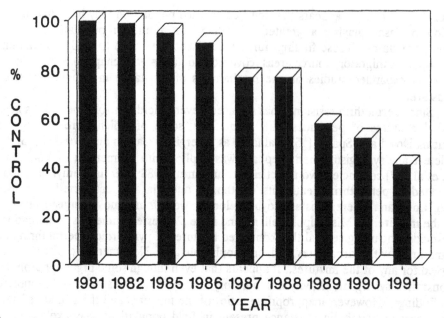

Figure 1. Control of soybean looper with permethrin (0.11 kg ai/ha) in small plot insecticide trials at various locations in Louisiana from 1981-1991. Data represents percent control from samples 1 to 7 days posttreatment.

not located near cotton. The variable results from state to state and locations within states present a dilemma for researchers, extension personnel and private pest management consultants who are charged with making recommendations for control.

McCutchen et al. [29] developed a glass vial technique for detecting resistance in tobacco budworm, Heliothis virescens (F.), larvae in cotton. We felt a similar procedure for the soybean looper would be even more applicable because insecticides are directed against populations comprised of all stages of larvae that must reach rather large populations before remedial measures are needed {Economic Threshold = 150 larvae per 100 sweeps -- Tynes and Boethel [30]}, as opposed to the situation with tobacco budworms where the economic threshold is low and control measures are directed at eggs and first instar larvae.

To develop a bioassay, Mink and Boethel [31] utilized two susceptible laboratory strains and eight populations of soybean looper collected in Louisiana soybean and cotton fields in 1990. Third through sixth instar larvae from these strains were exposed to permethrin residues of various concentrations in glass liquid scintillation vials to estimate concentration mortality lines. A discriminating concentration of 0.5 ug permethrin per vial was identified for detection of permethrin resistance in field populations. This concentration caused approximately 94-99% mortality of susceptible larvae, thereby killing a minimum number of resistant larvae. Results from each field strain indicated that LC_{50} values for third through fifth instar larvae were not significantly different which adds to the practicality of the technique for use by various clientele in the field. Also, a significant correlation between the vial bioassay results and topical bioassay results were found.

Several researchers have suggested that a discriminating dose is a more appropriate way to determine the proportion of resistant individuals in a population [32, 33, 34] than with dosage-mortality lines. Thus, it appears that the glass vial technique for soybean looper will be appropriate for both detecting resistance in the field (discriminating dose) and monitoring resistance (discriminating dose or a series of doses). During the 1990 growing season, entomologists in seven southeastern states [and one manufacturer of permethrin (FMC Corp.)] participated in a cooperative project utilizing vials prepared at Louisiana State University for this purpose. The data have confirmed what field trials and limited toxicological studies have indicated over the last few years--that soybean looper resistance levels vary depending on location, cropping system, and time (Mink and Boethel, unpublished).

In addition to being used to document control failures due to resistance, the vial technique may eventually become an aid in making decisions on the choice of adequate control measures. As part of the regional cooperative project, participants were asked to evaluate permethrin at the standard rate (0.11 kg ai/ha) on the same populations that were bioassayed. Preliminary results in Louisiana in 1990 and 1991 indicated that mortality at the discriminating dose was indicative of field control (Mink and Boethel, unpublished). Although the data from other states have not been compiled, a data base is being assembled that may prove useful in IPM programs. Because a relationship appears to exist between bioassay data and field control, a grower, consultant, etc. should be able to determine 24 hours after sampling a field whether pyrethroid insecticides are a viable option for management of soybean looper larvae.

Preliminary research on development of a discriminating dose for soybean looper adults has been completed at Louisiana State University, and field populations have been challenged with the dose (2.5 ug permethrin per vial) (Mink, unpublished). The data indicate resistance is present in the adults, and although this stage is not the target of insecticide applications in soybean, knowledge of the resistance levels in adults in various cropping systems may lead to an understanding of the role of cotton and migration in the pyrethroid resistance problem.

Role of Cotton in the Population Dynamics and Development of Resistance
Studies in Louisiana [35, 36] and Georgia [37] demonstrated that soybean fields in cotton-soybean agroecosystems exhibited larger populations of soybean loopers than soybean fields in ecosystems devoid of cotton. Although the initial explanation for this phenomenon was the decimation of natural enemies by the cotton insecticide regimes, subsequent research revealed that cotton nectaries provided a carbohydrate source for looper moths which resulted in increased fecundity [38, 39]. Thus, larger populations develop in cotton growing regions, and soybean looper adults frequent cotton fields where selection pressure with pyrethroids has been intense. In these areas, the loopers that appear on soybean may be the progeny of adults that survive pyrethroid applications on cotton, making within season selection highly probable.

It has been suggested that resistance in a population would first be detected and be most evident when large populations are encountered [40]. The data shown in Fig. 2 represent results of field insecticide performance trials conducted in the southern soybean producing states, with data from Louisiana detailed as to the ecosystem involved. It appears from these studies that lower percent control is not strongly linked to population size. Data from Louisiana demonstrate that percent control has decreased

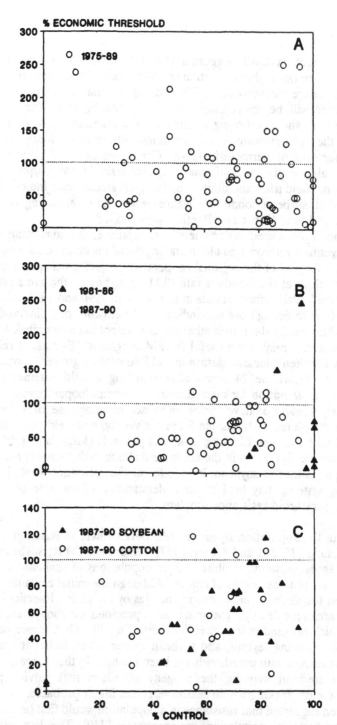

Figure 2. Relationship between soybean loopers population size and percent control with permethrin on soybean. A = data from 1975 to 1989 from southeastern states; B = Louisiana data from 1981-1986 before detection of resistance and 1987-1990 after detection of resistance; C = data since 1987 from cotton-soybean or soybean ecosystems in Louisiana.

since documentation of permethrin-resistance in 1987 (Fig. 2B) and support lower percent control during this time period in ecosystems containing cotton (Fig. 2C).

Soybean loopers develop on cotton, also. Surveys of Plusiinae populations in Louisiana from 1957 to 1963 revealed that >80% of the loopers collected on cotton were cabbage loopers, Trichoplusia ni, and >90% of the specimens found on soybean were soybean loopers [41]. Only 7% of the population on soybean were cabbage loopers, and soybean loopers accounted for 19% of the loopers on cotton. In Alabama, Canerday and Arant [42] also found mixed populations of these two species on cotton and soybean. Hensley et al. [41] cautioned that mixed populations occurred often enough to be of significance in control programs, and shortly after completion of these surveys, Chalfant [43] demonstrated that the two major species did not respond similarly to the same insecticides. The cabbage looper exhibited more tolerance; however, this was before the advent of the pyrethroids.

These surveys were conducted over 20 years ago, and it is reasonable to assume that the species composition on these two major crops has changed. The acreage of soybean increased dramatically over this time period, and insecticide use patterns on cotton, and to a lesser extent on soybean, shifted to pyrethroids. In Louisiana, incidental collections of loopers from cotton contained 85 and 100% soybean loopers in 1985 and 1988, respectively (Boethel, unpublished). In 1989, a cursory survey of untreated cotton and cotton treated with pyrethroids revealed some interesting results. In the untreated cotton, the percentage of soybean loopers ranged from <1% in August to >95% in September. However, in cotton treated with pyrethroids, the soybean looper comprised 100% of the loopers collected regardless of the date. If soybean loopers have become more common on cotton, and nectaried cotton cultivars grown throughout the South are an attractive food source for the migrant moths as they enter the Gulf Coastal States, then selection pressure with pyrethroids on larvae and adults on cotton contributing to resistant populations on soybean appears more likely. The need for current more intense surveys on the looper species on these two commodities is evident and should help solve this dilemma.

Migration and Gene Flow
The soybean looper appears to be incapable of diapause and is known to overwinter in southern Florida and southern Texas in the continental United States [8]. The lack of a diapause period means that soybean looper reproduces year round. As a result, the number of generations produced per year is increased, and larger numbers of generations have been correlated with increasing risk of resistance development [40].

Noctuid moths such as the soybean looper are generally strong fliers. The insect has been captured in light traps on oil platforms in the Gulf of Mexico as far as 66 to 100 miles from land [44, 45]. In addition to the reservoirs in Florida and Texas, Herzog [8] indicated that other sources for the annual migration of soybean loopers that invade the southern United States may be Central or South America or the islands of the Caribbean. If these areas are the source, insecticide usage there may result in development of resistance in the pest, which in turn presents a problem for soybean growers in the United States. The soybean looper has become troublesome on several commodities in Puerto Rico recently, and appears difficult to control with some of the insecticides currently being used on soybean in the United States. Both topical and glass vial bioassays conducted on populations collected on vegetables in Puerto Rico indicate high levels of permethrin-resistance (Mink, Boethel, and Gallardo, unpublished).

At this time, it is not known whether migratory populations originate in Puerto Rico but if it or some other country in the region is the source, the situation could parallel that seen in Florida and Georgia in the late 1970's. If the insecticide usage pattern on various commodities (i.e., vegetables) in these source areas within and outside the continental United States has changed from organophosphates and carbamates to pyrethroids, the appearance of pyrethroid-resistant soybean loopers in states along the Gulf Coast could be explained. Certainly, information on the responses of soybean looper populations from these suspected origins of migratory populations to currently used soybean insecticides would aid in determining if those populations are contributing to infestations of resistant soybean loopers on soybean.

The hypothesis linking migration and insecticide resistance is currently being considered under the assumption that return migration of resistant soybean loopers does not occur. Although permethrin was extremely effective when initially used, without return migration, the selection that has occurred and occurs each year in the southeastern states should not influence populations that arrive in subsequent years. Mitchell et al. [46] reported some evidence of return migration in Florida, but more recent reports make no mention of return migration [8].

Georghiou and Taylor [40] suggest that polyphagous insect pests tend to develop resistance more slowly than monophagous ones and along with Tabashnik [47] assume that immigrants are primarily susceptible. At this stage of our knowledge, it appears the soybean looper's ecology fits neither of these modes.

Of the two migrant lepidoptera problematic on soybean, the soybean looper would be considered polyphagous {82 hosts - Herzog [8]} relative to velvetbean caterpillar which confines its activity primarily to legumes. Insecticide resistance in the velvetbean caterpillar has not occurred, although it is as frequent a pest as soybean looper, and in some regions considered to have achieved key pest status [48]. This is further evidence that selection pressure on soybean, the major commodity attacked by the velvetbean caterpillar, has not been great. However, polyphagy may in fact enhance the potential for resistance development in species like soybean looper, as a result of inherently higher levels of certain enzymes, such as microsomal oxidases [40].

Refugia, that are free from insecticide pressure, generally allow polyphagous species to exhibit delays in development of resistance. In the case of the soybean looper, the individuals that appear on alternative hosts (as detailed earlier) are subject to intense selection pressure. Thus, the dilution of resistant genes through immigration and the haven of refugia for polyphagous species appears to be outside the assumptions for soybean looper and insecticide resistance development. In view of the potential seriousness of the current resistance situation, research on migration, inheritance of resistance, fitness of resistant populations, gene flow, etc. need immediate attention.

Pheromone Trapping

Beginning in 1990, a soybean pheromone trapping project that involves the cooperation of scientists in nine southeastern states was initiated to monitor the seasonal activity of soybean looper moths in the region [49]. It is hoped that after several years, patterns of migration may emerge. The lure used in the traps is not species specific, and all moths captured are identified by taxonomists at Mississippi State University. With the development of a discriminating dose for detection of pyrethroid resistance in adult soybean loopers, the network and methodology are in place to monitor pyrethroid resistance throughout the region. However, a more specific pheromone is needed to facilitate this research. Although not available commercially, a pheromone specific for the soybean looper has been identified [50].

Alternative Insecticides, Microbial Control, and New Chemistry

Other pyrethroids have been evaluated for soybean looper control since 1975 [4]. None have provided control superior to permethrin, at least at field rates considered economically competitive. Dowd and Sparks [51] found tralomethrin to be the most toxic of six pyrethroids (including cis- and trans-permethrin) tested in the laboratory. However, it along with lamba-cyhalothrin, cyfluthrin, cypermethrin, fenvalerate, and flucythrinate did not demonstrate greater or in some cases even equal, toxicity in field efficacy trials in Louisiana [16, 17, 26, 27, 52, 53]. Some of these same pyrethroids (plus esfenvalerate and fenpropathrin) demonstrated similar results in field trials in other southeastern states and Texas (i.e., Insecticide and Acaricide Tests Vol. 4-14). Felland et al. [7] found increased tolerances to tralomethrin and fenvalerate in permethrin-resistant soybean looper populations. Because target site insensitivity has been identified as a resistance mechanism, resistance to one pyrethroid confers resistance to all pyrethroids, thereby removing this entire class of insecticides as control alternatives [7, 54]. Even before documentation of permethrin resistance, the field data were a prelude to this conclusion.

Thiodicarb and methomyl have remained on the list of recommended insecticides for soybean looper control in the mid-south states. Since the 1970's, there has been reluctance to recommend methomyl because of the resistance problems that appeared in states further east [3]. Leonard et al. [6] found several strains of soybean loopers collected in Louisiana fields during 1987 and 1988 to be more tolerant to methomyl than a susceptible laboratory strain. However, in field trials, methomyl provided control of two of these strains that ranged from 63 to 83% which was comparable to control observed for permethrin. In Mississippi, Felland et al. [7] did not demonstrate methomyl resistance in laboratory bioassays and found field control to range from 69 to 73%, similar to the performance of permethrin. During the time these studies were being conducted, permethrin efficacy was decreasing compared to past performance while methomyl was providing adequate control. Data from Louisiana over the past two growing seasons support continued decreases in permethrin effectiveness (Fig. 1) but an increase in methomyl control into the 85-90% range in several tests [17, 18]. The reason for this increased activity cannot be explained except to speculate that if resistance is linked to a source population as proposed by Newsom et al. [3], then release of the pressure may be responsible. However, the elevated dosage response data seen in laboratory bioassays and the past history of field control problems make recommendations for control with methomyl tenuous.

The situation with thiodicarb parallels that for methomyl in the mid-South states and those further east. In Louisiana, thiodicarb has become the insecticide of choice for management of soybean looper, especially in regions where permethrin resistance has been common. Thiodicarb, at rates of 0.50 to 0.67 kg ai/ha, has consistently exhibited control in the range of 80-90% over the last several years [17, 18] which is a slight increase compared to performance in previous years (Boethel, unpublished). In general, control with thiodicarb in Mississippi has fallen in the range of that observed in Louisiana [55]. However, rates above 0.84 kg ai/ha do not cause similar population reductions in Georgia [10], Alabama [56], or South Carolina [57]. Thus, the pattern of decreased efficacy in soybean looper control from west to east across the region fits thiodicarb, as well as methomyl and permethrin. These findings may be an indication that the populations that invade each region originate from different sources.

Baseline dosage-mortality data on thiodicarb are limited at the present time but needed. The variable results already evident in the region and the expanding use of thiodicarb on cotton during the past several years make this insecticide the next possible candidate for resistance development by the soybean looper. Tobacco budworm resistance to thiodicarb was reported from populations collected on cotton in Louisiana in 1990 [58, 59]. Parallels between these two species in development of pyrethroid resistance make this development a concern for management of soybean looper, also.

Microbial insecticides containing B.t. have been recommended for soybean looper control in Louisiana and other southern states since the early 1970's. Until recently, they have not been as efficacious (primarily, not as rapid knockdown) or as economical as other insecticides. Since the onset of permethrin resistance, interest in these products has resurfaced in part due to the resistance problem but also the general interest in B.t.'s and their potential for manipulation through biotechnology.

Several commercially available B.t. compounds provide adequate control of soybean looper; however, because they are slow acting, usually 5 to 7 days are necessary before population reductions comparable to that seen with the more effective insecticides are obtained. This has prompted some to advocate downward adjustments in economic thresholds. Preliminary research in our laboratory has shown that larvae cease feeding almost immediately after exposure to B.t.--treated foliage (Mink, unpublished). Thus, adjustments of thresholds may be premature until further research is conducted but expanded educational programs may be necessary.

There has been some interest in using combinations of B.t. with other labelled insecticides for soybean looper control. Particular attention was given to mixtures with permethrin. In Louisiana, the data have not supported any advantage of these mixtures but have indicated that combinations with stomach poisons, such as thiodicarb, may have potential [18]. Data from other states have been variable (Insecticide and Acaricide Tests Vol 14-15). Certainly, mixtures of standard insecticides with B.t. products will increase the costs associated with management of the pest.

Jenkins et al. [60] reported that cotton cultivars with genes from B.t. for resistance to lepidopterous insects should be available in the near future. Because of the association of soybean looper with cotton, the selection pressure in larvae will surely increase when B.t. transgenic cotton appears. This will present an additional challenge for those involved in soybean IPM to preserve these microbial pesticides which currently are a viable alternative for insecticide resistant soybean loopers.

Beach and Todd [61] determined the toxicity of avermectin (MK936) to larva and adult soybean loopers in laboratory and field-cage studies. The compound was toxic to both stages and surviving adults that fed on the material exhibited reduced fertility and fecundity. Exposure of adults to treated foliage affected oviposition but the residual activity was limited to <72 hours after application. In 1991, we tested another avermectin product (MK244--Merck, Sharp, Dohme Research Laboratories, Three Bridges, N. J.) and found it highly effective against a population of pyrethroid-resistant soybean loopers in northeast Louisiana [62]. At 7 days posttreatment, MK244 demonstrated control equal to that of methomyl, thiodicarb, and another experimental insecticide, AC303,630 (American Cyanamid, Princeton, N. J.). The latter chemical was effective at rates lower than those being proposed for tobacco budworm control on cotton. This should make the product available at competitive prices but heavy use at higher rates on cotton could present problems for insecticide resistance management for the soybean looper. Both MK244 (avermectin) and AC303,630 (pyrrole) represent new chemistry, and although it will take several years to gather the data necessary for registration, they appear to offer alternatives for soybean looper management.

Other Management Tactics

There are few tactics other than insecticides that are available currently for management of soybean looper. Lambert [63] reviewed the status of host plant resistance to insects in soybean. Only three insect-resistant cultivars have been released since research began 20 years ago to develop resistance to foliar feeding insects. Only one of these, 'Lamar,' has potential for use in southern agriculture but when evaluated under production conditions, it exhibited levels of defoliation due to soybean looper that were similar to commercial cultivars. Several plant introductions have demonstrated resistance to soybean looper, and 35 insect-resistant breeding lines have been developed [63, 64, 65]. Therefore, the outlook for an insect-resistant cultivar is good but is hampered because the cultivars must also contain resistance to a myriad of diseases and nematodes found in the region in order to yield favorably. Once available, resistant cultivars will be a major management tool, ideally suited for a low value per acre commodity such as soybean.

Soybean cultivars in maturity groups V through VIII are adapted for growing in the latitudes that form the boundaries for the southern soybean growing region. The phenology of these cultivars place them at risk to soybean looper outbreaks that usually occur from mid-August to mid-September. Recently, cultivars (groups III and IV) that mature earlier in the season have been developed that are adapted to growing in the region. From an agronomic standpoint, they have advantages in that the pod-filling stage coincides with periods of favorable rainfall. If these cultivars mature early enough (by mid-August), they likely would escape attack by soybean looper and velvetbean caterpillar. The benefits of a management plan for these migratory lepidoptera based on early maturing cultivars must be balanced against the impact on other pests such as the bean leaf beetle, Ceratoma trifurcata (Forester) and southern green stink bug, Nezara viridula (Say). However, these cultivars would appear to be well suited for a management plan in the cotton growing areas in north Louisiana where the soybean looper has historically been the most important pest.

Because of the migratory behavior of the soybean looper, outbreaks of the pest are sporadic in time and space. This condition results in situations where natural enemies are not capable of regulating populations below economic thresholds. The problem appears more acute for parasitoids than predators.

Resurgence of soybean loopers following applications of insecticide was attributed to removal of natural biotic agents in South Carolina [66]. Brown and Goyer [67] found chemical exclusion of ground- or foliage-inhabiting predators resulted in larger populations of soybean loopers in tests conducted in Louisiana. Also, in Louisiana, Yanes [68] found that soybean looper developed larger populations and developed them earlier in the season in plots sprayed frequently with methyl parathion compared with plots that remained untreated. Collectively, these studies illustrate that the decimation of natural enemies with insecticides can lead to outbreaks or resurgence of soybean looper and that biological control agents provide a dampening effect on the population growth of the pest.

Shepard and Herzog [69] caution that it is usually difficult to separate the effects of pesticides on predators from effects on other natural enemies (parasitoids and entomopathogens). Numerous surveys to identify parasitoids and assess their impact on soybean looper populations have been conducted in the southeastern states {see Harper et al. [70] for review}. Among the most recent was a survey conducted during 1984 and 1985 in Louisiana [71]. The levels of parasitism varied from location to location and from year to year. However, the average seasonal parasitism at a site in south Louisiana was 64.5% in 1984, indicating that parasitoids played a role in regulating populations of the pest below the economic threshold. Daigle et al. [71] reported that native parasitoids and entomopathogens effectively utilize soybean looper as a host in Louisiana soybean fields despite its absence during much of the year, due to the migratory nature of the insect. Attempts at classical biological control of the soybean looper have been limited [72].

Harper et al. [70] reviewed the seasonal and geographical distribution of entomopathogens on soybean. Nomuraea rileyi (Farlow) and Entomophthora gammae (Weiser) are the major fungal pathogens of the soybean looper in the region, and natural epizootics commonly occur. However, economic damage frequently occurs before disease halts the pest population. Because of the lack of understanding of factors which cause initiation and spread of these diseases, outbreaks of entomopathogens are difficult to predict [69].

Newsom et al. [3] reported that a soybean looper virus (NPV) from Guatemala imported by Livingston and Yearian [73] and tested in Arkansas and Louisiana demonstrated good potential for control of soybean looper. Kuo [74] found the soybean looper NPV to persist in the soil and provide control a year after application. During 1989-90, this virus was still killing 38-63% of the soybean loopers in the fields where it was released in 1975-77, which was significantly greater than in nearby control fields (J. R. Fuxa, Dept. of Entomology, Louisiana State University, personal communication).

It is apparent that natural enemies are important in regulation of soybean looper populations. At this time, the best approach appears to be conservation of parasitoids and predators through selection of insecticides and timing of applications that are least detrimental. The recent findings concerning the soybean looper NPV are encouraging and should be pursued.

SUMMARY

Over the past 25 years, the soybean looper has developed resistance to several classes of insecticides to include DDT, the cyclodienes, organophosphates, carbamates, and most recently, the pyrethroids. These developments have been alarming in view of the infrequent usage of insecticides on soybean. The factors responsible relate to the pest's biology and ecology, which challenge the assumptions frequently made when attempting to model the development of insecticide resistance in insects.

Since documentation of pyrethroid resistance in 1987, research has been conducted to gather data necessary to formulate a resistance management plan. Progress has been made in 1) development of a rapid technique to detect and monitor pyrethroid resistance in larvae and adults, 2) determination of the mechanisms of pyrethroid resistance, 3) documentation of pyrethroid resistance in populations outside the continental United States that may be the source of migratory populations, 4) clarification of the role cotton plays in resistance development, 5) continued evaluation of insecticides currently registered as alternatives for permethrin, 6) examination of insecticide mixtures, synergists, and IGR's as alternatives, 7) evaluation of microbial insecticides as alternatives, 8) identification of experimental insecticides that are as efficacious as registered materials, and 9) organization of regional cooperative research efforts in resistance monitoring and pheromone trapping. More research in the areas of migration, genetics of resistance, gene flow, fitness of resistant populations, selection pressure on alternate host plants, and other management tactics (HPR, biological control, early cultivars) will be necessary to develop a comprehensive plan.

It is evident that a resistance management plan for the entire southern growing area would not be feasible at this time. The carbamates and permethrin currently are more efficacious in the mid-South states of Louisiana and Mississippi than in Alabama, Georgia, and South Carolina. This would support a regional approach. States such as Louisiana and Mississippi may need separate plans within each state for those areas where cotton and soybean are grown together as opposed to areas where soybean predominate. However, these states share common cotton-soybean ecosystems in the Mississippi Delta wherein a plan would logically cross state boundaries.

Also, it is evident that the soybean looper resistance problem cannot be addressed adequately without the cooperation of fundamental and applied scientists but more importantly, scientists working on cotton, vegetables, and in areas outside the United States. The research challenges that remain are great. Without the involvement of those outside the soybean research community, formulation and implementation of a resistance management plan will be formidable.

ACKNOWLEDGEMENTS

We thank J. B. Graves and J. A. Ottea for reviewing the manuscript and Ellen Plaisance for typing the manuscript. Approved for publication by the Director of the Louisiana Agricultural Experiment Station as manuscript No. 91-17-5565.

REFERENCES

1. Georghiou, G. P. and Mellon, R. B., Pesticide resistance in time and space. In Pest Resistance to Pesticides, eds. G. P. Georghiou & F. Sait, Plenum, N. Y., 1983, pp. 1-46.

2. Palazzo, R. J., Comparison of the responses of adults and larvae of five lepidopterous species to seven insecticides. M.S. Thesis, 1978, Louisiana State University, Baton Rouge, Louisiana.

3. Newsom, L. D., Kogan, M., Miner, F. D., Rabb, R. L., Turnipseed, S. G. and Whitcomb, W. H., General accomplishments toward better pest control in soybean. In New Technology in Pest Control, ed. C. B. Huffaker, Wiley, N. Y., 1980, pp. 51-98.

4. Chiu, P. S-B and Bass, M. H., Soybean looper control on soybeans, 1975-1976. Insecticide & Acaricide Tests., 1979, 4, 156-8.

5. Herzog, G. A., Performance of pyrethroids in the southeastern region of the cotton belt. Proc. Beltwide Cotton Conf., 1988, pp. 231-2.

6. Leonard, B. R., Boethel, D. J., Sparks, A. N., Jr., Layton, M. B., Mink, J. S., Pavloff, A.M., Burris, E. and Graves, J. B., Variations in response of soybean looper (Lepidoptera: Noctuidae) to selected insecticides in Louisiana. J. Econ. Entomol., 1990, 83, 27-34.

7. Felland, C. M., Pitre, H. N., Luttrell, R. G. and Hamer, J. L., Resistance to pyrethroid insecticides in soybean looper (Lepidoptera: Noctuidae) in Mississippi. J. Econ. Entomol., 1990, 83, No. 1, 35-40.

8. Herzog, D. C., Sampling soybean looper on soybean. In Sampling Methods in Soybean Entomology, eds. M. Kogan & D. C. Herzog, Springer-Verlag, N. Y., 1980, pp. 141-68.

9. Alford, A. R. and Hammond, A. M., Jr., Plusiinae (Lepidoptera: Noctuidae) populations in Louisiana soybean ecosystems as determined with looplure-baited traps. J. Econ. Entomol., 1982, 75, 647-50.

10. McPherson, R. M. and Herzog, G. A., Soybean looper control problems in Georgia. In Proc. Soybean Looper Resistance Workshop, eds. J. Hamer & H. Pitre, Orange Beach, Alabama, 1990, pp. 9-10.

11. Turnipseed, S. G. and Newsom, L. D., Chemical control of soybean insect pests. In Proc. of the World Soybean Reserch Conference IV, ed. A. Pascale, Buenos Aires, Argentina, 1989, V, pp. 1509-18.

12. Sparks, T. C., Development of insecticide resistance in Heliothis zea and Heliothis virescens in North America. Bull. Entomol. Soc. Am., 1981, 27, No. 3, 186-92.

13. Graves, J. B., Impediments to developing a resistance management plan for soybean looper. In Proc. Soybean Looper Resistance Workshop, eds. J. Hamer & H. Pitre, Biloxi, Mississippi, 1989, pp. 13-5.

14. Rose, R. L., Leonard, B. R., Sparks, T. C. and Graves, J. B., Enhanced metabolism and knockdown resistance in a field versus a laboratory strain of the soybean looper (Lepidoptera: Noctuidae). J. Econ. Entomol., 1990, 83, No. 3, 672-7.

15. Dowd, P. F. and Sparks, T. C., Inhibition of trans-permethrin hydrolysis in Pseudoplusia includens (Walker) and use of inhibitors as pyrethroid synergists. Pesticide and Biochemistry and Physiology, 1987, 27, 237-45.

16. Layton, B. and Boethel, D. J., Control of soybean looper on soybean in Louisiana, 1987. Insecticide & Acaricide Tests, 1988, 13, 286-7.

17. Mink, J. S., Boethel, D. J. and Burris, G., Control of soybean looper, 1988. Insecticide & Acaricide Tests, 1989, 14, 277-8.

18. Wier, A. T., Mink, J. S., Boethel, D. J., Leonard, B. R. and Burris, E., Control of soybean loopers on soybean, 1989. Insecticide & Acaricide Tests, 1991, 16, 229-31.

19. Taylor, J. D. and McPherson, R. M., Soybean looper control, 1988. Insecticide & Acaricide Tests, 1989, 14, 281.

20. McPherson, R. M. and Perry, C. E., Soybean looper control with foliar insecticides on soybean, 1989. Insecticide & Acaricide Tests, 1990, 15, 285.

21. McPherson, R. M. and Moss, R. B., Soybean looper and velvetbean caterpillar control, 1990. Insecticide & Acaricide Tests, 1991, 16, 219-20.

22. Buckelew, L. D. and Mack, T. P., Control of soybean looper, 1988. Insecticide & Acaricide Tests, 1989, 14, 272-3.

23. Leonard, B. R., Control of soybean insect pests--Macon Ridge II, 1990. Insecticide & Acaricide Tests, 1991, 16, 224.

24. Felland, C. M. and Pitre, H. N., Soybean looper control, 1988. Insecticide & Acaricide Tests, 1990, 15, 278.

25. Turnipseed, S. G., Heinrichs, E. A., da Silva, R. F. P. and Todd, J. W., Response of soybean insects to foliar applications of a chitin synthesis inhibitor TH6040. J. Econ. Entomol., 1974, 67, No. 6, 760-2.

26. Yanes, J., Jr. and Boethel, D. J., Evaluation of insecticides for control of soybean insect pests, 1984. Insecticide & Acaricide Tests, 1985, 10, 247.

27. Layton, B. and Boethel, D. J., Effects of insecticides on pest and beneficial insects in soybean, 1985. Insecticide & Acaricide Tests, 1986, 11, 344-5.

28. Layton, M. B., Review of product performance in large-plot extension demonstrations in Mississippi. In Proc. Soybean Looper Resistance Workshop, eds. J. Hamer & H. Pitre, Biloxi, Mississippi, 1989, pp. 22-3.

29. McCutchen, B. F., Plapp, F. W., Jr., Nemec, S. J. and Campanhola, C., Development of diagnostic monitoring techniques for larval pyrethroid resistance for Heliothis spp. (Lepidoptera: Noctuidae) in cotton. J. Econ. Entomol., 1989, 82, No. 6, 1502-7.

30. Tynes, J. S. and Boethel, D. J., Control of soybean insects. Louisiana Cooperative Extension Service, 1991, Pub. 2211.

31. Mink, J. S. and Boethel, D. J., Development of a diagnostic technique for monitoring permethrin resistance in soybean looper (Lepidoptera: Noctuidae) larvae. J. Econ. Entomol., 1992, **85**, (IN PRESS).

32. Roush, R. T. and Miller, G. L., Considerations for design of insecticide resistance monitoring programs. J. Econ. Entomol. 1986, **79**, 293-8.

33. ffrench-Constant, R. H. and Roush, R. T., Resistance detection and documentation. In Pesticide Resistance in Arthropods, eds. B. E. Tabashnik & R. T. Roush, Chapman and Hall, N. Y., 1990, pp. 4-38.

34. Plapp, F. W., Jr., Jackman, J. A., Campanhola, C., Frisbie, R. E., Graves, J. B., Luttrell, R. G., Kitten, W. F. and Wall, M., Monitoring resistance in the tobacco budworm (Lepidoptera: Noctuidae) in Texas, Mississippi, Louisiana, Arkansas, and Oklahoma. J. Econ. Entomol., 1990, **83**, No. 2, 335-41.

35. Burleigh, J. G., Population dynamics and biotic control of the soybean looper in Louisiana. Environ. Entomol., 1972, **1**, No. 3, 290-4.

36. Wuensche, A. L., Relative abundance of seven pest species and three predacious genera in three soybean ecosystems. M.S. Thesis, 1976, Louisiana State University, Baton Rouge, Louisiana.

37. Beach, R. M. and Todd, J. W., Comparison of soybean looper (Lepidoptera: Noctuidae) populations in soybean and cotton/soybean agroecosystems. J. Entomol. Sci., 1986, **21**, No.1, 21-5.

38. Jensen, R. L., Newsom, L. D. and Gibbens, J., The soybean looper: Effects of adult nutrition on oviposition, mating frequency, and longevity. J. Econ. Entomol., 1974, **67**, 467-70.

39. Beach, R. M., Todd, J. W. and Baker, S. H., Nectaried and nectariless cotton cultivars as nectar sources for the adult soybean looper. J. Entomol. Sci., 1985, **20**, No. 2, 233-6.

40. Georghiou, G. P. and Taylor, C. E., Factors influencing the evolution of resistance. In Pesticide Resistance: Strategies and tactics for management, ed. National Research Council, National Academy Press, Washington, D. C., 1986, pp. 157-69.

41. Hensley, S. D., Newsom, L. D. and Chapin, J. B., Observations on the looper complex of the noctuid subfamily Plusiinae. J. Econ. Entomol., 1964, **57**, 1006-7.

42. Canerday, T. D. and Arant, F. S., The looper complex in Alabama (Lepidoptera, Plusiinae). J. Econ. Entomol., 1966, **59**, 742-3.

43. Chalfant, R. B., Toxicity of insecticides against two looper species, Pseudoplusia includens and Trichoplusia ni, in the laboratory. J. Econ. Entomol. 1969, 62, No. 6, 1343-4.

44. Sparks, A. N., An introduction to the status, current knowledge, and research on movement of selected lepidoptera in southeastern United States. In Movement of Highly Mobile Insects: Concepts and methodology in research, eds. R. Rabb & G. Kennedy, University Graphics, North Carolina State University, Raleigh, North Carolina, 1979, pp. 382-5.

45. Lingren, P. D., Henneberry, T. J. and Sparks, A. N., Current knowledge and research on movement of the cabbage looper and related looper species. In Movement of Highly Mobile Insects: Concepts and Methodology in Research, eds. R. Rabb & G. Kennedy, University Graphics, North Carolina State University, North Carolina, 1979, pp. 394-405.

46. Mitchell, E. R., Chalfant, R. B., Greene, G. L. and Greighton, C. S., Soybean looper: Populations in Florida, Georgia, and South Carolina, as determined with pheromone-baited BL traps. J. Econ. Entomol., 1975, 68, No. 6, 747-50.

47. Tabashnik, B. E., Computer simulation as a tool for pesticide resistance management. In Pesticide Resistance: Strategies and tactics for management. ed. National Research Council, National Academy Press, Washington, D. C., 1986, pp. 194-206.

48. Kogan, M., Insect problems of soybeans in the United States. In Proc. of World Soybean Conference II, ed. F. T. Corbin, Westview Press, Boulder, Co., 1980, pp. 303-25.

49. Hamer, J. and Pitre, H., Southern regional soybean looper pheromone trapping project, 1990 Report, Mississippi State University, Mississippi State, Mississippi, 126pp.

50. Linn, Jr., C. E., Du, J., Hammond, A. and Roelfs, W. L., Identification of unique pheromone components for soybean looper moths, Pseudoplusia includens. J. Chem. Ecol., 1987, 13, No. 6, 1351-60.

51. Dowd, P. F. and Sparks, T. C., Relative toxicity and ester hydrolysis of pyrethroids in the soybean looper and tobacco budworm (Lepidoptera: Noctuidae). J. Econ. Entomol., 1988, 81, No. 4, 1014-8.

52. Ratchford, K. F., Insect control on soybeans, 1985. Insecticide & Acaricide Tests, 1986, 11, 347.

53. Layton, B. and Boethel, D. J., Control of soybean looper on soybean looper in Louisiana, 1986. Insecticide & Acaricide Tests, 1987, 12, 277.

54. Leonard, B. R., Sparks, T. C. and Graves, J. B., Insecticide cross-resistance in pyrethroid-resistant strains of tobacco budworm (Lepidoptera: Noctuidae). J. Econ. Entomol., 1988, **81**, 1529-35.

55. Layton, B. R., Hamer, J. and Smith, R., Control of soybean insects in Mississippi, 1990. Insecticide & Acaricide Tests, 1991, **16**, 222-3.

56. Smith, R. H., Soybean looper situation and demonstration results, 1989. In Proc. Soybean Looper Resistance Workshop, eds. J. Hamer & H. Pitre, Orange Beach, Alabama, 1990, p. 6.

57. Sullivan, M. J. and Chapin, J. W., Soybean looper resistance work. In Proc. of Soybean Looper Resistance Workshops, eds. J. Hamer & H. Pitre, Orange Beach, Alabama, 1990, p. 37.

58. Leonard, B. R., Burris, E., Graves, J. B. and Elzen, G., Tobacco budworm: Insecticide resistance and field control in the Macon Ridge region of Louisiana, 1990. In Proc. Beltwide Cotton Conf., 1991, pp. 642-8.

59. Ayad, H. M., Hope, J. H. and Sullivan, T. A., Toxicity of Larvin Brand thiodicarb to Helicoverpa virescens collected from six sites in the USA. In Proc. Beltwide Cotton Conf., 1991, p. 796-8.

60. Jenkins, J. N., Parrott, W. L., and McCarty, J. C., State of the art of host plant resistance to insects in cotton. Proc. Beltwide Cotton Conf., 1991, 627-33.

61. Beach, R. M. and Todd, J. W., Toxicity of avermectin to larvae and adult soybean looper (Lepidoptera: Noctuidae) and influence on larvae feeding and adult fertility and fecundity. J. Econ. Entomol., 1985, **78**, 1125-8.

62. Mink, J. S., Wier, A. T., Thomas, J., Boethel, D. J., Leonard, B. R. and Burris, E., Soybean looper control with selected experimental insecticides, 1991. Insecticide & Acaricide Tests, 1992, 17, IN PRESS.

63. Lambert, L., Status of host plant resistance to insects in soybeans. In Proc. of Soybean Looper Resistance Workshop, Orange Beach, Alabama, 1990, pp. 28-9.

64. Sullivan, M. J., Resistance to insect defoliators. In Proc. World Soybean Research Conference III, ed. E. Shibles, 1985, pp. 400-5.

65. Smith, C. M., Expression, mechanisms, and chemistry of resistance in soybean, Glycine max L. (Merr.) to the soybean looper, Pseudoplusia includens (Walker). Insect Science and Its Application, 1985, **6**, 243-8.

66. Shepard, M., Carner, G. R. and Turnipseed, S. G., Colonization and resurgence of insect pests of soybean in response to insecticides and field isolation, Environ. Entomol, 1977, **6**, No. 4, 501-6.

67. Brown, D. W. and Goyer, R. A., Effects of a predator complex on lepidotperous defoliators of soybean. Environ. Entomol., 1982, 11, 385-9.

68. Yanes, J., Jr., Impact of individual species and complexes of pest arthropod species on soybean grown on various row spacings. PhD. Dissertation, 1985, Louisiana State University, Baton Rouge, Louisiana.

69. Shepard, M. and Herzog, D. C., Soybean: Status and current limits to biological control in the southeastern U. S. In Biological Control in Agricultural IPM Systems, eds. M. A. Hoy & D. C. Herzog, Academic Press, N. Y., 1985, pp. 557-72.

70. Harper, J. D., McPherson, R. M. and Shepard, M., Geographical and season occurrence of parasites, predators, and entomopathogens. In Natural Enemies of Arthropod Pests in Soybean, ed. H. N. Pitre, South. Coop. Ser. Bull. 285, 1983, pp. 7-19.

71. Daigle, C. J., Boethel, D. J. and Fuxa, J. R., Parasitoids and pathogens of soybean looper and velvetbean caterpillar (Lepidoptera: Noctuidae) in soybeans in Louisiana. Environ. Entomol., 1990, 19, 746-52.

72. Boethel, D. J. and Orr, D. B., Pests of soybean. In Classical Biological Control in the Southern United States, eds. D. Habek, F. D. Bennett, & J. H. Frank. Southern Coop. Series Bull., 355, 1990, pp. 30-8.

73. Livingston, J. M. and Yearian, W. C., A nuclear polyhedrosis virus of Pseudoplusia includens, (Lepidoptera: Noctuidae). J. Invert. Path., 1972, 19, 107-12.

74. Kuo, S.-y., Persistence of nuclear polyhedrosis of soybean looper, Pseudoplusia includens (Walker) (Lepidotpera: Noctuidae) in the soybean ecosystem. M.S. Thesis, 1975, Louisiana State University, Baton Rouge, Louisiana.

EFFICACY OF BACULOVIRUSES AND THEIR IMPACT ON PEST MANAGEMENT PROGRAMS

JOE FUNDERBURK
North Florida Research and Education Center
University of Florida
Route 3 Box 4370
Quincy, Florida, USA

JIM MARUNIAK, DRION BOUCIAS, and ALEJANDRA GARCIA-CANEDO
Department of Entomology and Nematology
University of Florida
Gainesville, Florida, USA

ABSTRACT

A highly species-specific nuclear polyhedrosis virus isolated from velvetbean caterpillar larvae (VBC), Anticarsia gemmatalis, has been used widely as a biological control for VBC in Brazilian soybean. A single virus application of a similar Brazilian isolate significantly suppressed VBC populations for the remainder of the growing season in field trials in Florida. However, the virus had to be applied when larval populations were one half normal economic threshold densities to prevent yield loss under severe pressure. Studies were expanded to compare efficacy of the above-mentioned viruses with a less species-specific nuclear polyhedrosis virus isolated from Autographa californica larvae. All three viruses at two treatment rates provided significant suppression of the damaging middle and late instar VBC larvae. The impacts on nontarget pests and predators of each virus and a standard chemical insecticide treatment also were determined in this study. The chemical insecticide treatment did not affect populations of nontarget pests or predators, and tolerance of subeconomic VBC densities in the virus treatments did not enhance populations of predators. Consequently, there was little ecological advantage in our research for using baculoviruses for VBC control compared with the standard chemical insecticide. Baculoviruses may have greater commercial potential in the USA as biopesticides against other pests more difficult to control with chemical insecticides.

INTRODUCTION

The velvetbean caterpillar (VBC), Anticarsia gemmatalis, is a serious defoliator of soybean throughout much of the soybean growing region in South America and in the extreme southern soybean producing regions of North America. A nuclear polyhedrosis virus of VBC (AgNPV) was first isolated from dead larvae in Peru [1]. An isolate from the region of Campinas in Brazil was described by Allen and Knell [2]. Additional isolations from other regions have been made [3]. Some of the initial field trials that demonstrated its potential as a biopesticide for use in soybean integrated pest management (IPM) programs were conducted in Florida in the USA [4].

The chapter by Moscardi describes the success utilizing the AgNPV for controlling outbreak VBC populations in a soybean IPM program in Brazil. When applied against populations of early instar VBC larvae in Brazil, a wide range from low to high AgNPV doses results in >70% mortality [5]. A single application is sufficient to maintain VBC populations below damaging levels for the remainder of the growing season. Although the half life on soybean leaves is about 6 days, persistence apparently results from increase in diseased larvae directly in the field. By all indications, there are no substantial differences in virulence to different AgNPV isolates. Although the natural host is very susceptible to all isolates, the AgNPV is very specific, and other lepidopterous pests are infected only at very high doses.

While other lepidopterous species in soybean are occasional pests rather than perrenial pests like VBC, economic problems from occasional lepidopterous pests in the USA are more widespread and greater acreage is threatened. Further, outbreak populations of some occasional pests are more difficult to suppress with chemical insecticides, and fewer efficacious labeled products are available. Important occasional pests include the soybean looper (SBL), Pseudoplusia includens, and the corn earworm (CEW), Helicoverpa zea. Some baculoviruses are known to infect multiple species at low dosages, including the nuclear polyhedrosis virus of Autographa californica (AcNPV) [6].

Soybean is habitat for a wide range of natural enemies of arthropod pests. Kogan and Turnipseed [7] review and discuss the literature. Important generalist predators include spiders (Aranaea: Araneidae), bigeyed bugs (Hemiptera: Lygaeidae), damsel bugs (Hemiptera: Nabidae), and others. A fungal pathogen, Nomuraea rileyi, is an important natural disease of many lepidopterous pests. Parasitism of VBC larvae is low, but can be important for other pests. The impacts of beneficial organisms on pest populations are rarely well understood, but there is no doubt that indigenous natural enemies are very important in suppressing pest populations in all soybean production regions. Consequently, conservation and enhancement of natural enemy populations is a primary consideration in soybean IPM programs.

Efficacy against several pests, especially those difficult to control with chemical insecticides, undoubtedly would enhance commercial potential of baculoviruses in the USA. An objective of our research was to compare efficacy of two Brazilian AgNPV isolates, an AcNPV isolate, and a chemical insecticide standard against VBC, SBL, and CEW larvae. Using baculoviruses to control these pests in soybean IPM programs may enhance populations of beneficial species, thereby reducing the likelihood of outbreaks by nontarget pests. An additional objective was to compare the impacts of each treatment on populations of nontarget pests and predators.

MATERIALS AND METHODS

Soybean was grown on a Norfolk sandy loam soil at Quincy, Fla. USA. In 1990 and 1991, 'Braxton' soybean was planted on 7 and 10 June, respectively, in plots 15.5 by 4.6 m. Row spacing was 76 cm and seeding rate was about 35 viable seed per linear m of row. Experimental design was a randomized complete block each year. Plots were separated within blocks by 4 rows and between blocks by 4-m alleys. Standard herbicide treatments were applied similarly across all plots as needed. No fungicide applications were made to the plots in either year. No insecticide applications were made except for the virus and chemical insecticide treatments under investigation.

Treatments were the same each year, including an untreated control, diflubenzuron 25W (Uniroyal Chemical Co., Inc., Middlebury, Conneticut USA), and six baculovirus treatments. These consisted of plaque-purified AgNPV-2D (obtained from Jim Maruniak [8]), a commercial AgNPV preparation (obtained from Flavio Moscardi, Empresa Brasileira de Pesquisa Agropecuaria, Londrina, Brazil), and plaque-purified AcNPV-E2 (obtained from Max Summers, Texas A&M University, College Station, USA [9]), each applied at rates of 1.2×10^{10} and 1.2×10^{11} polyhedra per ha. Treatments were applied on 15 August in both years, when VBC larval populations were about 6 larvae per linear m of row. Both virus and chemical insecticides were applied with a 2-row, gas-pressurized backpack sprayer that was equipped with 3 nozzles per row. The amount of spray was about 234 l per ha.

Plaque-purified and prototype isolate AgNPV-2D was amplified in vivo by injecting the hemolymph of fourth- and fifth- instar VBC larvae with nonoccluded virus, and the plaque-purified and prototype isolate AcNPV-E2 was amplified similarly by injecting Trichoplusia ni larvae. Polyhedra were purified from the infected larvae, by using a modification of the protocols reported by Maruniak [10]. The Londrina AgNPV commercial preparation obtained as a dry powder was used in the experiments. Numbers of polyhedra in all preparations were quantified using a hemacytometer.

Densities of small (instars 1 and 2), medium (instars 3 and 4), and large (instars 5 and 6) VBC, CEW, and SBL larvae were estimated weekly in each plot, beginning each year one week following treatment application and ending near crop senescence. Nymphal and adult southern green stink bugs (Nezara viridula), bigeyed bugs (Geocoris punctipes), and damsel bugs (Raduviolus spp. and Nabis spp.) also were estimated on each sample date. Insect sampling was carried out as described by Kogan and Pitre [11]. All plots were sampled on each sampling date by beating the plants on both sides of the row onto a 0.9-m ground cloth placed between the rows. Two samples were taken per plot per sample date. Adjacent plants were searched at their base and the soil visually examined. To avoid unnecessary spread of virus between plots, untreated controls were sampled first, followed by the chemical insecticide treatment, the low virus rate treatments, and the high virus rate treatments.

The influence of virus and insecticide treatments on estimates were determined by analyses of variance. Data for each sample date of each growing season were analyzed separately. Significant ($P < 0.05$) treatment differences were separated by using Duncan's [12] multiple range test.

RESULTS AND DISCUSSION

Populations of VBC do not overwinter in northern Florida, and larvae usually first appear in soybean fields in July or August. Population densities reach the economic threshold of 12 larvae per linear m of row in late August or September. Previous experiments were conducted by us to evaluate the effect of date of application on efficacy of the AgNPV-2D against VBC larvae in soybean. In one experiment in which the untreated control was completely defoliated, application of AgNPV-2D when VBC density was about 6 per linear m of row significantly increased population suppression compared with later application when density was about 12 per linear m of row (Table 1) [13]. Both AgNPV-2D treatments gave significant suppression of larval populations for the remainder of the growing season. Diflubenzuron also significantly suppressed VBC larvae for the remainder of the growing season, unlike the other chemical insecticide treatments included in the experiment. By comparison with the diflubenzuron treatment, early application of AgNPV-2D prevented yield loss, while late application did not. Yields for the short-residual chemical insecticide treatments were similar to the untreated control.

The economic threshold is a pest or injury level established which indicates when a management action should be taken to prevent economic loss [14]. Because baculoviruses do not provide the rapid suppression typical of chemical insecticides, early application is needed in some situations to prevent economic loss. Our results in 1987 (Table 1) also indicated that levels of suppression of AgNPV-2D against VBC populations are less in Florida than in Brazil [5]. For these reasons, treatments were applied in our 1990 and 1991 experiments when population densities were about 6 per linear m of row. Chemical insecticide and baculovirus treatments were applied on the same day for experimental comparison purposes. Because of its long residual efficacy, diflubenzuron is the most commonly used chemical insecticide against VBC in soybean IPM programs. Therefore, it was included as the chemical standard in these experiments.

Estimates of the damaging medium and large VBC larvae did not reach the economic threshold density of 12 per linear m of row on any sample date. Densities were greatest on sample dates in early September both years, and treatment efficacy can best be summarized by examining sample estimates on these dates (Table 2). Populations of small larvae were significantly suppressed only in the diflubenzuron treatment. In 1990, in all baculovirus and chemical insecticide treatments estimates of both medium and large larvae were significantly lower than the untreated control with no other significant differences. In 1991, density estimates of medium larvae were significantly lower than the untreated control in the diflubenzuron and two baculovirus treatments. Density estimates of large larvae were very low, and there were no significant treatment differences.

Results from our field trials demonstrated that the Brazilian AgNPV isolates are efficacious against Florida populations of VBC, although levels of suppression were less than those reported against Brazilian populations [5]. As in Brazil, a single application suppressed populations for the remainder of the growing season. The AgNPV-2D, the commercial preparation of AgNPV from Londrina, and the AcNPV-E2 each demonstrated efficacy against VBC larvae at the treatment rates in the 1990 and 1991 trials; however, population densities were too low to obtain a good statistical

TABLE 1

Mean number of velvetbean caterpillar larvae per 25 sweeps on 4 dates and mean seed yield for selected baculovirus and chemical insecticide treatments in a soybean experiment conducted in 1987, Gadsden Co., Fla. (from Funderburk et al. [13]).

Treatment and Active Ingredient/ha	Mean No. VBC/25 Sweeps				Seed Yield kg/ha
	8 Sep	11 Sep	18 Sep	25 Sep	
Untreated Control	72 a[1]	58 a	11 b	56 bc	968 f
Diflubenzuron 25W 0.036 kg	13 bc	2 c	3 c	9 e	1998 ab
Methomyl L 0.280 kg	16 bc	3 c	178 a	90 a	962 f
Acephate 75S 0.841 kg	17 bc	5 c	177 a	70 ab	870 f
AgNPV-2D 1.6 x 10^{10} Polyhedra	65 a	42 b	110 b	21 de	1542 cde
AgNPV-2D[2] 1.6 x 10^{10} Polyhedra	64 a	51 ab	39 c	27 de	1739 bc

[1]Means in the same column followed by the same letter are not significantly different according to Duncan's [12] multiple range test (P<0.05). Letters are taken from analyses that include additional treatments not shown here.

[2]Treatment applied on 28 Aug. All other baculovirus and chemical insecticide treatments applied on 4 Sep.

TABLE 2

Mean number of small, medium, and large velvetbean caterpillar larvae in baculovirus and chemical insecticide treatments on 5 Sep 1990 and 4 Sep 1991 in soybean experiments, Gadsden Co., Fla. USA.

Treatment and Active Ingredient/ha		Mean No. VBC/1-m Row in 1990				Mean No. VBC/1-m Row in 1991			
		Small	Medium	Large	Total	Small	Medium	Large	Total
Untreated Control		30.7 a[1]	6.7 a	3.6 a	41.0 a	14.1 ab	5.5 a	1.0 a	20.6 a
Diflubenzuron 25W	0.036 kg	3.1 b	1.0 b	0.4 b	4.5 c	1.2 b	0.1 c	0.1 a	1.4 b
Londrina AgNPV	1.2×10^{10} Polyhedra	22.9 a	2.6 b	1.1 b	26.8 b	9.7 ab	2.2 bc	0.6 a	12.4 ab
Londrina AgNPV	1.2×10^{11} Polyhedra	20.2 a	1.8 b	0.0 b	22.0 b	10.3 ab	3.1 abc	0.7 a	14.1 ab
AgNPV-2D	1.2×10^{10} Polyhedra	21.8 a	2.6 b	0.6 b	25.0 b	9.9 ab	4.3 ab	1.1 a	15.3 ab
AgNPV-2D	1.2×10^{11} Polyhedra	19.1 a	1.5 b	0.0 b	20.7 b	9.8 ab	1.9 bc	0.4 a	12.1 ab
AcNPV-E2	1.2×10^{10} Polyhedra	27.1 a	2.0 b	0.1 b	29.2 b	21.2 a	3.4 ab	1.4 a	26.1 a
AcNPV-E2	1.2×10^{11} Polyhedra	20.7 a	1.2 b	0.0 b	21.9 b	9.2 ab	3.7 ab	1.0 a	14.0 ab

[1]Means in the same column followed by the same letter are not significantly different according to Duncan's [12] multiple range test ($P<0.05$).

comparison of individual treatments. Also, population densities of CEW, SBL, and other occasional lepidopterous pests remained very low (data not shown), and it was not possible to examine efficacy against these important pests. Populations of southern green stink bugs exceeded established economic thresholds in both the 1990 and 1991 field trials; however, there were no significant treatment differences in population estimates of nymphs or adults on any sample date (data not shown).

Populations of bigeyed bugs and damsel bugs were not much affected by any treatment. These predators were abundant on all sample dates, but density estimates of both were greatest in 19 September in 1990 and on 18 September in 1991 (Table 3). There were no significant treatment differences between estimates of damsel bug and bigeyed bug nymphs on any sample date either year. The only significant treatment differences for adult estimates occurred on the 18 September 1991 sample date. We can not conclude the reasons for these observed differences. Populations of the predators may have been directly affected by differences between treatments in VBC larval densities, or differences between treatments in the amount of leaf mass consumption may have indirectly affected developmental biology or dispersal behavior of the predators.

As previously discussed, indigenous natural enemies are very important in suppressing pest populations in soybean, and conservation and enhancement of natural enemy populations is a primary consideration in soybean IPM programs. Deleterious impacts resulting from management practices frequently result in resurgence of target pests and induced outbreaks of secondary pests [7]. However, tolerance of subeconomic VBC larval populations resulted in little increase in populations of damsel bugs or bigeyed bugs in our studies. Further, the standard chemical insecticide treatment demonstrated no negative impacts on populations of bigeyed bugs and damsel bugs.

Another important natural enemy of VBC and occasional lepidopterous pests in soybean is N. rileyi. Normally, epizootics on VBC larvae occur after an outbreak already has occurred. Previous research by Moscardi et al. [4] demonstrated that treatment of AgNPV at numerous rates retarded or inhibited the development of epizootics in Florida soybean. However, additional research is needed to determine if tolerance of subeconomic densities of VBC can result in an enhancement of N. rileyi in soybean and in a subsequent increase in mortality to later populations of occasional lepidopterous pests.

Populations of SBL and CEW remained very low in our plots, and efforts to evaluate field efficacy of AgNPV-2D, a commercial AgNPV preparation, and AcNPV-E2 against these occasional pests were unsuccessful. Efficacy against these pests, as well as VBC, would enhance commercial potential of baculoviruses in soybean in the USA. Populations of SBL in the Southeast USA recently have become resistant to many labeled insecticides [15]. The pest does not overwinter in the region, and resistant populations are believed to have developed as a result of exposure to heavy pressure from insecticides in overwintering areas of extreme southern Florida and the Caribbean. We urge additional research to evaluate field efficacy of baculoviruses against SBL in soybean. Considerable research of baculovirus effects on ecology of natural enemies when applied against SBL in soybean has been conducted. Ruberson et al. [16] is a recent publication, which also provides a review of previous research. The impacts of baculovirus treatment on populations of nontarget pests and natural enemies compared with standard chemical insecticides used against SBL have not been adequately

TABLE 3

Mean number of damsel bug and bigeyed bug nymphs and adults in baculovirus and chemical insecticide treatments on 19 Sep 1990 and 18 Sep 1991 in soybean experiments, Gadsden Co., Fla. USA.

Treatment and Active Ingredient/ha		Mean No. Damsel Bugs/1-m Row				Mean No. Bigeyed Bugs/1-m Row			
		1990		1991		1990		1991	
		Nymphs	Adults	Nymphs	Adults	Nymphs	Adults	Nymphs	Adults
Untreated Control		1.7 a¹	1.5 a	1.8 a	0.76 bc	1.7 a	2.3 a	1.5 a	1.2 a
Diflubenzuron 25W	0.036 kg	1.0 a	2.1 a	2.3 a	0.9 bc	1.4 a	1.8 a	1.8 a	0.0 c
Londrina AgNPV	1.2 x 10¹⁰ Polyhedra	1.5 a	1.1 a	2.8 a	1.5 ab	1.2 a	1.2 a	2.3 a	0.7 b
Londrina AgNPV	1.2 x 10¹¹ Polyhedra	1.7 a	1.4 a	2.5 a	1.2 abc	1.1 a	0.7 a	2.1 a	0.0 a
AgNPV-2D	1.2 x 10¹⁰ Polyhedra	1.0 a	1.5 a	3.2 a	1.5 ab	0.7 a	1.7 a	0.7 a	0.7 b
AgNPV-2D	1.2 x 10¹¹ Polyhedra	0.4 a	1.7 a	4.2 a	1.1 abc	0.7 a	2.5 a	1.4 a	0.1 c
AcNPV-E2	1.2 x 10¹⁰ Polyhedra	1.1 a	2.0 a	3.2 a	2.1 a	0.9 a	1.5 a	1.8 a	0.0 c
AcNPV-E2	1.2 x 10¹¹ Polyhedra	1.1 a	1.7 a	3.1 a	0.3 c	1.8 a	2.0 a	1.2 a	0.0 c

¹Means in the same column followed by the same letter are not significantly different according to Duncan's [12] multiple range test ($P<0.05$).

evaluated.

Populations of VBC are easily suppressed with numerous available chemical insecticides, and application of the standard chemical insecticide for VBC in soybean resulted in no undesirable nontarget effects to soybean IPM programs in our research. Commercial potential of baculoviruses in the USA at the present time may be greater for difficult-to-control pests in soybean or other crops, as recent commercial interest in and research on baculoviruses against Spodoptera spp. may indicate.

REFERENCES

1. Steinhaus, E.A., and Marsh, G.A., Report of diagnosis of diseased insects, 1951-1961. Hilgardia, 1962, 33, 349-90.

2. Allen, G.E., and Knell, J.D., A nuclear polyhydrosis virus of Anticarsia gemmatalis: I. Ultrastructure, replication, and pathogenecity. Fla. Entomol., 1977, 60, 233-40.

3. Moscardi, F., Use of viruses for pest control in Brazil: The case of the nuclear polyhedrosis virus of the soybean caterpillar, Anticarsia gemmatalis. Mem. Inst. Oswaldo Cruz, 1989, 84, 51-6.

4. Moscardi, F., Allen, G.E., and Greene, G.L., Control of the velvetbean caterpillar by nuclear polyhedrosis virus and insecticides and impact of treatments on natural incidence of the entomopathogenic fungus Nomuraea rileyi. J. Econ. Entomol., 1981, 74, 480-5.

5. Moscardi, F., and Correa-Ferreira, B.S., Biological control of soybean caterpillars. In World Soybean Research Conference III: Proceedings, ed. R. Shibles, Westview Press, Inc., Boulder, Colorado, 1985, pp. 703-11.

6. Maruniak, J.E., and Summers, M.D., Autographa californica nuclear polyhedrosis virus phosphoproteins and synthesis of intracellular proteins after virus infection. Virology, 1981, 109, 25-34.

7. Kogan, M., and Turnipseed, S.E., Ecology and management of soybean arthropods. Annu. Rev. Entomol., 1987, 32, 507-38.

8. Johnson, D.W., and Maruniak, J.E., Physical map of Anticarsia gemmatalis nuclear polyhedrosis virus (AgMNPV-2) DNA. J. Gen. Virol., 1989, 70, 1877-83.

9. Smith, G.E., and Summers, M.D., Analysis of baculovirus genomes with restriction endonucleases. Virology, 1978, 89, 517-27.

10. Maruniak, J.E., Baculovirus structural proteins and protein synthesis. In The Biology of Baculoviruses, ed. R.R. Granados and B.A. Federici, CRC Press, Inc., Boca Raton, Florida, 1986, Vol. I, pp. 129-46.

11. Kogan, M., and Pitre, Jr., H.N., General sampling methods for above-ground populations of soybean arthropods. In Sampling Methods in Soybean Entomology, ed. M. Kogan and D.C. Herzog, Springer-Verlag, Inc., New York, 1980, pp. 30-60.

12. Duncan, D.B., Multiple range and multiple F tests. Biometrics, 1955, 11, 1-42.

13. Funderburk, J., Brown, A., Maruniak, J., and Boucias, D., Velvetbean caterpillar control in soybean, 1987. Insect. Acar. Tests, 1988, 13, 281-2.

14. Pedigo, L.P., Hutchins, S.H., and Higley, L.G., Economic injury levels in theory and practice. Annu. Rev. Entomol., 1986, 31, 341-68.

15. Leonard, B.R., Boethel, D.J., Sparks, Jr., A.N., Layton, M.B., Mink, J.S., Pavloff, A.M., Burris, E., and Graves, J.B., Variations in response of soybean looper (Lepidoptera: Noctuidae) to selected insecticides in Louisiana. J. Econ. Entomol., 83, 27-34.

16. Ruberson, J.R., Young, S.Y., and Kring, T.J., Suitability of prey infected by nuclear polyhedrosis virus for development, survival, and reproduction of the predator Nabis roseipennis (Heteroptera: Nabidae). Environ. Entomol., 20, 1475-9.

APPENDIX

This is Florida Agricultural Experiment Station Journal Series No. R-02169.

USE OF VIRUSES AGAINST SOYBEAN CATERPILLARS IN BRAZIL

FLÁVIO MOSCARDI AND DANIEL RICARDO SOSA-GOMEZ
EMBRAPA-CNPSo
Caixa postal 1061, 86001 Londrina, Paraná, Brazil

ABSTRACT

Among the lepidopterous insects associated with soybeans in Brazil, the velvetbean caterpillar, Anticarsia gemmatalis, is the most important and widespread defoliator, seconded by species of Plusiinae, such as Pseudoplusia includens and Rachiplusia nu, which are of lesser importance but usually occur associated with populations of A. gemmatalis. Research with a nuclear polyhedrosis virus of A. gemmatalis (AgNPV) has led to a wide-scale use of the pathogen at farmer level, estimated in ca. 1,000,000 hectares annually. Most of its production has been carried out by estate organizations and farmer cooperatives, but at present four private companies are commercializing the AgNPV as a biological pesticide. NPVs of the Plusiinae complex are also being developed as a microbial insecticide, through a cooperative effort between Brazilian, Argentinian, and Uruguaian researchers. Research developments with the AgNPV, progress in its use at farmer level, and the work being carried out with baculoviruses of other soybean insects are discussed.

INTRODUCTION

In Brazil, soybean is cultivated on ca. 9.0 million hectares, ranging from temperate zones in the south to sub-tropical and tropical areas in Central and Northeast regions. Key insect pests are represented by lepidopterous defoliators and pod sucking pentatomids. Among the pentatomids, Nezara viridula, Piezodorus guildinii, and Euchistus heros are the predominant species, while the velvetbean caterpillar, Anticarsia gemmatalis, is by far the most important and widespread among the defoliators (1). A complex of Plusiinae, especially Pseudoplusia includens and Rachiplusia nu, usually occurs in association with A. gemmatalis populations, assuming higher importance in the South. The soybean shoot borer, Epinotia aporema, is also generally considered of secondary importance, except in some regions in the South, where it may reach high populations. Other lepidopterous insects of occasional or restricted occurence are species of Spodoptera, such as S. eridania, S. frugiperda and S. latifascia, Omiodes indicata, Urbanus proteus and various geometrids, among others.

A soybean integrated pest management (IPM) programme, established in Brazil in 1977 (2) has led to a three-fold reduction in insecticides usage on the crop (3,4). Among the IPM tactics employed, considerable emphasis has been given to biological control, either through strategies which allow their conservation or their employment for inoculative and inundative releases on soybeans. Several entomopathogens are associated with soybean insects, especially fungi and viruses (5). The fungus Nomuraea rileyi is the most important naturally-occurring agent on defoliating caterpillars, and usually maintains A. gemmatalis populations below the economic injury level in seasons with extended periods of high humidity (5). Its occurrence together with Entomophthoralean fungi and parasitoids generally maintain Plusiinae species under control in most regions. Viruses, mainly those belonging to the Baculovirus group, are often found on soybean lepidopterous insects in Brazil and other South American countries (Table 1), but usually at low natural incidences.

TABLE 1

Viruses associated with soybean lepidopterous insects in South America

Host species	Type of virus*	Reference
Anticarsia gemmatalis	NPV, IV	(6,7,8,9)
Epinotia aporema	GV	(10)
Pseudoplusia includens	NPV	(11)
Rachiplusia nu	NPV	(12)
Spodoptera eridania	NPV	(10)
Spodoptera frugiperda	NPV, GV	(13)
Spodoptera latifascia	NPV	(10)
Urbanus proteus	NPV	(10)

* NPV= nuclear polyhedrosis; GV= granulosis virus; IV= iridescent virus

Among these viruses, the nuclear polyhedrosis virus of Anticarsia gemmatalis (AgNPV) has been the most studied. It was first found in Brazil, in 1972, in diseased larvae collected in the region of Campinas, State of São Paulo (6). Later it was reported in other regions (7,8). Early results with this pathogen, obtained in small field-plot tests, indicated a high potential for its use as a viral insecticide against A. gemmatalis (7,14). Beginning in 1979, the National Soybean Research Center (CNPSo) of the Brazilian Organization for Agricultural Research (EMBRAPA), started a research programme directed to the development of the AgNPV as a microbial pesticide. Nowadays the AgNPV is widely used in Brazil and, to some extent, in other South American countries, and represents one of the most significant programmes, on a worldwide basis, aiming at control of insect pests with viruses of field crops (15,16,17,18). More recently, similar efforts were started towards the use of NPV against the Plusiinae complex.

The objective of this paper was to review the research developments with the AgNPV and other baculoviruses, the progress in AgNPV use as a microbial insecticide, as well as the future prospects for utilization of entomopathogenic viruses of soybean insect pests in Brazil and in other South American countries.

1. RESEARCH DEVELOPMENTS AND USE OF AgNPV
1.1. Field efficacy

Timing of AgNPV applications, based on A. gemmatalis population intensity and age composition, is necessary to guarantee adequate control of the insect and to avoid soybean defoliation above the economic injury level. Field cage studies showed that economic damage will be avoided when the virus is applied against populations of up to 20 small larvae per meter of row, when compared to populations of healthy larvae (Fig. 1). Field spraying when the majority of larvae are under 1.5cm is important, since larger larvae are much less susceptible to the virus, added to the fact that they can inflict higher damage to soybean than small ones (17).

Taking into account these conditions for AgNPV application, different field trials have shown that at a dosage of 50 larval equivalents (LE) per hectare, or ca. 1.0×10^{11} polyhedron inclusion bodies (PIB) per hectare, a control level of 80% or higher can be attained, with defoliation maintained at acceptable levels (17). In these trials, soybean yields in AgNPV-treated plots were not significantly different from those obtained in insecticide-treated plots. These results were confirmed in large farmer fields in different soybean producing regions (17).

1.2. AgNPV epizootiology

The half life of AgNPV on soybean leaves is ca. 6-7 days for crude and formulated preparations, compared to ca. four days for purified preparations of the pathogen (15,17). However, since treated insects start dying on the 5th day after application, there is a continuous reposition of the pathogen on the crop, resulting in a sufficient amount of inoculum to contaminate larvae hatched after the 6th day post application. As a consequence, usually a single application of the AgNPV controls the insect during the entire season (15,17). A major proportion of the virus is deposited in the soil, a natural reservoir of the pathogen, where it can persist for long periods. In heavy-clay soils of northern Paraná State, the AgNPV retained ca. 40% of its activity in no till soybean fields after one year of its deposition in the soils, compared to ca. 13% in a paired field cultivated under a conventional tillage system (17). After two years, virus activity was ca. 26% and ca. 8%, respectively, in the two systems. Collection of A. gemmatalis larvae in those areas, during the soybean growing season showed that the virus present in the soil was able to infect natural populations of the insect. As in other virus-host systems, predators and parasitoids play an important role in the AgNPV epizootiology, since these agents have been shown to act as passive carriers of the disease inside treated areas and between soybean fields (15,17).

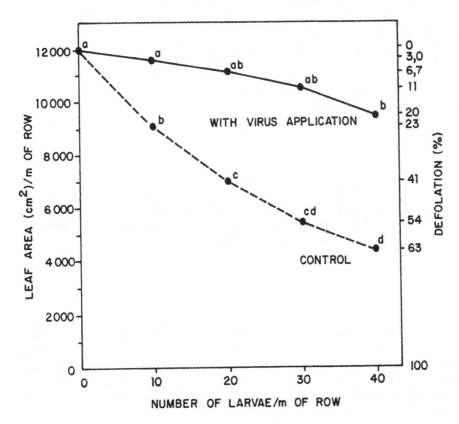

Figure 1. Leaf area and percentage of defoliation of soybean
artificially infested with _Anticarsia_ _gemmatalis_
larvae at different population levels, in the
presence or abscence of AgNPV application

1.3. Production and formulation

AgNPV production can be accomplished by different methods (17),
including: a) infection of larvae, continuously reared on artificial diet;
b) release of laboratory-reared larvae on soybeans treated with the virus,
either in glasshouses or in field screen cages; c) release of laboratory
obtained adults in large screen cages, with AgNPV application on resulting
larval populations; and d) application against naturally-occurring
populations of A. gemmatalis, followed by collection of AgNPV dead larvae.
In all cases, dead larvae are collected from the 7th to 10th day after
inoculation and frozen for further processing as a formulation. The less
expensive method, however, has been the production on naturally occurring

larval populations during the soybean growing season. In the 1990/91
season this procedure alone has allowed CNPSo-EMBRAPA to produce more than
3,000kg of infected larvae in a period of 20 days, enough for the
treatment of more than 150,000 hectares.

Studies towards development of a simple standardized formulation of
the AgNPV (17) have led to a wettable powder, obtained through air drying
and milling of a kaolin-based AgNPV slurry. This formulation has been
shown to provide adequate control of A. gemmatalis populations at dosages
varying from 1.0 x 10^{11} to 2.0 x 10^{11} polyhedron inclusion bodies per
hectare (Table 2), and has been routinely used by soybean farmers since
the 1986/87 season.

TABLE 2

Average number of large larvae (LL) (<1.5cm) of Anticarsia gemmatalis per
two meters of soybean row and percent control (PC) obtained in field plots
treated with a wettable powder (kaolin based) formulation of the AgNPV and
with a non formulated virus

Treatments	Dosage/ha	Day after application			
		7		10	
		LL	PC	LL	PC
Formulated virus	5 x 10^{10} PIB	10.2	74.0	1.8	64.5
Formulated virus	1 x 10^{11} PIB	8.5	78.4	0.8	87.1
Formulated virus	2 x 10^{11} PIB	6.3	84.0	0.8	88.7
Non formulated virus	1 x 10^{11} PIB	8.0	79.6	1.0	83.9
Control	–	39.3	–	6.2	–

1.4. Specificity and safety tests

Laboratory studies in the USA indicated low susceptibility to the
AgNPV by other noctuid species, such as Pseudoplusia includens , Heliothis
spp., and Spodoptera spp. (19). Similar trials conducted in Brazil (20)
showed that this virus was able to infect other lepidopterous species such
as Bombyx mori (silkworm), Chlosine lacinia saundersii (sunflower
caterpillar), Spodoptera latifascia (pod worm), and Trichoplusia ni
(cabbage looper), only at very high dosages, while the natural host
(A. gemmatalis) presented high susceptibility to the pathogen (120 to
250,000 - fold differences as compared to the other species).

Safety tests to vertebrates have been conducted with the kaolin-based,
wettable powder formulation (ca. 1.0 x 10^{10} PIB/g) by exposing rats, rabbits
and guinea pigs to the AgNPV in different ways (Table 3). In all of these
tests, no adverse effects were observed on the test animals.

TABLE 3

Type of tests conducted with the AgNPV on vertebrates

Test animal	Type of AgNPV administration	Observation	Dosage
Albino Rat	"per os"	oral toxicity	2000mg/kg
Albino Rat	dermal	dermal toxicity	4000mg/kg
Albino Rat	dermal	dermal irritation	non specified
Guinea Pig	intradermic	dermal sensitivity	10 inoculations of 0.1ml of a 1:1000 dilution in a physiological solution
Albino Rat	eye	eye irritation	non specified

Source: Technological Institute of Paraná (TECPAR), Brazil, 1991
        (unpublished)

1.5. Mixtures of AgNPV with pesticides

A study involving mixtures of AgNPV with eight insecticides and two post emergence herbicides, at 2% in a aqueous suspension of the pathogen, showed that virus activity was not affected by the pesticides (Moscardi, unpublished). Furthermore, mixtures with low rates of insecticides (1/4 to 1/8 of the recommended dosages) were efficient in reducing A. gemmatalis populations which had surpassed the threshold for virus use. Some of these data are shown in Table 4, where the effect of low rates of insecticides mixed with the AgNPV provided adequate control of the insect, as well as avoided high defoliation levels, when compared to the recommended dosage of profenophos, to the use of virus alone, to the control, and to the respective insecticides at the reduced rates.

For larval populations above the threshold for virus application, these mixtures have worked well because of the knock down effect of the insecticides, even at reduced rates, with the virus acting on the remaining population. This allows reposition of the AgNPV in soybean fields, resulting in a positive interaction between chemicals and the AgNPV. This tactic is being further evaluated and may prove useful for situations where the AgNPV can not be used alone, based on current recommendations for virus application.

1.6. Ongoing research with the AgNPV

Studies are underway towards evaluating the possibility of development of resistance by populations of A. gemmatalis to the NPV, as well as changes in virulence of the pathogen, when passed sequentially through laboratory and field populations of the insect. Initial data, obtained with NPV-exposed and non exposed insect populations, collected in widespread geographical regions of Brazil, have shown that all were highly susceptible to the virus, in spite of some differences detected among the

TABLE 4

Percent control of Anticarsia gemmatalis larval populations and percent defoliation in field plots of soybean treated with AgNPV, with low rates of insecticides and in mixtures with the pathogen

Treatment	Percent Control 1/			Percent defoliation
	Days after application			
	Three	Seven	Twelve	
Profenophos 100 g.a.i./ha	98.0	99.3	98.0	5.0
Virus 60 LE/ha	11.0	93.1	96.7	25.0
Profenophos 30 g.a.i./ha	80.9	80.9	93.4	15.0
Profenophos (30g) + virus	70.6	97.5	95.4	10.0
Carbaryl 50 g.a.i./ha	61.4	53.8	53.9	28.0
Carbaryl (50g) + virus	56.5	92.5	94.7	13.0
Bacillus thuringiensis 125g PC/ha	90.5	68.2	47.4	10.0
B. thuringiensis (125g) + virus	85.7	87.3	86.8	8.0
Diflubenzuron 3.3 g.a.i./ha	87.5	98.1	100	10.0
Diflubenzuron (3.3g) + virus	81.1	99.0	100	8.0
Control	–	–	–	65.0

1/ Calculated in relation to control plots, which registered an average of 97.8, 69.2, and 15.2 larvae per two meters of row at the 3rd, 7th, and 12th days after application, respectively. Economic injury level is 40 larvae per two meters of row or 30% defoliation (vegetative stages) or 15% defoliation (after flowering).

populations. On the other hand, selection pressure on laboratory populations has shown that A. gemmatalis became increasingly resistant to the NPV after a few generations (J.R. Fuxa, Louisiana State University, USA, pers. communication; A. Abot, F. Moscardi and D.R. Sosa-Gomez, unpublished data). Even though genomic variations have been detected in AgNPV isolates, obtained yearly through applications against field populations of A. gemmatalis (J.E. Maruniak, University of Florida, USA, pers. communication), their virulence has not greatly varied in initial laboratory bio-assays (F. Moscardi and D.R. Sosa-Gomez, unpublished data). When available, conclusive results on these aspects will be extremely important, considering that the AgNPV has been extensively used in Brazil since the 1982/83 season.

1.7. Progress in AgNPV use at farmer level

The use of AgNPV in Brazil started as a pilot programme at farmer level, conducted during the 1980/81 and 1981/82 soybean seasons, when the virus was shown to provide adequate control of A. gemmatalis larval populations as compared to paired insecticide-treated and non-treated areas (15). In the 1982/83 season the AgNPV was applied in ca. 2,000ha. In the 1984/85 season a substantial increase in the treated area occurred (ca. 200,000ha) after the implementation of regional production units, established in existing laboratories of research institutions, universities and farmer cooperatives, especially in the States of Paraná and Rio Grande do Sul. AgNPV-treated area kept increasing afterwards reaching ca. 1,000,000ha in the 1989/90 season, corresponding to ca. 8-10% of the total area cultivated with soybeans in Brazil (18).

Up to 1985, the AgNPV was used at farmer level as crude preparations. With the development of a kaolin-based wettable powder formulation, by EMBRAPA-CNPSo, this type of preparation was predominantly used since the 1986/87 season. In 1989, a liquid formulation of the AgNPV was registered in Brazil by a private industry (AGROGGEN) under the trade name MULTIGEN. Unfortunately, possibly due to inadequate quality control procedures, this product did not provide dependable control of A. gemmatalis at farmer level, and its production was discontinued in 1990. Currently, the Organization of Farmers Cooperatives of the Parana State (OCEPAR), and four other private companies are producing and commercializing the wettable powder formulation developed by EMBRAPA-CNPSo, which has been successfully employed in different regions of Brazil. The AgNPV has been used to some extent in other South American countries, especially in Paraguay, where the pathogen has been applied in over 60,000ha in the 1990/91 soybean season.

2. STUDIES WITH BACULOVIRUSES OF OTHER SOYBEAN INSECTS

Research with baculoviruses of other soybean insects is quite recent, and has concentrated mainly on NPVs of Plusiinae species and on the granulosis virus (GV) of E. aporema. Laboratory studies with a NPV of P. includens (21) have pointed out this virus as highly virulent to the

insect and as capable of drastically reducing the foliar consumption of
larvae infected in the first three instars. Since Plusiinae species,
mainly P. includens and R. nu, usually occur associated to populations of
A. gemmatalis in Brazil, the use of a single virus isolate to control this
insect complex has been attempted, by sequential passages of an Autographa
californica NPV (AcNPV) through A. gemmatalis and P. includens (22,23).
After five passages through P. includens, the resulting AcNPV variant was
more virulent to this species than a NPV isolated from this insect.
Passages through A. gemmatalis also resulted in increased virulence to this
species; however, in both cases, the multiplication of the AcNPV in one
species did not result in isolates virulent to the other, except for the
Fl AcNPV isolate in P. includens, which was also highly active on R. nu,
but lost virulence to the later species after further passages through
P. includens. Due to the importance of Plusiinae, specially R. nu, on
soybeans, sunflower, and flax in Argentina, Brazil (South), and Uruguay,
a joint project is presently underway involving research institutions of
these countries, aiming at developing NPVs of Plusiinae as microbial
insecticides. Efforts have been mainly directed to methods of production,
formulation, and field application.

The use of GV against the shoot borer, E. aporema, would be desirable
in areas of high incidence of the insect, since it usually occurs in
early vegetative stages of soybeans and is currently controlled solely by
applications of broad-spectrum chemical insecticides, which delay field
colonization by natural enemies and, therefore, make further IPM practices
more difficult on the crop. This GV has been investigated under laboratory
conditions, and has been shown to be highly virulent to different larval
instars of E. aporema at various temperature conditions (Kastelic and
Moscardi, unpubl. data). These results are to be confirmed at field
conditions in the next soybean seasons.

3. FUTURE PERSPECTIVES FOR BACULOVIRUS USE IN SOYBEANS

The interest in the use of viruses for insect control has greatly
increased in Brazil and other South American countries with the success
achieved with the NPV of A. gemmatalis at farmer level. The utilization of
this microbial insecticide, currently estimated at 1,000,000ha annually,
is expected to increase substantially in the next soybean seasons, mainly
because the increasing demand for this pathogen is bound to be gradually
satisfied by the four private industries which are presently
commercializing the AgNPV. The participation of these industries is
considered important for improvement of current production and formulation
methods, which ultimately will result in a better product reaching the
farmer. Current research projects on AgNPV epizootiology, on genetic
stability, and on possibilities of resistance to the NPV by populations of
A. gemmatalis, will certainly provide invaluable information for the
continuity of an adequate wide-scale use of the pathogen.

Despite the present efforts towards developing viral insecticides for
the control of P. includens, R. nu, and E. aporema, further studies on

production, formulation, and field application methods are necessary before they can be efficiently employed at farmer level. Although the studies conducted so far indicate that it is difficult to obtain a single virus isolate highly active to the main soybean lepidopterous insects (A. gemmatalis, P. includens and R. nu), this strategy should be further pursued, since these species usually occur simultaneously in most soybean regions of Brazil.

REFERENCES

1. Gazzoni, D.L. and Oliveira, E.B., Soybean insect pest management in Brazil. I. Research effort; II. Program implementation. In Proceedings International Workshop Integrated Pest Control in Grain Legumes, ed. P.C. Matteson, EMBRAPA, Departamento de Difusão de Tecnologia, Brasília, 1983, pp. 312-25.

2. Kogan, M., Turnipseed, S.G., Shepard, M., Oliveira, E.B. and Borgo, A., Pilot insect pest management program for soybean in Southern Brazil. J. Econ. Entomol., 1970, 70, 659-63.

3. Finardi, C.E. and Souza, G.L., Ação da extensão rural no manejo integrado de pragas da soja, Associação de Crédito e Assistência Rural do Paraná, Curitiba, 1980, 16pp.

4. Moscardi, F., Soybean IPM programme in Brazil, Proc. FAO/UNEP/USSR Workshop on Integrated Pest Management, Kishinev, Moldavia, 12-15 June 1990. (In press).

5. Moscardi, F., Microbial control of insect pests in grain legume crops. In Proceedings International Workshop Integrated Pest Control in Grain Legumes, ed. P.C. Matteson, EMBRAPA, Departamento de Difusão de Tecnologia, Brasília, 1983, pp. 189-223.

6. Allen, G.E. and Knell, J.D., A nuclear polyhedrosis virus of Anticarsia gemmatalis: I. Ultrastructure, replication, and pathogenicity. Florida Entomologist, 1977, 60, pp. 189-223.

7. Carner, G.R. and Turnipseed, S.G., Potential of a nuclear polyhedrosis virus for control of the velvetbean caterpillar in soybean. J. Econ. Entomol., 1977, 70, pp. 608-10.

8. Corso, I.C., Gazzoni, D.L., Oliveira, E.B. and Gatti, I.M., Ocorrência de poliedrose nuclear em Anticarsia gemmatalis na região sul do Brasil. An. Soc. Entomol. Bras., 1977, 6, pp. 312-14.

9. Sieburth, P.J. and Carner, G.R., Infectivity of an iridescent virus for larvae of Anticarsia gemmatalis. J. Invertebr. Pathol., 1987, 42, 49-53.

10. Moscardi, F., Correa-Ferreira, B.S., Hoffmann-Campo, C.B., Oliveira, E.B. and Boucias, D.G., Ocorrência de entomopatógenos em lepidópteros que atacam a cultura da soja no Paraná. In Nono Congresso Brasileiro

de Entomologia, Sociedade Brasileira de Entomologia, Londrina, 1984, pp. 143.

11. Hoffmann, C.B., Foerster, L.A. and Newman, G.G., Incidência estacional de doenças e parasitas em populações naturais de Anticarsia gemmatalis e Plusia spp. em soja. An. Soc. Entomol. Bras., 1979, 8, 115-24.

12. Diez, S.L., Gamundi, J.C., Mendez, J.M. and Saccone, A.M., Evaluacion de factores de mortalidad (entomopatogenos y parasitos) en poblaciones de Rachiplusia nu (Guenee) (Lep.: Noctuidae) en el cultivo de soja. In Segundo Congreso Argentino de Entomologia, La Cumbre, Cordoba, Argentina, 1991, pp. 69.

13. Moscardi, F. and Kastelic, J.G., Ocorrência de virus de poliedrose nuclear e virus de granulose em populações de Spodoptera frugiperda atacando soja na região de Sertaneja, PR. In Resultados de Pesquisa de Soja 1984/85, EMBRAPA-CNPSo, Londrina, 1985, pp. 128, (Documentos, 15).

14. Moscardi, F., Allen, G.E. and Green, G.L., Control of the velvetbean caterpillar by nuclear polyhedrosis virus and insecticides and impact of treatments on the natural incidence of the entomopathogenic fungus Nomuraea rileyi. J. Econ. Entomol., 1981, 74, 480-85.

15. Moscardi, F. and Correa-Ferreira, B.S., Biological control of soybean caterpillars. In III Proceedings of Soybean Research Conference, ed. R. Shibles, Westview Press, London, 1985, pp. 703-11.

16. Moscardi, F., Production and use of entomopathogens in Brazil. In Proceedings of Conference on Biotechnology, Biology Pesticides and Novel Plant-Pest Resistance for Insect Pest Management, ed. D.W. Roberts and R.R. Granados, Boyce Thompson Institute, Ithaca, 1989, pp. 53-60.

17. Moscardi, F., Use of viruses for pest control in Brazil: the case of the nuclear polyhedrosis virus of the soybean caterpillar, Anticarsia gemmatalis. Mem. Inst. Oswaldo Cruz, 1989, 84, 51-6.

18. Moscardi, F., Development and use of soybean caterpillar baculovirus in Brazil. In 5th Proc. Int. Colloquium on Invertebr. Pathol. and Microbial Control, Soc. Invertebrate Pathology, Adelaide, 1990, pp. 184-87.

19. Carner, G.R., Hudson, J.S. and Barnet, O.W., The infectivity of a nuclear polyhedrosis virus of the velvetbean caterpillar for eight noctuidae hosts. J. Invertebr. Pathol., 1979, 33, 211-16.

20. Moscardi, F. and Corso, I.C., Ação de Baculovirus anticarsia sobre a lagarta da soja (Anticarsia gemmatalis) e outros lepidópteros. In Segundo Seminário Nacional de Pesquisa de Soja, EMBRAPA-CNPSo, Londrina, 1981, pp. 51-57.

21. Chiaravalle V., W.R., Biologia comparada de Pseudoplusia includens (Walker, 1857) (Lep.: Noctuidae) em dietas naturais e artificiais e efeito de um virus de poliedrose nuclear na sua mortalidade e no consumo da área foliar de soja. Universidade de São Paulo, Piracicaba, 1988, pp. 164. (Tese de Mestrado).

22. Morales C., L., Avaliação do potencial de um VPN de Autographa californica, como agente biológico no controle de Pseudoplusia includens, Anticarsia gemmatalis e Rachiplusia nu. UNESP, Jaboticabal, 1990, pp. 126. (Tese de Mestrado).

23. Morales C., L., Moscardi, F. and Gravena, S., Potencial de uso de baculovirus de Autographa californica no controle de Pseudoplusia includens [Chrysodeixis includens] e Anticarsia gemmatalis. Pesquisa Agropecuaria Brasileira. (In press).

MANAGEMENT OF INSECT PESTS OF SOYBEAN IN BRAZIL AND OTHER
LATIN AMERICAN COUNTRIES

JOHN FISHER
Shell International Chemical Co. Ltd.,
Shell Centre, London, U.K.

HORACIO de ALMEIDA
Shell Brasil SA, Sao Paulo, Brazil

ABSTRACT

The use of crop protection agents to maintain pest populations below
economic injury levels is described. The importance of regular crop
scouting and the availability of unambiguous guidelines which are
appropriate to the insecticide to be used is stressed.

INTRODUCTION

Within the Latin American region in general and Brazil in
particular a wide variety of phytophagous arthropods has been recorded
from soybean even though it is not a native plant. One list (1) runs to
85 species although not all those included cause significant damage
consistently and on a wide scale. Some species remain locally
important while others reach pest proportions for limited time spans or
only in certain years. Rational explanation for such occurrences is
not always forthcoming since for many species we lack understanding of
the complex balance between forces acting for and against population
increase. Changes in pest status have been reviewed (2) and two
examples from Brazil deserve mention. One is the severe outbreaks of
the whitefly Bemisia tabaci in Parana and Sao Paulo in 1972/73 which
have not been repeated even though endemic populations remain in cotton
and beans. The other is the increasing importance of the pentatomid
bugs Piezodorus guildinii and Euschistus heros which have supplanted
Nezara viridula in warmer areas. In this instance reduced levels of
parasitism and lower levels of susceptibility (but not resistance) to
most common insecticides have been cited. Brazilian and other Latin
American growers therefore face an array of endemic arthropods which
are or could be serious pests of soybean together with the threat of
accidental introductions from the presumptive centre of origin of this
crop (The Orient) or indeed other parts of the world. The recent
arrival of the cotton boll weevil (Anthonomus grandis) and the horn fly
(Haematobia irritans) in Brazil serve to emphasise this point.

INSECT ATTACK

Insect attack on soybean can reduce crop stand, yield quantity and yield quality. For this reason it is helpful to group pests according to the part of the plant attacked, the damage caused and the period during the crop cycle when attacks occur. An overall listing of major pests for the Latin American region is as follows (2, 3, 4).

TABLE 1
Feeding niches for major pests of soybean

Niche	Pest	Damage
Roots	Agrotis	cutworm
	Peridroma	cutworm
	Cerotoma	roots
Lower stem	Elasmopalpus	stemborer
Leaves	Anticarsia	
	Spodoptera	defoliators
	Plusia	
	Pseudoplusia	
	Estigmene	
	Omiodes	
	Semiothisa	
	Hedylepta	leafroller
	Caliothrips	damage to
	Sericothrips	young plants and flowers, virus vector.
	Cerotoma	defoliators of
	Epicauta	young plants
	Diabrotica	
	Bemisia	leaf damage, sooty moulds, virus vector.
	Tetranychus spp	leaf damage, defoliation.
Upper stems/ terminal buds	Epinotia	prevents further growth
Pods/seeds	Nezara	
	Piezodorus	seed feeders
	Euschistus	
	Acrosternum	
	Etiella	pod borers
	Maruca	

Agricultural systems are influenced by economic, technical, social and political factors in a highly complex manner so that the use of crop protection agents cannot logically be considered in isolation. Commercial growers remain in business by satisfying the demands of their customers in respect of price and quality and often in addition the demands of their bank manager. In responding to a pest attack a farmer should take account of the following:

i)   the nature of the attack and the potential
     damage.
ii) the range of control measures known to him.
iii) his own objectives-whether formulated personally
     or externally imposed-and his attitude to risk.

Crop protection agents can be used either prophylactically or
curatively. The former is pre-emptive and can be made without reference
to pest levels as in the case of seed treatments. Curative treatments
presuppose the existence of a rational basis for decision making to
avoid over-reaction to either pest presence or crop damage at levels
below the economic injury level. This is provided by reliable sampling
methodology and the economic or action threshold which links:

i)   Intensity of attack.
ii) Pest density and damage.
iii) Effectiveness of applied control measure.
iv) The eventual price per unit weight of yield.

In addition farmers may well be influenced by questions of health
and safety and also environmental impact. The sum total of information
and advice relating to these topics is covered by product stewardship.
Hazardous materials must be supplied in suitable packaging that is
correctly labelled for use by operators equipped with protective
clothing and properly calibrated sprayers. Users must also be aware of
environmental impact and the need to avoid contamination of water
courses and other non-target areas while spraying and when disposing of
empty containers. Suppliers must maintain a high profile programme of
information and practical training appropriate to their products.
In respect of soybean a considerable body of knowledge exists in regard
of sampling methodology and the relationship between pest levels and
damage(5) while numerous attempts have been made to produce predictive
models. Not all this information is however in a suitable form for
assimilation by the non-specialist. Farmers need clear instructions on
sampling at the lowest possible cost compatible with the accuracy
needed, together with clearly stated guidelines for decision taking.
This will cover pest identification and instar (often a crucial factor)
together with reliable guidelines for estimating percentage
defoliation. This is clearly an area where companies involved in
providing crop protection agents can actively cooperate with official
bodies and universities and at the same time promote sound principles
of product stewardship.

In Brazil soybean is grown in Sao Paulo, Parana, Rio Grande do Sul,
Mato Grosso do Sul and Mato Grosso, Goias, Sta. Catarina and Minas
Gerais, being planted in October/December and harvested in March/May of
the following year. The time scale of crop development is as follows,
and is broadly similar in all parts of the region.

TABLE 2

Growth of Soybean

Period	Days	Milestones
	0	planting
Germination	0--10	
Establishment	10--50	
	50	first flower
Flowering	50--70	
Pod formation	60--80	
Grain filling	80--100	
	100	50% yellow leaves
Maturation	100--120+	
	120+	harvest

Growers are advised to scout their crop at weekly intervals as a minimum and more frequently as pest levels increase. Pests are dislodged from the crop onto a sampling sheet placed between the rows, except for Epinotia where plants are examined at random. Results from several sampling points are averaged before making a decision. In addition to pest insects samples usually contain predators and parasites. In addition to spiders these are predatory bugs and beetles which may be closely related to the pests themselves. Beneficials are not counted or included in the decision-making but their presence is clearly desirable. Threshold values which have been in common use in Brazil are given below.

TABLE 3

Threshold values for Brazil

Pests	Crop Stage	Action Threshold
Defoliating caterpillars	1) before flowering	40 larvae > 15mm per sample 30 per cent defoliation
	2) after flowering	40 larvae > 15mm per sample 15 per cent defoliation
Epinotia	to pod formation	30 per cent terminals attacked
Plant bugs	from pod formation to maturity	4 insects > 5mm per sample

In Mexico, where a wider variety of caterpillars and other pests can occur, more detailed guidelines are given for the critical period of flowering and pod formation(5).

TABLE 4

Action thresholds for defoliators in Mexico

Percentage defoliation	Number of larvae >15 mm per metre	
	8--14	15--18
0--20	continue sampling	preventive spray
20-30	spray	spray
over 30	spray immediately during flowering and pod formn. during pod filling sample on next two days, spray if population remains above threshold.	spray

In all countries in the region the need for accurate estimation of defoliation is stressed. It is widely accepted that the inexperienced show a consistent tendency to overestimate. This can be corrected both by the provision of illustrated guidelines and encouraging the collection of leaves at random from lower, middle and upper thirds of plants selected at random for estimation of an approximate average. However, the basis for action remains larval numbers either from point sampling as in Brazil or from using a sweep net.

Broadly similar guidelines for foliar pests are available throughout the Latin American region. In addition to defoliators which cause obvious loss of leaf area others such as Tetranychus can cause leaf drop. The whitefly Bemisia can reduce photosynthetic efficiency both directly and by the growth of sooty mould on excreted honeydew, as well as acting as the vector of soybean crinkle and dwarf mosaic viruses. Thrips can damage foliage and perhaps more importantly flowers as well as vectoring the tobacco ringspot virus. While guidelines are available for subsidiary pests, advice on their control is often restricted to the selection of a product for a major pest with an appropriate spectrum. The presence of several pest species all at levels just below their individual thresholds also remains problematical. Expected weather patterns can influence the decision in such cases since spider mites, thrips and whitefly populations are capable of explosive increases in hot, dry conditions while rain will have adverse effects and also promote spread of pathogens such as Nomuraea rileyi. Defoliation as an index permits the summation of damage by major pests but correct identification remains important when counter-measures are considered since different dose rates or even different products may be needed. These can be either chemically or biologically based. Preparations based on Bacillus thuringiensis are included in the former category. Other approaches include the deliberate use of epizootics such as <u>Baculovirus anticarsia</u> (dealt with elsewhere in this symposium) and the periodic mass release of egg parasites such as <u>Trichogramma</u> spp. during the period 30-65 days after germination.

Thresholds for Heteroptera are four insects > 0.5cm long per metre of row for grain crops reducing to two insects for seed crops because of the deleterious effects of attack on germination levels. Correct identification of pests is important since <u>Piezodorus</u> and <u>Euschistus</u> are less susceptible than <u>Nezara</u> to many commonly used insecticides. When thresholds are exceeded a wide range of materials is available for use. The following table is illustrative without claiming to be exhaustive.

TABLE 5

Dose rates for commonly used insecticides

<u>Anticarsia</u> g A.I./ha	Insecticide	<u>Nezara</u> Piezodorus g A.I./ha
***	Baculovirus	NO
***	B.thuringiensis	NO
210	carbaryl	800
70	thiodicarb	NO
87.5	endosulfan	437.5
15--25	diflubenzuron	NO
500	fenitrothion	500
150	methamidophos	300
150	monocrotophos	200
210	parathion methyl	480
100	profenophos	NO
400	trichlorfon	800
30	fenvalerate	NO
25	permethrin	50
2	bifenthrin	NO
7.5	cyfluthrin,	15
3.5	lambdacyhalothrin	7.5

*** dose rate depends on the preparation being used

An overview of the relative amounts of different products used in Brazil in 1990 can be deduced from market analysis.

Table 6

Per cent market share for Brazil based on area treated

Product/type	percent share	
monocrotophos	28.6	
methamidophos	12.2	
other OP's	8.9	
TOTAL OP's		49.7
endosulfan	18.4	18.4
permethrin	24.2	
other pyrethroids	4.3	
TOTAL PYRETHROIDS		28.5
TOTAL CARBAMATES		0.8
TOTAL ACYLUREAS		2.2
others		0.4
		(100)

Given the wide use patterns of monocrotophos, which is classified as an extremely hazardous material, recent work in Brazil aimed at reducing dose rates while maintaining knockdown and persistence of effect is of interest. One approach has been to evaluate mixtures of monocrotophos with common salt. Since higher rates are needed to control Heteroptera attention was focussed on these species.

TABLE 7

Effect of common salt on performance of monocrotophos
against Heteroptera

Treatment	Dose g A.I./ha	Nezara			Piezodorus		
		percentage efficacy/days after treatment					
		3	5	7	3	5	7
monocrotophos	120	92	84	89	88	78	83
	150	100	94	96	92	87	83
plus salt	100+ 0.5%	100	97	96	96	87	99
	120+ 0.5%	100	100	100	100	97	97
endosulfan	437.5	84	87	92	90	84	83
salt alone	0.5%	20	20	20	31	35	31

Percentage efficacy was calculated using Abbott's formula.

These results demonstrate the feasibility of reducing the dose rate of monocrotophos from the commercial recommendation of 200 g a.i./ha and therefore reducing the environmental impact of this product. No phytotoxicity was noted in this trial and further work is clearly justified.

Several pyrethroids can also be used against pest combinations, as well as endosulfan and carbaryl. All of these products share with monocrotophos and methamidophos the attribute of fast action so that very little extra defoliation occurs while the pest population is being brought under control. This is not the case for other more selective materials such as acylureas which are ingestion toxicants which kill at the subsequent moult. When these products are applied at the existing thresholds unacceptable levels of defoliation can occur.

TABLE 8

Control of Anticarsia using acylureas

Treatment	Dose	percentage control				percentage defoliation
	g A.I./ha	days after trtmnt				+ 16 days
		4	5	5	7	
flufenoxuron	7.5	69	80	83	96	24
	9.0	78	90	98	95	25
	10.0	74	90	86	95	22
teflubenzuron	7.5	73	91	90	96	20
	9.0	77	82	88	91	19
	10.5	77	91	96	94	21
diflubenzuron	25	72	88	96	96	21
monocrotophos	150	99	99	99	100	13
control	---	--	--	--	--	48

The increase in percentage defoliation over monocrotophos is a measure of the damage caused before adequate levels of control are achieved. Clearly such materials need to be applied much earlier in the infestation when control of early larval instars which are of shorter duration would be more efficient. This is a deficiency which is shared by the baculovirus which typically requires about eight days to achieve 80-90 per cent mortality against Anticarsia. Growers wishing to use acylureas against lepidopterous and coleopterous defoliators possibly in conjunction with mass release of Trichogramma are therefore at a disadvantage unless they can be provided with a new threshold that accommodates for the extra defoliation.

A final consideration is that of cost. Accurate and strictly comparable figures are unobtainable so that what follows must be regarded as an approximation. However even taken as an overview the figures are illuminating. An overall estimate of costs for insect and weed control agents plus their application for Brazil for 1990 gives a figure of 6.50 % of total costs. Rather more detailed analyses are available from Colombia (7) and (8).

TABLE 9

Costs of crop protection in Colombia 1990

	item	cost as per cent of total
Example 1	insecticides	2.89
	applications	1.38
	herbicides	6.96
	applications	1.21
	technical assistance	1.98
	TOTAL	14.42
	(MACHINERY AND TRANSPORT	24.66)
	(FINANCE AND ADMIN.	32.42)
Example 2	insecticides	2.89
	applications	1.59
	Trichogramma	0.42
	releases x 3	0.09
	herbicides	3.31
	applications	1.68
	technical assistance	0.95
	TOTAL	10.93
	(MACHINERY AND TRANSPORT	28.0)
	(FINANCE AND ADMIN.	35.10)

Without over-agile use of these figures the argument can be sustained that other items on the balance sheet such as the examples included above will be of greater concern to the grower than costs for insect control of five per cent or below. It still remains vitally necessary however, that growers continue to receive and to put into practice the highest possible standards of advice on product use and product stewardship. It is the firmly held view of the authors that this is best achieved by cooperation between government and university workers and the crop protection industry.

REFERENCES

1. Smith, J.G., Pests of soybean in Brazil. In Pests of Grain Legumes:Ecology and Control, eds. S.R.Singh, H.F. van Emden and T. Ajibola Taylor, Academic Press, London, 1978, pp. 167-177.

2. Kogan, M., Turnipseed, S.G., Ecology and management of soybean arthropods. Ann. Rev. Entomol. 1987, 32, 507-538.

3. Sifuentes, J.A., Plagas de la Soya y Su Control en Mexico. Mex. Inst. Nac. Invest. Agric. Foll. Divulg. No.70, 1978.

4. Garcia, F., Plagas de la Soya y Su Manejo. In La Soya en la Agroindustria 30 Anos. Instituto Colombiano Agropecuario. 1990.

5. Sampling Methods in Soybean Entomology, eds. M. Kogan and
   D.C. Herzog,
   Springer-Verlag, New York, 1980.

6. Anon., Guia Tecnica para el Cultivo de la Soya Parte IV.
   Agromundo 1991, 6, 22-34.

7. Alvarez, G., Murcia, P., Costos de Produccion Cultivo Soya
   Semestre 90A. El Algodonero Edicion No. 215, 1991, 13-15.

8. Anon., Costos de Produccion Soya-Semestre B 1990. Evaluacion
   Agricola No. 1, 1991, 10-13.

THE EFFECTS OF A PYRETHROID, LAMBDA-CYHALOTHRIN, ON NATURAL PEST CONTROL IN BRAZILIAN SOYBEANS

JENNIFER WHITE, RICHARD BROWN, ANILDO BETTENCOURT AND CARLOS SOARES

Ecology and Soil Science Section, ICI Agrochemicals,
Jealott's Hill Research Station, Bracknell,
Berkshire, UK. RG12 6EY

ABSTRACT

The effects of lambda-cyhalothrin on soybean pests and their natural enemies was investigated in 1990-91 in southern Brazil. Lambda-cyhalothrin (3.75 g ai/ha) and a standard monocrotophos (120 g ai/ha) were applied to control the velvetbean caterpillar Anticarsia gemmatalis. Crop and ground dwelling arthropods, percentage parasitism and soybean leaf area index were monitored throughout the season and survivorship curves were constructed for Anticarsia.

Anticarsia populations were effectively controlled by both treatments. Following lambda-cyhalothrin applications, transient effects were observed on crop dwelling predators but not on ground dwelling arthropods or parasites. Following monocrotophos treatments, effects were observed on crop and ground dwelling predators but not on parasites. Typically, effects lasted 24 days and natural enemies recovered at the time of subsequent Anticarsia infestations, which did not reach economic threshold levels. There was no evidence for a reduction in the potential for natural control of subsequent infestations of Anticarsia or other late season pests.

INTRODUCTION

Lambda-cyhalothrin (PP321, KARATE) is a broad spectrum synthetic pyrethroid insecticide used to control a wide range of pests, including aphids, Coleoptera and Lepidoptera in a variety of crops worldwide. Small plot trials in Southern Brazil have shown lambda-cyhalothrin to give good control of the key soybean pests the velvetbean caterpillar Anticarsia gemmatalis Hübner and the green stink bug Nezara viridula Linnaeus [1]. Natural enemies are considered important in the control of these pests [2]. In Brazil, insecticides are generally applied twice to soybean during the season, once in the late vegetative stages, against defoliators such as Anticarsia and again at the end of the season to control pod damage by the stink-bug complex. Early season applications against

Anticarsia have caused concerns regarding the possible disruption of natural control of subsequent generations of Anticarsia, secondary defoliator pests and late season pests. There is evidence for up to 80% reduction in the abundance of natural enemies five days after applications of some broad-spectrum insecticides [3]. However, few longer term data, assessing the consequences of such effects for the natural control of following pests on large plots, are available.

Due to lambda-cyhalothrin's broad-spectrum nature, much work has been done to investigate the effects of this compound on natural enemies. Substantial laboratory and field data exist for temperate cereals [4, 5] but only limited data are available for tropical crops such as soybean. A field study was therefore set up in southern Brazil during the 1990-91 soybean season to investigate the effects of lambda-cyhalothrin on the key soybean pests and their natural enemies over a complete season.

MATERIALS AND METHODS

The trial was located in the municipality of Passo Fundo in the state of Rio Grande do Sul in Southern Brazil, in direct drilled soya (Variety BR4). The site, which had been managed using direct drilling (plantio direto) for eight years, was divided into four replicate blocks of three plots, lambda-cyhalothrin, untreated control and monocrotophos, each plot greater than 1 hectare in area. In early February, applications of lambda-cyhalothrin as KARATE 5EC, 3.75 g ai/ha and monocrotophos as AZODRIN 400, 120 g ai/ha, were made using a tractor mounted boom sprayer, to control Anticarsia. Spray deposition on lambda-cyhalothrin plots, at both crop canopy and soil level, was measured by residue analysis of filter paper squares mounted on supports in the crop at crop canopy and soil level. These were placed in the crop just before spraying, collected within an hour of application and held frozen until analysis for lambda-cyhalothrin residues. It was intended to make a second insecticide application in early March to control the stink bug complex, but stink bugs remained below economic threshold levels throughout the season, so a second application was not made.

Throughout the season, regular estimates of the leaf-area index of the crop were made using point quadrats. Arthropods were sampled pre and post-treatment, weekly samples being taken between mid January and mid April. In each plot, crop dwelling arthropods were sampled by taking ten 1 meter crop-beating samples and ground dwelling arthropods from 10 pitfall traps containing 4% formalin. In addition, Anticarsia larvae were collected before and after treatment and reared in the laboratory to determine the percentage parasitism.

Anticarsia larvae were separated into instars based on head capsule width and larval length [6] and then separated into cohorts, each new infestation was considered the beginning of a new cohort. To investigate mortality from sprays and natural enemies, a life table was then constructed for each cohort and survivorship curves constructed using the logarithms of the lx values from the life tables (lx = number of individuals surviving out of a starting number of 1000) [7]. The mortality rates of each cohort, given by the slope of the linear component of the survivorship curves, were compared between treatments. In cohort II on treated plots, numbers of fifth and sixth instar larvae were very low (less than 5 per cohort), and there were more fifth than sixth instar larvae. This appeared as a negative mortality rate and so created a bias in the analysis. The data for sixth instar larvae on the treated cohort

II were therefore removed from the analysis. All other time series data were analysed using a 2-way ANOVA, to detect differences between the treatments over time.

RESULTS

During 1990-91, the state of Rio Grande do Sul was seriously affected by drought, rainfall at the trial site being only 65% that of a typical year. As a result the crop remained poorly developed throughout the season and the crop canopy was not closed at spraying. This resulted in a high percentage (56%) of the spray reaching ground level (Table 1). The leaf area index of the crop was approximately half that of a normal year and significant drought stress was evident 14-35 days after treatment (DAT), when the crop appeared to be wilting (Figure 1). Following spraying the leaf area on the lambda-cyhalothrin plots were significantly higher (p<0.05) than the control for the majority of the season. This was reflected in the yield which was significantly greater on lambda-cyhalothrin than on control plots (Table 2).

TABLE 1
Spray deposition at crop and soil level
(calculations based on nominal concentration)

Replicate	Crop Level Deposition		Soil Level Deposition	
	g ai/ha	% deposition	g ai/ha	% deposition
1	3.75	100	2.19	58
2	3.25	87	1.75	47
3	3.88	103	2.13	57
4	3.88	103	2.31	62
Mean	3.69	98	2.09	56

TABLE 2
Soybean Yield (kg/ha)

Replicate	Lambda-cyhalothrin	Untreated Control	Monocrotophos
1	1381	1145	1142
2	1003	898	1100
3	924	574	601
4	396	303	232
Mean	926 (a)	730 (b)	769 (ab)

Numbers with different letters are significantly different at 5% (LSD).

Figure 1: Mean Soybean Leaf Area Index

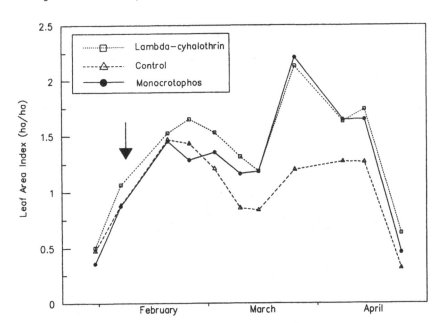

The first infestation of <u>Anticarsia</u> was sprayed in early February at a density of 8 larvae per metre of crop row. At these pest densities relative to the poor crop development, chemical treatment was probably necessary although levels were below the economic damage threshold (EDT) of 20 larvae per metre [2]. At spraying, the majority of the larvae were second instars and both lambda-cyhalothrin and monocrotophos gave over 90% control for 14 DAT (Figure 2). Untreated populations reached densities of 19 larvae per meter 10 DAT, after which there was a drought induced crash of the control populations, numbers of all instars falling simultaneously. In late February, twenty-one days after treatment, a second <u>Anticarsia</u> infestation was observed on the treated plots, numbers falling again 28 DAT. This was not observed on the control plots which were by this stage heavily defoliated (>60%) (Figure 1), but other fields on the farm also experienced a second <u>Anticarsia</u> infestation in late February.

The <u>Anticarsia</u> populations were divided into three cohorts, cohort I (all pre-treatment), cohort II, the sprayed cohort and cohort III the re-invading cohort which appeared on the treated plots only. Survivorship curves for cohorts II and III are given in Figures 3 and 4 respectively. In cohort II, the mortality rates for lambda-cyhalothrin and monocrotophos were significantly greater than that of the control (p<0.01) and there was weak evidence for the mortality rate for lambda-cyhalothrin being greater than that for monocrotophos (p=0.06). In cohort III, the mortality rates for lambda-cyhalothrin and monocrotophos were significantly greater than that of the control in cohort II (p<0.01), but not different to one another (p=0.09).

The total number of ground dwelling predators, as collected in pitfall traps, and the total number of crop dwelling predators, as collected by crop-beating, are shown in Figures 5 and 6 respectively. Ground dwelling predator samples included ants, spiders, staphylinid and

124

Figure 2. <u>Anticarsia</u> Larvae Remaining Expressed as a Percentage of the Larvae Present Before Spray

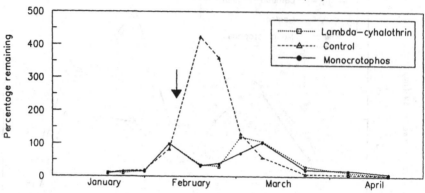

Figure 3. Survivorship Curve Cohort II

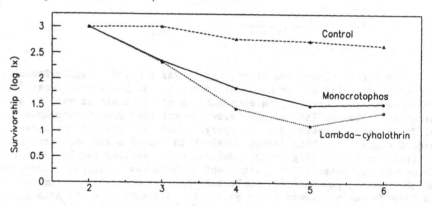

Figure 4. Survivorship Curve Cohort III

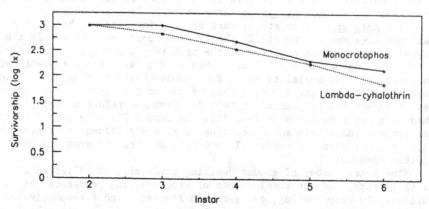

Figure 5: Total Ground Dwelling Predators (mean no. per trap per day)

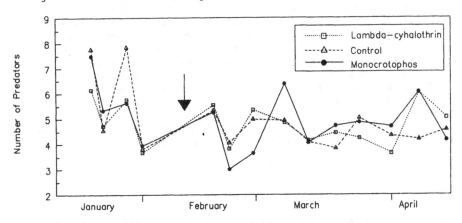

Figure 6: Total Crop Dwelling Predators (per 10 crop-beating samples)

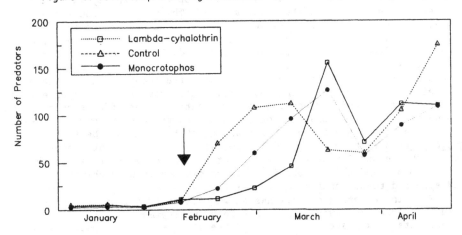

Figure 7: Percentage Parasitism

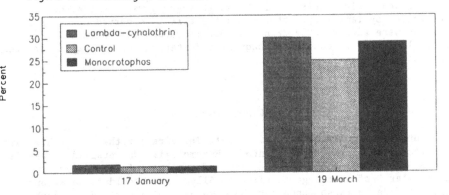

Date Anticarsia Sampled from Field

carabid beetles, while crop dwelling predators included predatory hemipteran, chrysopid larvae, spiders and carabid beetles. There were no differences in the abundance of ground dwelling predators between control and lambda-cyhalothrin plots following treatment but abundance on monocrotophos plots is reduced between 12 and 24 DAT ($p<0.05$). Crop dwelling predators were not very active at the time of spraying, but activity was reduced for 24 DAT following lambda-cyhalothrin treatment and 17 DAT following monocrotophos treatment, compared to the control ($p<0.05$), after this time, it was similar on all treatments. Field collection and laboratory rearing of _Anticarsia_ larvae showed there were no differences between the treatments with between 25 and 30% parasitism on the treated and control plots 40 DAT (Figure 7).

DISCUSSION

Good pest control was achieved with both treatments and despite sprays being applied at below the EDT, a yield advantage was observed on the lambda-cyhalothrin plots, possibly as the crop was unusually susceptible to pest damage because of the water-stress. The poor crop condition at spraying resulted in high spray deposition at crop and ground level, probably causing above average exposure of natural enemies to the sprays. Generally, effects were greatest on the crop dwelling predators where the spray deposition was highest. Ground dwelling predators and parasites were unaffected following lambda-cyhalothrin treatment, although there were effects on crop dwelling predators. There was no evidence of effects on parasites following monocrotophos applications, but effects on crop and ground dwelling predators were observed. Effects were, however, transient and natural enemy abundance was increasing at the time of the second _Anticarsia_ infestation. _Anticarsia_ mortality rates on treated plots during the second infestation were greater than mortality rates on the control plots and this second infestation did not require chemical control and appeared to be controlled naturally.

Work in temperate cereals has shown similar transient effects on natural enemies following lambda-cyhalothrin treatments. Manipulative experiments on key species, suggested the effects on some species were due to behavioral responses such as migration out of the plot or reduced activity and for other species were due to mortality [5]. Further work on soya pest and natural enemy interactions is being conducted.

In conclusion, this trial showed _Anticarsia_ was satisfactorily controlled by lambda-cyhalothrin and the effects observed on natural enemies were sufficiently transient not to reduce the potential for natural enemies to control subsequent infestations of _Anticarsia_ or other late season pests.

ACKNOWLEDGEMENTS

The authors acknowledge Sr. Luis Graeff Teixeira for the use of his farm, the lecturers of the Departomento de Entomologia, Universidade Federale do Rio Grande do Sul (UFRGS) for advice and the loan of facilities, in particular Professor Rogerio Pires da Silva, the Faculdade da Agronomia, Universidade do Passo Fundo for the loan of facilities and researchers at EMBRAPA-CNPSo for their advice, particularly Decio Luiz Gazzoni.

The authors also thank Osvaldo dos Santos Lima, Gerson Roberto Miola

and Cesar Antonio Michel for help in the field, Maria Angelica Heineck,
Vera Lucia Eick and Lucia Regina Gomes for help in the laboratory, Alberi
Jardim for applying the chemical, Francisco Ely a Silva for all his help
and Eddie McIndoe and Peter Chapman for the statistical analysis.

REFERENCES

1. Belarmino, L.C., Link, D., Costa, E.C., da Silva, M.T.B., Cardosa,
   I. C., Lorini, I., Fontes, L.F., da Silva, R.F.P. and Bertoldo, N.G.
   Efficiency of Insecticides in Soybean in Brazil: I-Anticarsia
   gemmatalis Hübner 1818 (Lepidoptera, Noctuidae, Erebinae).
   Proceedings of the XII International Plant Protection Congress, Rio
   de Janeiro, Brazil. 1991.

2. Gazzoni, D., Oliveira, E.B., Corso, I.C., Ferreira, B.S.C., Villas
   Boas, G.L., Moscardi, F. and Panizzi, A.R. Manejo de pragas da soja.
   Circular tecnica 5, EMBRAPA - CNPSo, Londrina PR. 1988.

3. EMPRAPA-CNPSo Pesquisa com inseticidas em soja: summario dos
   resultados alancados entre 1975 e 1987 por Edilson Bassoli de
   Oliveira e outros. Londrina, PR. 1988.

4. White, J.S., Everett, C.J., & Brown, R.A. Lambda-cyhalothrin:
   Laboratory and Field Methods to Assess the Effects on Natural
   Enemies. Proceedings of the Brighton Crop Protection Conference -
   Pests and Diseases. Vol 3, pp 969-974. 1990.

5. Brown, R.A., Everett, C.J., White, J.S. How does an Autumn Applied
   Pyrethroid Affect the Terrestrial Arthropod Community? In Field
   Methods for the Study of Environmental Effects of Pesticides, M.P.
   Greaves, P.W. Greig-Smith & B.D. Smith (eds). BCPC Monograph no. 40
   pp 137-145. 1988.

6. Heineck, M.A. Influencia de 3 cultivares de soja sobre consumo
   folhar, desenvolvimento e fecundidade de Anticarsia gemmatalis.
   Porto Alegre, UFRGS. Dissertacão de Mestrado. 1989.

7. Southwood T.R.E. Ecological Methods. 2nd Ed., Chapman and Hall,
   London. pp 524. 1978.

MANAGEMENT OF NEMATODE PROBLEMS ON SOYBEAN IN THE UNITED STATES OF AMERICA

R. D. RIGGS

Arkansas Agricultural Experiment Station
Fayetteville, Arkansas, USA

ABSTRACT

Nematode management may be nematicidal, resistance, cultural, or biological. Before the 1970s the prices received for soybeans was too low to justify the use of nematicides. During the 1970s and 1980s 1,2-dibromo-3-chloropropane and ethylene dibromide were effective and were used to some extent on soybean. Both are banned from use in the USA now. Resistance is an effective and economical method of nematode management. However, resistance is not available for many of the nematodes discussed here. In addition, growers tend to overuse resistance when it is available and often the resistance becomes ineffective because of shifts in the nematode population. The life of a resistant cultivar can be extended by proper management such as use in rotations. Cultural practices that can be effective in managing nematode population levels include rotation, subsoiling, modification of planting date, proper fertilization, and irrigation. Care must be taken to use the proper tactic at the right time. The right time may be determined by the nematode involved or environmental conditions. For example a nematode such as *Pratylenchus penetrans*, with a broad host range may be difficult to manage with rotations because nonhosts may be difficult to find. If one is found it may not be profitable for growers. Biological management occurs naturally in many soils. Research aimed at utilization of these natural enemies has been ongoing for many years. The types of nematodes discussed here are the more difficult ones to manage biologically. Because they constantly move from one place to another they must be trapped before a fungus can infect them. Fungi with trapping mechanisms occur in the soil but so far they cannot be applied in such a way as to be effective in reducing population levels of plant parasitic nematodes enough to prevent yield suppression.

INTRODUCTION

Plant parasitic nematodes are ubiquitous on soybean in the United States (Fig. 1). Soybean cyst nematodes, *Heterodera glycines* Ichinohe are widespread where soybean is grown, and root-knot nematodes, *Meloidogyne* spp. occur in sandy soils in which soybean is grown. Reniform nematodes, *Rotylenchulus reniformis* Linford and Olivera, are more prevalent in states that border the Gulf of Mexico and the Atlantic Ocean [1]. Lance nematodes, *Hoplolaimus* spp., are widespread in the south but different species may have limited distribution. Lesion nematodes, *Pratylenchus* spp., are widespread but a particular species may have limited distribution. Sting nematodes, *Belonolaimus* spp., generally occur in deep sandy soils in southern states. Other nematodes such as dagger nematodes, *Xiphinema* spp., stubby root nematodes, *Paratrichodorus* spp. and *Trichodorus* spp., sheath nematode, *Hemicycliphora* spp., ring nematodes, *Criconemella* spp., and pin nematodes, *Paratylenchus* spp. occur sporadically and only infrequently cause obvious damage. Stunt nematodes, *Tylenchorhynchus* spp., *Quinisulcius acutus,* and *Merlinius* spp., and spiral nematodes, *Helicotylenchus* spp., *Rotylenchus* spp. and *Scutellonema* spp. are widespread in soybean growing areas in the United States of America but seldom reach population levels that reduce soybean yields.

The management of soybean cyst and root-knot nematodes will be covered in later discussions. Therefore, this discussion will be devoted to other nematodes that parasitize soybean.

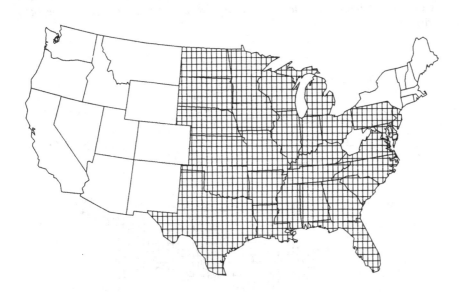

Figure 1. Map of the 48 contiguous states of the United States of America with cross-hatched area showing where soybean is grown.

RENIFORM NEMATODES
Rotylenchulus reniformis

Reniform nematode was considered a problem on soybean in the southern USA by 1973 when Rebois [2] investigated the effects of soil temperature on its infectivity and development on resistant and susceptible soybean cultivars. It was associated with damaged soybean plants in Arkansas as early as 1978. It is known to occur in Alabama, Arkansas, California, Florida, Georgia, Louisiana, Mississippi, South Carolina, Texas, (Fig. 2) and the commonwealth of Puerto Rico [3, 4].

Above-ground symptoms of *R. reniformis* infection include stunting and chlorosis. The root system also is stunted and, under severe conditions, may become necrotic. When infected roots are carefully extracted from the soil, the nematode egg masses can be detected because of the soil particles adhering to them. The damage to the roots occurs when the vermiform, immature female penetrates the root. She becomes oriented perpendicular to the longitudinal axis of the root with the posterior portion of the body outside the root [5]. The nematode feeds on an endodermal cell to initiate a syncytium. The syncytium is comprised of endodermal and parenchymal cells which coalesce when walls between them dissolve [6]. The nematode feeds in the syncytium located adjacent to the xylem vessels. Presumably the nematodes draw nourishment that would normally support plant growth. In a growing season *R. reniformis* can reach very high population levels.

Nematicides have been used to manage *R. reniformis* population levels on soybean but they have not been very effective (Birchfield, pers. comm.). In addition few nematicides are available and they are too expensive to use on soybean at the present grain prices.

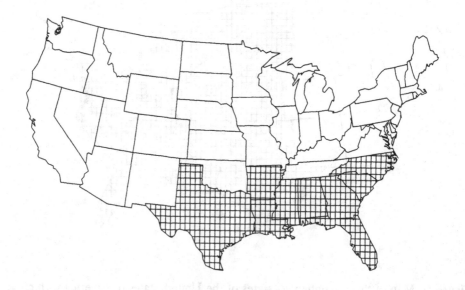

Figure 2. Cross-hatched states are known to have infestations of reniform nematode on soybean.

Several soybean cultivars have resistance to *R. reniformis*. There is an association between resistance to *H. glycines* and resistance to *R. reniformis*, however, the association does not hold with all *H. glycines*-resistant cultivars. Breakdown of resistance to *R. reniformis* has not been reported as with *H. glycines*, and rotation is an effective management tactic. Grasses appear to be poor hosts of *R. reniformis* and planting grass for 2 years reduces the population density enough to allow good soybean growth. Nonhost crops are limited so rotation options are few.

LANCE NEMATODES
Hoplolaimus spp.

Species of lance nematodes that have been associated with soybean are *H. columbus* Sher, *H. galeatus* (Cobb) Thorne and *H. macrostylus* Robbins [7]. *H. columbus* is found in Alabama, Georgia, Louisiana, Mississippi, North and South Carolina, and Texas [4] (Fig. 3) and is the most destructive of the three species. *H. galeatus* is widespread but does not appear to damage soybean. *H. magnistylus* is known to occur in Arkansas, Mississippi and Tennessee [7]. It has been shown to cause damage to soybean in the greenhouse but not in the field (Riggs, unpub.).

Hoplolaimus columbus penetrates into the root tissue and migrates through the cortex leaving a trail of damaged cells. The damaged cells die, the cortex sloughs off and lateral roots are not produced. This results in chlorosis, stunting, and reduced pod production [8].

Management of lance nematodes should not be difficult. Nematicides reduce nematode population densities and may provide an economic return [1]. Foster,

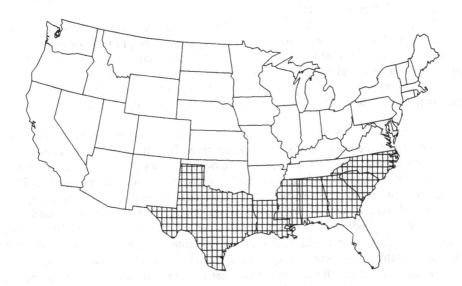

Figure 3. Cross-hatched states have infestations of Columbia lance nematode (*H. columbus*) on soybean.

Centennial, Coker 317, and Coker 368 are tolerant to *H. columbus*; therefore they yield better even when nematodes are present than do intolerant cultivars [1, 9]. Tolerance was expressed better in subsoiled than in non-subsoiled plots [9]. Coker 488, Deltapine 506, Kirby, Ring Around 680, and Young also sustained high yields on *H. columbus*-infested soil [10]. Reduction in nodulation of intolerant cultivars may be a factor in the lower yield of these cultivars [11].

In *H. columbus*-infested soil high rates of nematicides (1,3-D, aldicarb, or fenamiphos) did not always result in better yield of Coker 317 or Coker 156 soybean [12]. Subsoiling resulted in as high a yield as any nematicide treatment at one location, and only slightly lower at another. At a third location no treatment had higher yields than the check. Schmitt and Bailey [13] stated that "achieving high and profitable yields of soybean in soils infested with *H. columbus* will require more than simply applying an effective nematicide."

LESION NEMATODES
Pratylenchus spp.

Lesion nematodes that have been found on soybean include *P. agilis* Thorne and Malek, *P. alleni* Ferris, *P. brachyurus* (Godfrey) Filipjev and Schuurmanns Stekhoven, *P. coffeae* (Zimmerman) Filipjev and Schuurmans Stekhoven, *P. crenatus* Loof, *P. hexincisus* Taylor and Jenkins, *P. neglectus* (Rensch) Filipjev and Schuurmans Stekhoven, *P. penetrans* (Cobb) Filipjev and Schuurmans Stekhoven, *P. pratensis* (de Man) Filipjev, *P. scribneri* Steiner, *P. vulnus* Allen, and *P. zeae* Graham. Lesion nematodes are widely distributed in the USA, though a given species may be limited in distribution [4]. Individual species may have a large number of hosts and the group as a whole a has broad host range [14]. The host range includes many weeds that serve as reservoirs to prevent the usefulness of some control procedures [15].

Lesion nematodes penetrate into and migrate through the root tissue. They destroy cells by feeding on them and simply by migrating through them. Generally they destroy enough cells in local areas to cause lesions, thus the common name. The lesions may increase in size and coalesce to form extensive necrotic areas on the roots. The root tissue damage may result in chlorosis and stunting of the shoots or in extreme cases plants may die. Soybean cultivars may vary in their susceptibility to various *Pratylenchus* spp. [16, 17].

Management of populations of *Pratylenchus* species may be difficult. Little has been reported on the use of nematicides of any kind. Schmitt (pers. comm.) and Riggs (unpub.) obtained higher yield from soybean growing in *Pratylenchus*-infested soil when they used DBCP, but it is no longer available. Aldicarb application resulted in a 25% increase in yield because of a reduction in the population level of *P. scribneri* [18].

In soil infested with *P. brachyurus*, Essex outyielded Forrest in untreated soil, however, Forrest yields were increased 150% by carbofuran treatment [19]. Nematicide treatment of *P. brachyurus*-infested soil and planting a sensitive soybean cultivar demonstrated that yield increases (even near 100%) may not be profitable [19].

Rotations can be effective but care must be taken to determine the species of *Pratylenchus* in a particular field. *Pratylenchus hexincisus*, *P. scribneri* and *P. penetrans* increase on corn and soybean and *P. hexincisus* on oat [20]. *P. neglectus* increased more on corn or soybean than on cotton or peanut [21]. *P. agilis* does not reproduce well on

Essex soybean. Hosts were I. O. Chief maize> Marglobe tomato> Williams soybean> Essex soybean [22]. *Pratylenchus brachyurus* can be managed by a maize-soybean rotation in North Carolina [23] but not in Florida [24].

The Forrest cultivar is resistant to *P. scribneri* [16] but is very susceptible to *P. brachyurus* [25]. Essex appears to be tolerant to *P. brachyurus* [25]. Williams 79 and Fayette supported different amounts of population level increase [18].

Planting date affects winter survival of *P. brachyrus* and the subsequent damage to soybean. Wheat during the winter reduced the population more than winter fallow [26]. Early planting dates also affect the yield of the current crop [26].

STING NEMATODES
Belonolaimus spp.

At least two species of sting nematodes, *Belanolaimus* spp., are associated with soybean. *B. longicaudatus*, the species of greater importance, is found in states along the Atlantic Seaboard from Virginia to Florida and in Alabama [4]. *B. nortoni* is not as destructive and is found in the river valleys of Arkansas, Kansas, Louisiana, New Jersey, Oklahoma, and Texas [4]. They are always found in very sandy soil.

Belonolaimus spp. are ectoparasites that feed near the tips of roots. They have long stylets with which they can penetrate deeply into the root tissue. Lesions are formed that extend into the stele and consist of a cavity surrounded by injured cells [27]. The root damage results in stubby roots and lack of feeder roots. The above-ground symptoms are chlorosis, stunting and wilting which are nondiagnostic.

Crop rotations and nematicides are effective in managing sting nematodes. Non-hosts that are acceptable to growers may be difficult to find. Tobacco and watermelon are good nonhosts but may not fit in the cropping system of most growers. Fenamiphos gives excellent control of *B. longicaudatus* as does aldicarb (Schmitt, pers. comm.), both resulting in significant yield increases. However, the cost of the nematicide may be prohibitive, depending on the market price for soybeans.

OTHER NEMATODES

Criconemella ornata [28], *C. macrodora, Helicotylenchus pseudorobustus, H. dihystera, Rotylenchus* spp., *Scutellonema bradys, S. brahhyurum, Quinisulcius acutus, Paratrichodorus minor, Paratylenchus projectus, P. tenuicaudatus, Tylenchorhynchus canalis, T. claytoni, T. ewingi, T. goffarti, T. martini,* and *Meiodorus hollisi, Xiphinema americanum, X. chambersi,* and *X. rivesi,* [13, 29, Schmitt, pers. comm.] have been reported to inhabit the rhizosphere of soybean. However, only in few instances has damage to soybean been attributed to any one of these species. Little time has been spent in attempting to manage these nematodes on soybean.

REFERENCES

1. Schmitt, D.P., and Imbriani, J.L., Management of *Hoplolaimus columbus* with tolerant soybean and nematicides. Ann. App. Nematol. (Supp. J. Nematol.), 1987, 1, 59-63.

2. Rebois, R.V., Effect of soil temperature on infectivity and development of *Rotylenchulus reniformis* on resistant and susceptible soybeans, *Glycine max.* J. Nematol., 1973, 5, 10-3.

3. Heald, C.M., and Robinson, A. F., Survey of Current distribution of *Rotylenchulus reniformis* in the United States. J. Nematol. Supp., 1990, 22, 695-9.

4. Schmitt, D.P. and Noel, G.R., Nematode parasites of soybeans. pp. 13-59. In W. R. Nickle (ed.) Plant and Insect Parasitic Nematodes. Academic Press. New York. 1984.

5. Rebois, R.V., Epps, J.M. and Hartwing, E.E., Correlation of resistance in soybeans to *Heterodera glycines* and *Rotylenchulus reniformis*. Phytopathology, 1970, 60, 695-700.

6. Rebois, R.V., Madden, P.A. and Eldridge, B.J., Some ultrastructural changes induced in resistant and susceptible soybean roots following infection of *Rotylenchulus reniformis*. J. Nematol., 1975, 7, 122-39.

7. Robbins, R.T., Description of *Hoplolaimus magnistylus* n. sp. (Nematoda: Hoplolaimidae). J. Nematol., 1982, 14, 500-06.

8. Lewis, S.A., Smith, F.A. and Powell, W.M., Host-parasite relationships of *Hoplolaimus columbus* on cotton and soybean. J. Nematol., 1982, 8, 141-45.

9. Weiser, G.C., Mueller, J.D. and Shipe, E.R., Differential nodulation of soybean cultivars in the presence of *Hoplolaimus columbus*. Soy. Genet. Newsl., 1988, 15, 121-3.

10. Mueller, J.D., Schmitt, D.P., Weiser, G.C., Shipe, E.R. and Musen, H.L., Performance of soybean cultivars in *Hoplolaimus columbus*. Ann. App. Nematol. (Supp. to J. Nematol.), 1988, 2, 65-9.

11. Weiser, G.C., Mueller, J.D. and Shipe, E.R., Response of tolerant and susceptible soybean cultivars to columbia lance nematode. Soy. Genet. Newsl., 1988, 14, 260-2.

12. Ross, J.P., Nusbaum, C.J. and Hirschmann H., Soybean yield reduction by lesion, stunt, and spiral nematodes. Phytopathology, 1967, 57, 463-4 (Abstr.).

13. Schmitt, D.P., and Bailey, J.E., Chemical control of *Hoplolaimus columbus* on cotton and soybean. J. Nematol. Supp. (Ann. App. Nematol.), 1990, 22, 689-94.

14. Goodey, J.B., Franklin, M.T., and Hooper, D.J., T. Goodey's The nematode parasites of plants catalogued under their hosts. 3rd ed. Commonw. Agr. Bur., 1965.

15. Manuel, J.S., Reynolds, D.A., Bendixen, L.E. and Riedel, R.M., Weeds as hosts of *Pratylenchus*. Ohio Agric. Res. Devel. Cent. Res. Bull., **1123**, 1980.

16. Acosta, N., Malek, R.B., and Edwards, D.I., Susceptibility of soybean cultivars to *Pratylenchus scribneri*. J.Agric. Univ. Puerto Rico, 1979, **63**, 103-110.

17. Zirakparvar, M.E., Susceptibility of soybean cultivars and lines to *Pratylenchus hexencisus*. J. Nematol., 1981, **14**, 217-20.

18. Lawn, D.A., and Noel, G.R., Field interrelationships among *Heterodera glycines*, *Pratylenchus scribneri*, *Helicotylenchus pseudorobustus*, *Paratylenchus projectus*, and *Tylenchorhynchus martini*. J. Nematol., 1986, **18**, 98-106.

19. Koenning, S.R., and Schmitt, D.P., Control of *Pratylenchus brachyurus* with selected nonfumigant nematicides on a tolerant and a sensitive soybean cultivar. Ann. App. Nematol. (Supp. to J. Nematol.), 1987, **1**, 26-8.

20. Ferris, V.R., and Bernard, R.L., Population dynamics of nematodes in fields planted to soybeans and crops grown in rotation with soybeans. I, The genus *Pratylenchus* (Nemata: Tylenchida). J. Econ. Ent., 1967, **60**, 405-10.

21. Johnson, A.W., Dowler, C.C. and Hauser, E.W., Crop rotation and herbicide effects on population densities of plant parasitic nematodes. J. Nematol., 1975, **7**, 158-68.

22. Rebois, R.V., and Golden, A.M., Pathogenicity and reproduction of *Pratylenchus agilis* in field microplots of soybean, corn, tomato, or corn-soybean cropping systems. Plant Disease, 1985, **69**, 927-9.

23. Koenning, S.R., and Schmitt, D.P., Integrated pest management of *Pratylenchus brachyurus* on soybean. Nematropica, 1986, **16**, 237-44.

24. Kinloch, R A., and Lutrick, M.C., The relative abundance of nematodes in an established field crop rotation. Soil crop Sci. Soc. Fla. Proc., 1975, **34**, 192-94.

25. Schmitt, D.P., and Barker, K.R., Damage and reproductive potentials of *Pratylenchus brachyurus* and *P. penetrans* on soybean. J. Nematol., 1981, **13**, 327-32.

26. Koenning, S.R., Schmitt, D.P. and Barker, K.R., Influence of planting date on population dynamics and damage potential of *Pratylenchus brachyurus* on soybean. J. Nematol., 1985, **17**, 428-34.

27. Standifer, M.S., The pathologic histology of bean roots injured by sting nematodes. Plant Dis. Rep., 1959, **43**, 983-6.

28. Barker,K.R., Schmitt, D.P., and Campos, V.P., Response of peanut, corn, tobacco, and soybean to *Criconemella ornata*. J. Nematol., 1982, **14**, 576-581.

29. Robbins, R.T., Riggs, R.D., and Von Steen, D., Results of annual phytoparasitic nematode surveys of Arkansas soybean fields, 1978-1986. Ann. Appl. Nematol. (J. Nematol. Supp.), 1987, **1**, 50-5.

MANAGEMENT OF THE SOYBEAN CYST NEMATODE, *HETERODERA GLYCINES*, IN SOYBEANS

LAWRENCE D. YOUNG
Agricultural Research Service,
United States Department of Agriculture
605 Airways Blvd., Jackson, TN, 38301, USA

ABSTRACT

The soybean cyst nematode, *Heterodera glycines* Ichinohe, was first
described in Japan. It also occurs in Canada, The People's Republic of
China, Columbia, Indonesia, Korea, the Soviet Union, and in 26 states
within the United States. Males are necessary for reproduction, and most
of the eggs are retained within the dead female body or cyst where they
can survive for many years. The primary methods of reducing yield damage
caused by the soybean cyst nematode are rotation of soybean, *Glycine max*
(L.) Merr., with nonhost crops, use of nematicides, and planting resistant
cultivars. Primary research on management of *H. glycines* involves
development and deployment of resistant cultivars. This is necessary
because of the great genetic variability in populations of the nematode for
reproduction on cultivars with different sources of resistance. Long-term
management of the nematode requires integration of control tactics.

INTRODUCTION

The soybean cyst nematode, *Heterodera glycines* Ichinohe, was first

described as a new species in 1952 [1]. However, the organism occurred in

Japan prior to 1915 where it was known as *Heterodera schachtii* Schmidt [2].

It was first found in the United States (U.S.) in North Carolina in 1954

and has been confined mostly to the southern states for many years, but

current infestations extend to 26 states including the northern areas of

soybean production. The nematode is also known to occur in Canada, The

People's Republic of China, Columbia, Indonesia, Korea, and the Soviet

Union [2]. In this discussion of the management of the nematode, its

biology will be described, followed by consideration of current control

practices and their merits.

THE SOYBEAN CYST NEMATODE

Life Cycle

After egg deposition, the nematode changes through four juvenile stages and undergoes four molts to become an adult [3]. The first-stage juvenile develops within the egg and after molting emerges from the egg as a second-stage juvenile. The second stage juvenile moves through the soil, invades the root, and induces a syncytium as a feeding site, primarily in the vascular tissue. After feeding begins, the juvenile enlarges and becomes sedentary. The nematode molts three more times as it enlarges to become an adult. The fourth-stage male juvenile ceases feeding and reverts to an elongate form which allows it to be mobile in the soil. Males are necessary for reproduction and are frequently observed in the gelatinous matrices secreted by females. The lemon-shaped, adult female changes color from white to yellow to brown as it matures to become a cyst. Most of the 200-600 eggs produced by the female are retained within the cyst except for a few which are deposited in a gelatinous matrix outside the vulva. The eggs can survive within the cyst for at least seven years [3].

The soybean cyst nematode is spread by agents which move soil such as farm machinery, soil around nursery stock, wind, and water. Blackbirds can ingest cysts while feeding in infested fields and eggs that pass through the intestinal tract of the bird remain viable [4]. The nematode can also survive for several months as cysts in soil peds which are dry soil particles mixed with uncleaned seed [5]. Planting of the contaminated seed may spread the nematode to noninfested fields.

The nematode develops best between 24 and 30 C [6]. Optimum temperature for hatch is 24 C but hatch occurs at 20 to 30 C. Few, if any, soybean cyst nematodes develop above 35 C, and degeneration of juveniles in roots becomes more frequent when temperatures exceed 24 C. During summer months the nematode takes four weeks to complete its life cycle from penetration of roots to egg deposition. Diapause is probably induced by declining temperature in the fall [7].

Nematode Variability

The soybean cyst nematode has a high degree of genetic variability expressed through differential reproduction on soybean cultivars. Four races of the nematode were described originally [8], but 16 races were recently characterized [9]. Occurrence of only 12 of these races has been reported. This variability creates problems in the development and deployment of cultivars resistant to the nematode.

Economic Importance

The soybean cyst nematode often reduces yield by 10 to 50% of expected production, although nearly 100% of production can be lost. There is evidence that most of the yield reduction is caused by infections occurring within 6 weeks of planting [10]. Usually yield reduction is greater in sandy soils than in soils with low sand content [11]. However, yield reduction of 50% has been measured in soils containing only 10-15% sand and more than 60% silt. Irrigation does not appreciably reduce the damage caused by the nematode [12].

CONTROL

Most of the issues involved in management of the soybean cyst nematode revolve around the use of resistant cultivars. Due to the great genetic variability of the nematode, frequent planting of resistant cultivars can lead to loss of most cultivars with the same source of resistance if there is a race shift within the nematode population. Two strategies proposed to limit this risk are to reduce use of resistant cultivars and to rotate cultivars with different sources of resistance. Although much of the recent research involves resistant cultivars, other control practices must also be considered. These alternative control practices and using resistant cultivars reduce the number of nematodes that invade and establish feeding sites within the plant roots. Crop yield is improved because the relationship between soybean yield and number of nematode eggs is generally linear [2].

Crop Rotation

The soybean cyst nematode has a limited host range, therefore, rotation of nonhost crops with soybean is an effective means of reducing the number of nematodes in the soil. This was the earliest control measure used and is still used where corn and grain sorghum are frequently grown. Usually, two consecutive years of nonhost crops are needed between soybean crops to reduce nematode populations to nondamaging levels [2]. An additional year of nonhost crop may be beneficial if the beginning soil population of nematodes is extremely high. In contrast, one year of nonhost crops may be sufficient if the nematode population density is low or if it is parasitized by microbial antagonists. In North Carolina, the nematode has been managed with crop rotation combined with cultural practices such as planting early-maturing cultivars late in the growing season and planting soybean with no tillage [13]. Planting late in the growing season allows the juveniles to hatch, but because a host is not present, the nematode dies. No-till planting is reported to inhibit nematode survival, but this inhibition may be partially associated with late planting effects.

Nematicides

Few nematicides are available for control of the soybean cyst nematode, and the available ones give inconsistent results. Nematicides are seldom used because of the cost factor and their failure to raise the yield of susceptible cultivars above that of resistant cultivars (Table 1). In conjunction with resistant cultivars, nematicides are often used to control nematodes such as *Meloidogyne* species that occur together with *H. glycines* in fields planted with soybean [14].

TABLE 1
Yield and number of cysts associated at harvest with resistant 'Bedford' soybean and susceptible 'Forrest' soybean grown with and without nematicides in soil infested with soybean cyst nematode race 14

Cultivar	Nematicide	Seed yield (kg/ha)	Cysts in soil (no./liter)
Bedford	None	2,984 a	144 c
Forrest	Phenamiphos	2,431 b	636 b
	Aldicarb	2,312 bc	532 b
	DBCP	2,227 cd	552 b
	Carbofuran	2,183 cde	524 b
	Ethoprop	2,031 e	736 b
	None	2,072 de	1,148 a

C.V. = 9%.
Means in same column followed by same letter are not significantly different (Duncan's Multiple Range Test, P = 0.05).
Reprinted from Epps et al. [15].

Resistant Cultivars

Host resistance is often extremely effective in suppressing soybean cyst nematode populations and increasing yields (Table 2). 'Pickett' was the first cultivar released in the U.S. with resistance to the soybean cyst nematode. This cultivar is resistant to races 1 and 3 of the nematode. The 'Bedford' cultivar was the first cultivar with resistance to race 14 (then considered race 4), and 'Cordell' was the first cultivar resistant to race 5. The 'Hartwig' cultivar will soon be available, and it is thought to be resistant to all races of the nematode. Although there are many cultivars available with resistance to the soybean cyst nematode, only four sources of resistance ('Peking', PI 88788, PI 90763, and PI 437654) have been used.

TABLE 2
Yield and cyst count at harvest for soybean cultivars grown in fields infested with soybean cyst nematode race 3

Cultivar	Cysts/liter	Yield (kg/ha)
Forrest	0	3,425
Pickett 71	5	2,900
Centennial	0	3,300
Tracy	1,640	2,080
LSD (0.05)		436

'Tracy' is susceptible to race 3; other cultivars are resistant.
Reprinted from Hartwig [17].

As an example of the contribution of resistant cultivars in suppressing yield losses to the nematode, the value of soybeans not lost to the soybean cyst nematode due to planting of the race 3 resistant cultivar 'Forrest' was estimated to exceed 400 million dollars during a six-year period [16].

The primary problems associated with planting of resistant cultivars have been the lack of resistance to all races of the nematode and race shifts within nematode populations due to the selection pressure exerted by the cultivars upon the nematode. The first problem is expected to be solved, at least temporarily, with the release of the 'Hartwig' cultivar which has putative resistance to all races of the nematode. Reducing selection pressure on the nematode population through use of host resistance requires less frequent planting of resistant soybean cultivars. Planting susceptible cultivars or alternative crops can be considered a control cost if it is more profitable to grow resistant soybean.

Some experiments have focused on shifting the predominant genotype of the nematode from one race to another. In North Carolina, an attempt was made to shift race 2 to race 1 in the field so that available resistant cultivars could be used again [13]. This strategy was based on greenhouse research in which mixtures of race 1 and race 2 became predominantly race 1 when susceptible cultivars were grown. However, the attempts to do this in the field were unsuccessful. Francl and Wrather [18] rotated 'Forrest' soybean with 'Bedford' soybean in an unsuccessful attempt to shift the ability of the nematode population to reproduce on 'Bedford'. They concluded that 'Bedford' and 'Forrest' did not have mutually incompatible reactions with their respective selected nematode populations. This incompatibility was considered necessary for success in rotating cultivars with different sources of resistance. We have attempted to prevent shifts of the nematode genotypes by alternating cultivars with different sources of resistance. In our greenhouse, breeding line J82-21 (resistance from 'Peking' and PI 90763) is mutually incompatible with 'Bedford' (resistance from 'Peking' and PI 88788) for race 14 reproduction. However, in a field infested with race 14, rotating these two soybeans has not consistently prevented a shift for more reproduction on 'Bedford' compared to growing either J82-21 or a susceptible cultivar in continuous monoculture (Table 3). The genetics of the nematode for parasitism of soybean is not adequately understood to manipulate genotypes of the nematode by rotating sources of resistance.

Reduction of selection pressure on soybean cyst nematode populations has been attempted by alternating the resistant cultivar with susceptible cultivars and nonhost crops. Young and Hartwig [19] reported that an original race 9 population reproduced less on resistant cultivar 'Bedford' when susceptible cultivars were rotated with it than when 'Bedford' was grown in a continuous monoculture (Table 4). Rotating corn (*Zea mays* L.), a nonhost of the nematode, with 'Bedford' delayed the increase in ability of the nematode population to reproduce on 'Bedford'. Blending 'Bedford' with susceptible cultivars effectively maintained nematode reproduction at a low, acceptable level for several years, but reproduction on 'Bedford' increased dramatically after 11 years of planting the blend. In another study, neither rotating 1 year of susceptible cultivar 'Tracy M' with 2 years of the resistant cultivar 'Centennial' nor blending the two cultivars

TABLE 3
Relative reproduction of *Heterodera glycines* on *Glycine max* cv. 'Bedford',
a resistant cultivar, grown in the greenhouse in soil from field plots
planted with soybeans with different sources of resistance.

Cropping	Relative reproduction[*]					
sequence	1985	1986	1987	1988	1989	1990
FFFFFF	8	35	18	15	32	24
BBBBBB	18	67	68	68	89	135
JJJJJJ	14	10	10	14	18	22
BBJJBB	16	78	18	27	36	62
JJBBBJ	17	27	28	49	63	60
CEBCEB	11	20	46	17	33	77
EBCEBC	15	19	68	38	81	91
BCEBCE	13	13	14	31	33	51
LSD (0.05)	NS	NS	39	26	30	47

Each letter or number designates the soybean grown for 1 year: F =
'Forrest', B = 'Bedford', E = 'Essex', J = J82-21, and C = corn. 'Forrest'
and 'Essex' are susceptible, 'Bedford' is resistant, and J82-21 is
moderately resistant to race 14 which was the original race in this field.
[*]Number of cysts on 'Bedford' expressed as a percentage of cysts on 'Essex'
soybean 35 days after planting in soil collected from field plots.

for a 10-year period, limited the ability of the nematode population
(originally race 3) to reproduce on 'Pickett 71' [20], another race 3
resistant cultivar, at the conclusion of the test.

A rotation of resistant cultivar, nonhost crop, and susceptible
cultivar may merely slow the shift of races. After 6 years of such a
rotation, the reproductive ability of the nematode (originally race 14) on
'Bedford' was significantly higher following both 'Bedford' and a nonhost
crop than a continuous monoculture of a susceptible cultivar (Table 3).
Additional data are needed to determine if the trend for a race shift
occurs in this rotation.

CONCLUSION

Management of the soybean cyst nematode requires integration of control
practices to allow soybean producers versatility in crop production and to
achieve the best economic returns for their farming operation. Although
rotating nonhost crops, susceptible soybean, and resistant soybean is not
totally effective in preventing race shifts within the nematode population,

TABLE 4

Relative reproduction of *Heterodera glycines* on *Glycine max* cv. 'Bedford', a resistant cultivar, grown in soil from field plots exposed to different cropping sequences.

Cropping sequence	Relative reproduction[*]							
	1980	1982	1984	1985	1986	1987	1988	1989
FFFFFFFFFF	8	12	3	4	11	8	23	20
BBBBBBBBBB	29	86	42	87	57	87	116	100
MMMMMMMMMM	10	16	11	24	32	34	38	77
BCBCBCBCBC	12	52	20	48	50	49	59	88
FEBFEBFEBF	6	28	7	14	14	22	32	46
EBFEBFEBFE	12	8	7	15	8	37	22	31
BFEBFEBFEB	36	31	4	11	20	8	30	44
LSD (0.05)	22	39	13	40	34	30	28	54

Each letter indicates the soybean cultivar grown for 1 year. F = 'Forrest', B = 'Bedford', E = 'Essex', M = a blend of 70% 'Bedford' and 30% 'Forrest', and C = corn. 'Bedford' is resistant, and 'Forrest' and 'Essex' are susceptible.
[*]Number of cysts occurring on 'Bedford' soybean expressed as a percentage of the number of cysts on 'Essex' 35 days after planting in the greenhouse in the soil collected from field plots.
Data on relative production on 'Bedford' soybean were not obtained in 1979, the first year of the experiment, and data for 1981 and 1983 were not significantly different.
Reprinted from Young and Hartwig [19].

it is one of the best strategies for stabilizing soybean production. The strategy also allows maximization of the longevity of resistant cultivars for control of this nematode. The research in North Carolina [13] demonstrates that control practices can also be integrated successfully by utilizing practices other than resistant cultivars. The best combination of practices will depend on the crops that the producer can grow most successfully.

REFERENCES

1. Ichinohe, M. 1955. Studies on the morphology and ecology of the soybean nematode, *Heterodera glycines*, in Japan. Rep. Hokkaido Nat. Agri. Exp. Stat. 48:1-64.

2. Riggs, R. D., and Schmitt, D. P. 1989. Soybean cyst nematode, In Sinclair, J. B., and Backman, P. A., eds., Compendium of Soybean Diseases. 3rd. ed. APS Press, St. Paul, MN.

3. Slack, D. A., Riggs, R. D., and M. L. Hamblen. 1981. Nematode control studies in soybeans: Rotations and population dynamics of soybean cyst and other nematodes. Ark. Agri. Exp. Stat. Rep. Ser. 263:1-36.

4. Epps, J. M. 1969. Recovery of soybean cyst nematode (*Heterodera glycines*) from the digestive tract of blackbirds. J. Nematol. 3:417-419.

5. Epps, J. M. 1969. Survival of the soybean cyst nematode in seed stocks. Plant Dis. Rep. 53:403-405.

6. Alston, D. G., and Schmitt, D. P. 1988. Development of *Heterodera glycines* life stages as influenced by temperature. J. Nematol. 20:366-372.

7. Hill, N. S., and Schmitt, D. P. 1989. Influence of temperature and soybean phenology on dormancy induction of *Heterodera glycines*. J. Nematol. 21:361-369.

8. Golden, A. M., Epps, J. M., Riggs, R. D., Duclos, L. A., Fox, J. A., and Bernard, R. L. 1970. Terminology and identity of infraspecific forms of the soybean cyst nematode (*Heterodera glycines*). Plant Dis. Rep. 54:544-546.

9. Riggs, R. D., and Schmitt, D. P. 1988. Complete characterization of the race scheme for *Heterodera glycines*. J. Nematol. 20:392-395.

10. Wrather, J. A., and Anand, S. C. 1988. Relationship between time of infection with *Heterodera glycines* and soybean yield. J. Nematol. 20:439-442.

11. Koenning, S.R., Anand, S. C., and Wrather, J. A. 1988. Effect of within-field variation in soil texture on *Heterodera glycines* and soybean yield. J. Nematol. 20:373-380.

12. Young, L. D., and Heatherly, L. G. 1988. Soybean cyst nematode effect on soybean grown at controlled soil water potentials. Crop Sci. 28:543-545.

13. Schmitt., D. P. 1991. Management of *Heterodera glycines* by cropping and cultural practices. J. Nematol. 23:348-352.

14. Rodriguez-Kabana, R., Weaver, D. B., Robertson, D. G., King, P. S., and Carden, E. L. 1990. Sorghum in rotation with soybean for the management of cyst and root-knot nematodes. Nematropica 20:111-119.

15. Epps, J. M., Young, L. D., and Hartwig, E. E. 1981. Evaluation of nematicides and resistant cultivar for control of soybean cyst nematode. Plant Dis. 65:665-666.

16. Bradley, E. B., and Duffy, M. 1982. The value of plant resistance to soybean cyst nematode: A case study of Forrest soybeans. NRE Staff Report No. AGES820929. U. S. Dep. Agri. Econ. Res. Serv., Washington, DC.

17. Hartwig, E. E. 1981. Breeding productive soybean cultivars resistant to the soybean cyst nematode for the southern United States. Plant Dis. 65:303-307.

18. Francl, L. J., and Wrather, J. A. 1987. Effect of rotating 'Forrest' and 'Bedford' soybean on yield and soybean cyst nematode population dynamics. Crop Sci. 27:565-568.

19. Young, L. D., and Hartwig, E. E. 1992. Cropping sequence effects on soybean and *Heterodera glycines*. Plant Dis. 76:78-81.

20. Young, L. D., and Hartwig, E. E. 1988. Selection pressure on soybean cyst nematode from soybean cropping sequences. Crop Sci. 28:845-847.

MANAGEMENT OF ROOT-KNOT NEMATODES IN SOYBEAN

ROBERT KINLOCH
University of Florida,
Agriculture Research And Education Center,
Jay, Florida 32565, U.S.A.

ABSTRACT

The use of resistant cultivars has been the mainstay of root-knot nematode management in soybean. Genes which impart a quantitative resistance to Meloidogyne incognita have been incorporated into several cultivars of maturity groups VI - VIII. Though some yield losses are experienced by these cultivars when grown in heavily infested soil, the additional use of nematicidal soil treatment is not generally practiced. Management by crop rotation is complicated by the extensive host ranges of the causal organisms, which also include M. arenaria, and M. javanica. However, the declining economics of soybean production over the last decade has encouraged growers individually to adopt rotational cropping practices. This should favor the management of the latter two species, against which resistant cultivars have not been adequately developed.

INTRODUCTION

Root-knot nematodes are pests of soybean in the warmer regions of the crop's distribution. They are often major yield limiting factors where soybean cultivars of maturity group VI and above are grown, especially in drought sensitive soils. Three species of root-knot nematode are major pests of soybean: Meloidogyne incognita, M. arenaria, and M. javanica. Though the first mentioned species is the most widely distributed and has received the most management attention, M. arenaria, predominantly race 2 (1), has been increasing in importance, especially in the U.S.A.

This has been primarily due to the development and extensive
planting of <u>M</u>. <u>incognita</u> resistant soybean cultivars (2), most of
which are highly susceptible to <u>M</u>. <u>arenaria</u>. In a few restricted
sites in north Florida and south Georgia in U.S.A., <u>M</u>. <u>javanica</u>
is a pest of soybean where it has been grown in association with
tobacco (3). A fourth species, <u>M</u>. <u>hapla</u>, is of considerably less
consequence to soybean production in the cooler more northerly
regions of the U.S.A.

Life History

<u>M</u>. <u>incognita</u>, the Southern root-knot nematode, is an obligate
plant parasite. It is an indigenous species throughout most of
the world where soil temperatures range within 20 - 30 C and
where sufficient soil moisture is present for host growth. Its
very extensive host range among cultivated crops has intensified
the pest's importance and has made management by crop rotation a
problematic option. The nematode is active when soil temperatures
are above 15 C and sufficient moisture is present to allow the
nematode to move through the soil. In the absence of a host
plant, or in winter when soil temperatures are below optimum, the
nematode survives in the egg stage or as a vermiform second stage
infective juvenile. The latter is the predominant survival stage
where soils are moist and temperatures are above 15 C.
Penetration of the root is at the growth meristem and the
nematode will enter plants that may not prove to be hosts, in
which case the nematode will emigrate or succumb. Within a
suitable host, development to an adult by molting through second,
third, and fourth juvenile stages takes twenty to thirty days
depending on temperature. Thus, several overlapping generations
are produced through a soybean season. The host response to
feeding by the nematode is the development of multinucleate
giants cells from which the pest gains its sustenance. There is
an accompanying swelling of the root tissue around the feeding
site. In heavy infestations these are manifest by abnormal root
swellings (galls) and the dislocation of the root's conducting
system. The consequent impediment of water and nutrient flow

through the plant causes it to wilt readily and display general symptoms of nutrient deficiency. Foliar symptoms are not usually evident until the latter half of the growth season. Root-knot galling is not readily noticeable in the first two months of the soybean's growth, however, if the infection is substantial, it is clearly evident at flowering. M. incognita is a parthenogenic species, though adult males may be produced under adverse environmental conditions such as over-infestation of the root late in the crop season. The swollen pear-shaped female may produced in excess of 200 eggs which are deposited to the root surface within a protective gelatinous matrix. At soybean maturity, the eggs may not hatch if the soil temperature and moisture are low and the root-knot population will over-winter in this stage. Normally, in regions where the major root-knot nematodes are serious pests of soybean, soil temperature and moisture conditions at soybean maturity are sufficient for egg hatch and the nematode survives the winter as second stage infective juveniles. Population densities remain quite stable through the winter months, but if host plants are absent, they will decline as the soil warms in the spring. Thus the mid-winter period is an optimum time to sample soil for root-knot nematode infestations since an early spring may bring soil populations to near undetectable levels as the soybean planting season approaches (4).

Distribution
In the U.S.A. root-knot nematode disease of soybean is prevalent in Louisiana, Mississippi, Alabama, Florida, Georgia, South Carolina, and North Carolina. Acknowledgement of the severity of the disease did not become apparent until the late 1960s when considerable surveys were being conducted to determine the distribution of the soybean cyst nematode, Heterodera glycines, an introduced species and a federally quarantined pest at that time. During the 1960s, there were increasing plantings of the soybean cultivar "Pickett" (maturity group VI) which was resistant to the soybean cyst nematode but very susceptible to M.

incognita. Since soybean had become a major cash crop of the
region there was no incentive to adopt rotational cropping
regimes as part of a root-knot nematode management program, and
chemical control strategies were questionably effective, even
where root-knot nematode infestations were light.

CHEMICAL CONTROL

The introduction of a M. incognita resistant cultivar, "Bragg"
(maturity group VII), in 1963 and its widespread acceptance
through much of the southeast U.S.A. into the next decade, was
accompanied by an increased use of nematicides, particularily the
soil fumigant DBCP. Resistance in Bragg to root-knot nematode is
quantitative (5) and, when the cultivar was grown in severe root-
knot nematode infestations, sufficient juveniles invaded the
roots and developed to maturity such that the use of a nematicide
was an economic practice. The standard method was to apply DBCP
fumigant at or immediately before planting a resistant cultivar
via one chisel per row injected at 15 cm below the seed at a rate
of 10 liters per hectare. The non-fumigant organophosphate and
carbamate nematicides were considerably less effective than DBCP
and were not widely used for root-knot nematode management (6).

TABLE 1

Yields (kg/ha) of Meloidogyne incognita resistant (R) and
susceptible (S) soybean cultivars grown in nematicide treated
infested soil in Florida, U.S.A. in 1972.

Nematicide	Bragg (R)	Pickett (S)
DBCP 12 1/ha	1984 a	1392 b
Carbofuran 3.4 kg/ha	1412 b	599 c
Ethoprop 3.4 kg/ha	1345 b	558 c
Fensulfothion 3.4 kg/ha	1197 b	713 c
None	1217 b	552 c

Yields are means of four replicates. Means followed by the same
letter are not significantly different ($P = 0.05$) according to
Duncan's Multiple Range Test.

DBCP remained the standard nematicide for root-knot nematode management on soybean until 1978 when it was removed from use because of environmental concerns (7). A similar fate befell its widely used replacement, ethylene dibromide, in 1981 (7). This soil fumigant was as effective and as economical as DBCP. It was chisel-injected in a similar manner but at the rate of 15 liters per hectare. Thus in the 1980's there were no effective nematicides for soybean pest management.

RESISTANCE

During the 1970s considerable progress had been achieved in improving the resistance to M. incognita in new soybean cultivars. The root galling response is a practicable symptom for comparing soybean germplasms for their resistance to root-knot nematode. It is an especially valuable marker since yield of soybean has been shown to be linearily and negatively related to the amount of root-knot galling (8). Replicated entries of breeding lines, grown either in infested fields or in potted cultures in a glasshouse, are rated according to a six-place scale depending on their amount of galling after 2 - 3 months exposure to the nematode (9). There are no cultivars that consistently escape galling, but those considered resistant to M. incognita normally have less than 5 % of their root surface galled when grown for a full season in soil heavily infested with the nematode (8).

As testimony to the importance of resistant cultivars in root-knot nematode management in soybean, many breeding programs both in the private sector and in public institutions have been involved in their development and the resistance genes have been incorporated into cultivars in several maturity groups (10). Advances in breeding for M. arenaria resistance have not been as successful. Though some cultivars perform considerably better than others in heavy soil infestations of this nematode, yield losses are generally in excess of 60% (11). At best, they may be considered as moderately susceptible cultivars and should only be grown in soil that has been rotated for two or more years with non-hosts of this pest.

TABLE 2

Comparative yield and galling responses of soybean cultivars grown in non-infested, <u>Meloidogyne</u> <u>incognita</u>-infested, and <u>M.</u> <u>arenaria</u>-infested soils in Florida, U.S.A. Data are averages of observations taken over three years.

	Non-infested	Galling	Yield	(% yield loss)	
	kg/ha	M. incognita		M. arenaria	
Kirby*	2,735	0.5	2,392 (13%)	2.2	1,195 (56%)
Coker 6738*	3,279	0.8	2,478 (24%)	2.1	1,244 (62%)
Coker 686*	3,129	1.7	2,230 (29%)	2.9	964 (69%)
Coker 6727*	3,241	1.2	2,479 (24%)	2.1	978 (70%)
Hartz 6385*	3,068	1.0	2,145 (21%)	2.3	921 (70%)
Cobb*	3,000	1.5	2,164 (28%)	3.4	177 (94%)
Gordon*	3,133	0.6	2,338 (25%)	2.0	948 (70%)
Centennial*	3,272	0.7	2,550 (22%)	3.2	715 (78%)
N K S69-96	3,370	2.6	1,100 (67%)	3.5	326 (90%)

* Cultivars resistant to <u>M.</u> <u>incognita</u>
Galling on a scale of 0 = 0%, 0.2 = < 5%, 1 = 5-25%, 2 = 26-50%, 3 = 51-75%, and 4 = > 75% root-surface galled.

ROTATION

Monoculturing of soybean was generally practiced in the southern states of U.S.A. through the 1970s, despite evidence of the effectiveness of single year rotations with maize and <u>M.</u> <u>incognita</u> resistant soybean cultivars without need of nematicidal soil treatments (12). The practice was maintained by the availability of resistant cultivars, effective nematicides, and strong market conditions for the crop. The current practice of rotating soybean with other crops has arisen more out of economic neccessity rather than as a conscientious effort to manage root-knot disease. A weakening of the soybean commodity market has forced growers to plant more economically attractive crops such as cotton and peanut or to leave hectarage temporarily out of production. Cotton is a host of <u>M.</u> <u>incognita</u>, and, though resistant soybean cultivars could yield adequately following cotton, residual nematode infestations in soil after soybean would negatively affect cotton yields. Alternatively, a rotation of soybean and cotton would be beneficial for the management of M. arenaria which is not a parasite of cotton. Peanut is not a host of either M. incognita or M. arenaria race 2 and would seem to be an ideal rotational crop with soybean. However, other soil pathogens, such as Sclerotium rolfsii and Rhizoctonia solani which these crops have in common, prohibit adoption of this as a management practice.

REFERENCES

1. Hartman, K.M. and Sasser, J.N., Identification of Meloidogyne species on the basis of differential host test and perineal pattern morphology. In An Advanced Treatise On Meloidogyne Vol II, ed. J.N. Sasser, United States Agency for International Development, Raleigh, North Carolina, 1985, pp. 69-78.

2. Kinloch, R.A., The control of nematodes injurious to soybean. Nematropica, 1980, 2, 141-153.

3. Garcia, R. and Rich, J.R., Root-knot nematodes in north central Florida soybean fields. Nematropica, 1985, 15, 43-48.

4. Kinloch, R.A., The relationship between soil populations of Meloidogyne incognita and yield reduction of soybean in the Coastal Plain. Journal of Nematology, 1982, 14, 162-167.

5. Hinson, K., Breeding for resistance to root-knot nematodes, In World Soybean Research Conference III: Proceedings, ed. R. Shibles, Westview Press, Boulder and London, 1985, pp. 387-393.

6. Kinloch, R.A., Response of soybean cultivars to nematicidal treatments of soil infested with Meloidogyne incognita. Journal of Nematology, 1974, 6, 7-11.

7. Johnson, A.W. and Feldmesser, J., Nematicides - a historical review. In Vistas On Nematology, ed. J.A. Veech, The Society of Nematologists, Hyattsville, Maryland, 1987, pp. 448-454.

8. Kinloch, R.A., Hiebsch, C.K. and Peacock, H. A., Evaluation of soybean cultivars for production in Meloidogyne incognita-infested soil. Annals of Applied Nematology, 1987, 1, 32-34.

9. Kinloch, R.A., Screening for resistance to root-knot

nematodes. In Methods For Evaluating Plant Species To Plant-Parasitic Nematodes, ed. J.L. Starr, The Society of Nematologists, Hyattsville, Maryland, 1990, pp. 16-23.

10. Hussey, R.S., Boerma, H.R., Raymer, P.L. and Luzzi, B.M., Resistance in soybean cultivars from maturity groups V-VIII to soybean cyst and root-knot nematodes. Journal of Nematology, 1991, 23(4S), 576-583.

11. Kinloch, R.A., Hiebsch, C.K. and Peacock, H. A., Galling and yields of soybean cultivars grown in Meloidogyne arenaria-infested soil. Journal of Nematology, 1987, 19, 233-239.

12. Kinloch, R.A., Soybean and maize cropping models for the management of Meloidogyne incognita in the Coastal Plain. Journal of Nematology, 1986, 18, 451-458.

SOYBEAN DISEASE MANAGEMENT: CHEMICAL AND BIOLOGICAL CONTROL IN TEMPERATE REGIONS

PAUL A. BACKMAN and BARRY J. JACOBSEN
Auburn University, Department of Plant Pathology
Auburn, Alabama 36849 USA

ABSTRACT

The soybean germplasm we utilize today was primarily derived from the cool temperate region of northeastern China. When soybeans are planted in similar regions around the world, disease severities and losses are typically lower than they are in warm temperate, subtropical and tropical locations. Diseases are usually less severe in cooler climates because of freeze-thaw cycles in the winter, cooler mean temperatures, and/or shorter summers that reduce the number of reproductive generations, and/or lower humidity that reduces the number of infection events for foliage-infecting pathogens. Management is difficult in warm temperate regions requiring an integration of genetic defenses, cultural practice modifications, and chemical control practices. Improvements in disease management will be made by providing farmers with decision-making aids, and by researching disease induced losses that are presently poorly understood.

INTRODUCTION

The domesticated crop we know as soybean *Glycine max* (L.) Merr. has its origins in northeastern China. This cool, wet, temperate origin probably indicates why the soybean has done so well when grown in cool temperate regions around the world. Initially, the soybean was separated from its adapted pathogens by being relocated to other continents. Over time, many of these pathogens made their way to these new production centers. Disease induced losses have remained relatively low in most temperate production centers and are higher in areas with tropical and subtropical environments.

Soil borne pathogens including Phytophthora rot, charcoal rot, and brown stem rot, and the soybean cyst nematode have caused the greatest losses in temperate regions. As the range of soybean adaptation has been expanded to lower latitudes with their humid environments, foliage and stem diseases, viruses, seed pathogens, and root knot nematode have been identified as limiting factors. In addition, as the hard freezes of winters in north China or the midwestern United States are replaced with the mild winters of the southern United States, or the total lack of winter in the tropics, the ability of pathogen inoculum to survive and infect the next crop greatly increases. In mild climates, increased inoculum at planting results from volunteer soybeans or alternative hosts surviving on a year-round basis, from vectors surviving, and/or a higher rate of inoculum survival resulting from fewer freeze-thaw cycles during the winter. Generally, higher inoculum levels at planting indicate more intensive management will be required during the growing season. Compounding this problem, warmer mean temperatures and longer growing seasons result in more generations per year for fungi, nematodes and insect vectors.

ROLE OF ENVIRONMENT

Using the United States as a model, it is interesting to see how disease severity, and even the mix of diseases present, can vary greatly between the northern and southern production zones within the temperate region. From figure 1, it is quite apparent that the more northern regions of the United States typically are affected by cooler air masses originating in the polar regions. During the winter,

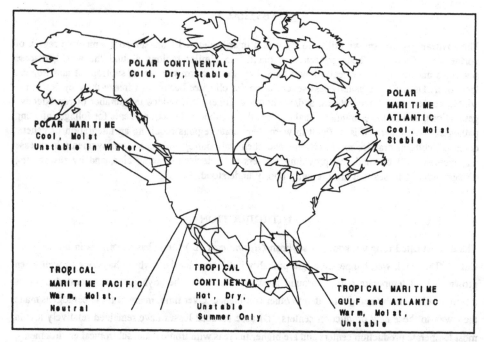

Figure 1. Air mass trajectories and source regions provide very different environments for soybean growing regions in the United States.

these air masses provide the killing frosts and the freeze-thaw cycles that suppress inoculum capable of infecting either foliage or roots. In the summer, warm weather lasts for a shorter time, is generally less humid, and often the nights are cooler. The end result is less primary inoculum to establish initial infections, shorter growing seasons, and lower mean temperatures that limit the number of cycles that multicycle organisms are capable of completing.

In the southern United States, conditions are decidedly different. Polar air enters this region only infrequently during the winter. During the summer, this area is almost exclusively under tropical influence with frequent convective thunderstorms, and generally elevated relative humidities compared to those in more northerly areas. Figure 2 illustrates the mean thunderstorm days during the month of July for four Southeastern states. Areas near the Gulf of Mexico and the Atlantic coast have a much higher incidence of thunderstorms than areas that are more inland. Since infection with the vast majority of foliage, pod and stem diseases occurs only when plants are wetted with rainfall, heavy dew, or fog, many more infection events occur near the ocean waters than inland; disease severity has been directly related to the number of infection events occurring during critical infection periods [2]. If soybeans are planted even more southerly, for example in Puerto Rico, the disease spectrum and intensity would change even more. Here, there are no killing frosts so diseases like soybean rust occur, with inoculum originating on soybeans or alternative hosts which can grow year-long.

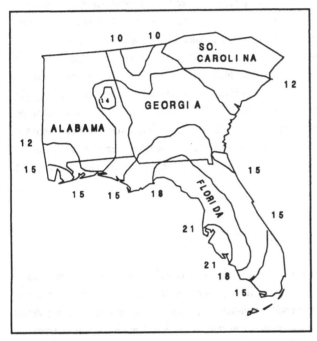

Figure 2. Mean thunderstorm days in July for the southeastern United States [1].

In this environment, humidity is omnipresent, with frequent rainfall that allows diseases such as Rhizoctonia aerial blight to become important. In the continental U.S., aerial blight is seen primarily in the most southerly growing areas and only then if soybeans are grown near rivers, swamps, or flooded rice fields that provide the necessary leaf wetness.

The distribution and severity of soil borne diseases and nematodes are affected also by climatic conditions. Sclerotinia white mold is only found in the coolest production areas, while *Sclerotium rolfsii* causes damage only in the warm temperate to tropical production zones. Nematodes cause the greatest damage where soils are light, and where there are long growing seasons that allow for more cycles per year. However, the warm winter soil temperatures in the southern U.S. allow for more biological control activity against soybean cyst nematodes, with the result that rotations of 1-2 years are adequate to prevent losses while rotations of 3-4 years are needed in more temperate areas. Losses to storage fungi are much higher in warm humid environments, thus storage parameters must be more closely managed than in northern climates where low temperatures reduce storage fungi growth for 3-5 months each year.

GENETIC RESISTANCE

The most economical and efficient defense against plant diseases is to incorporate resistance to the pathogens into high yielding cultivars. In the United States, diseases such as powdery mildew, stem canker, bacterial pustule, frogeye leaf spot, Phytophthora root rot, brown stem rot, and several virus and nematode induced diseases are all managed primarily by resistance. Resistance to *Phomopsis spp.* (pod and stem blight and seed decay) presently is being evaluated. Rapid development of new races in response to race specific resistance is often a problem in Phytophthora rot, soybean rust, frogeye leaf spot, and downy mildew, since numerous races of the causal pathogens are known. Where resistance is not absolute, or is easily defeated by the development of new races, cultivar rotation has emerged as a means of suppressing effective inoculum and preserving resistance [3]. Many of the genes for resistance mentioned above have not been utilized in many countries. Frequently their cultivar selections are the result of reselections made from a very few lines originally introduced. This lack of diversity can demonstrate itself when a new disease arrives. For example, Brazil presently has severe problems with stem canker, as for similar reasons, has Thailand with bacterial pustule.

CULTURAL PRACTICES

The second tactic available to soybean farmers is to alter cultural practices to suppress disease development. Rotation is one of the most frequently used tactics both in the United States and throughout the temperate zone. Rotation with a non-host crop results in decomposition of crop debris containing bacterial and fungal inoculum before the next soybean crop. Crop rotation is effective for pathogens that are poor saprophytes. These pathogens die out or are debilitated when the host debris is decayed and they are incapable of generating enough energy to produce infective propagules. Rotation also reduces the number of sclerotia that can survive until the next soybean crop. For pathogens such as *Sclerotium rolfsii* or *Cylindrocladium crotalariae*, effective rotations may require 3 or 4 years of nonhosts before disease potential is sufficiently reduced to grow soybean safely again. One year rotations with nonhosts can greatly suppress foliage and stem pathogens and can be effective in reducing cyst nematode severity. Rotation is useful in reducing inoculum of soybean cyst and root knot nematodes since non-host crops do not allow for second stage juveniles of those nematodes to complete their life cycle.

Tillage is another cultural control tactic that can aid in suppression of disease. Deep plowing of soil (particularly just after harvest) buries debris and accelerates the decomposition process, and for pathogens that are dispersed by wind or splashing rain, it effectively prevents their movement to the infection court. In the United States, many farmers have adopted no-till farming systems in order to reduce soil erosion. However, this practice has compromised some of the benefits of tillage, particularly in the control of diseases that can utilize the previous crop's debris for plant-to-plant spread or for production of primary inoculum. Deep plowing can reduce damage from *Sclerotinia sclerotiorum* by burying sclerotia deep enough to prevent apothecia from developing and then releasing ascospores above the soil surface.

Use of pathogen-free seed with high vigor can reduce the amount of primary inoculum in the field and also can improve the rate of emergence and seedling growth, thereby providing an escape from soilborne seedling diseases. Most certification programs rely on field inspections to assure genetic purity, and some also monitor virus and fungal diseases. Harvesting as soon after maturity as possible reduces fungal seed infections, further assuring high quality seed. Many seedsmen evaluate seed quality by measuring germination under both normal and stress conditions and also by measuring fungal seed infections. High quality seed in temperate climates can be characterized as having at least 85% germination in warm tests, 70% germination in the Ames cold test, and with less than 10% of the seed with visible disease. If seed lots do not meet these criteria the addition of seed treatment fungicides will often improve performance so that they may still be used for planting purposes. Many farmers who grow soybeans for seed in the U.S. utilize fungicide applications during the R_6 growth stage to reduce seed borne fungal infections and also to improve germination when planted [4].

Water management can prevent saturated soils and thus reduce damage from *Pythium* and *Phytophthora*. Maintaining adequate moisture with irrigation can reduce damage from charcoal rot and brown stem rot. Optimal soil pH, fertility (particularly potassium and phosphorus nutrition), seeding rates, herbicide selection, weed control, and method of cultivation are all factors that can have important effects on disease severity. Planting to coordinate the bloom to pod maturation with the dry season always has been an important strategy in the tropics. This strategy can also be useful in temperate zones if the farmer can select appropriate planting dates and maturity groupings to coordinate reproductive stages with drier months.

CHEMICAL CONTROL

Chemicals for disease control are finding increased usage in soybean production. Seed treatments are used to improve seedling emergence, reduce primary inoculum for midseason diseases, and prevent diseases such as stem canker from becoming established in previously clean fields. However, the amount of seed treated in the U.S. still remains less than 15%. Metalaxyl can be applied as a spray or as a granule into the seed furrow in order to control *Phytophthora* or *Pythium* root rots. It also effectively controls *Phytophthora* root rot when used as a seed treatment in combination with tolerant varieties. Fields with a history of these diseases are the best candidates for treatment, with large yield increases often reported. However, some farmers apply metalaxyl on a prophylactic basis without a previous history of the disease, and still frequently report improved yields. These increases may well be the result of chronic damage that often has been related to pythiaceous fungi.

Foliage-applied fungicides commonly have been used to increase yields and improve seed quality by controlling anthracnose, Cercospora leaf blight, frogeye leaf spot, Rhizoctonia aerial blight, Septoria brown spot, pod and stem blight and purple stain. Where warm wet conditions prevail during growth stages R_1 through R_5, yields may be suppressed by 10-15% or more with serious reductions in seed quality. Timely applications of benzimidazole fungicides, chlorothalonil, or triphenyltin hydroxide can recover up to 80% of losses induced by these diseases. Yield increases are due primarily to larger seed size, but control of anthracnose also can increase seed numbers by reducing numbers of seedless pods.

In the United States, disease control with foliage-applied fungicides had an early history of erratic results. During those early years, fungicides were applied at the R_3 and R_5 growth stages without regard to environment. Later, as epidemiologic principles were applied, applications were made only when conditions favored yield-damaging infections. These conditions were defined to farmers by a series of rules, check lists, or point systems that resulted either in a decision to apply fungicides or to wait to see if infection conditions would develop later [5]. For advanced farmers these rules could be imbedded into a comprehensive computerized soybean pest management model [6]. Applications of fungicides during vegetative growth stages have been shown to reduce stem canker incidence and severity and to reduce latent infections of stem and pod infesting fungi.

Techniques for the application of fungicides for control of foliage, stem, and pod diseases are well developed, frequently relying on aircraft for quick delivery at critical growth stages when conditions favor infection. However, environment can seriously reduce the efficiency of disease control if suboptimal conditions exist at the time of application. Optimal conditions occur from dawn until midmorning, and from late afternoon through darkness. Conditions are optimal when temperature is low and humidity is high, and when wind is less than 7 kph, thus early to mid morning and late afternoon are favored times for aerial spraying. Unfortunately, when farmers are faced with the need to apply fungicides to many hectares and with only a few hours per day of suitable weather, they often compromise and apply during suboptimal conditions. Applications made during the heat of the afternoon in the southern U.S. deposit only 35-40% of the active ingredient on the crop surface, compared to >90% if sprayed by the same aircraft during the morning hours. Efficient disease control will not occur with these reduced rates of active ingredient. Use of ground spray equipment allows farmers to apply the chemicals at any time, but for large farms several sprayers may be needed to assure timely application. Narrow row widths also limit the use of ground sprayers.

BIOLOGICAL CONTROL

Biological control is presently the subject of a great deal of research, and though it has reached commercialization in some crops, this has not been the case for soybean. Numerous countries around the world are trying to reduce the amount of pesticides applied. "Green Laws", as they are called, mandate reductions and do not make allowances for lack of research, or lack of beneficial results on a crop. In the future we will almost certainly see this type of legislative impact on soybean pest control practices. Results from peanut indicate that seeds treated with the A-13 strain of *Bacillus subtilis* predictably develop higher yields due to a healthier root system, enhanced root growth, and improved nutrient status [7]. These results led to the registration of this organism for biological control, *Rhizobium* nodulation enhancement, and growth stimulation for use in peanuts, common beans and cotton under the trade names Quantum 4000 and Epic (Gustafson, Inc., Plano, Texas). Preliminary research with *Bacillus megaterium* applied to soybean seed has shown similar responses [8]. Control of soil borne diseases always has been difficult with chemicals. These data indicate that there is potential for colonizing the growing root with beneficial organisms that protect the root from disease-causing organisms.

Biological control of pod, stem, and foliage diseases of soybean is at an even more embryonic level. Data from other crops indicate that applied organisms have a difficult time becoming established on leaf surfaces because they must compete with indigenous organisms for space they already occupy, plus the harsh wet-dry cycling, UV radiation, and heat can decimate populations of an applied antagonist [9]. Recent research has examined the use of insoluble, selective food-bases such as chitin to provide a stable food supply for the antagonist and to provide improved niches for its growth and development [10]. Disease control has been achieved on peanut, tomato, potato, and apple, indicating that this technology has good potential for transfer into soybean production systems. Biological control of nematodes and *Sclerotium rolfsii* has been shown to be possible by altering soil microbiology by the addition of soil amendments. While amounts required at this time preclude practical application, the research has identified key microbes that may be useful for biological control.

INTEGRATED DISEASE MANAGEMENT

Management of soybean diseases can rarely be accomplished in the long term by only one control method. Disease management must be placed within the context of agronomic practices, weed and insect control, and economics. Integrated pest management, which utilizes economic injury levels, economic thresholds, scouting, record keeping, and pest mapping, allows for a planned, economically sound approach to crop management. Further, it is becoming increasingly important to identify pathogens to the race level. Integrated pest management programs must be adapted to local situations, and several have been developed, including some that use computer software.

One such program is the Auburn University Soybean Integrated Pest Management Model (AUSIMM), developed to predict the profitability of soybean pest management practices in the southeastern United States. The model begins with a ranking of about 120 cultivars according to maturity group, relative productivity, and susceptibility to nematodes and diseases. Potential yields of optimal cultivars are calculated, given the location, soil type, planting date, rotation, and pest histories of individual fields. Submodels are utilized later for management of nematodes, plant diseases, and insect pests. These integrate data on soybean plant growth, population dynamics of major pests, pesticide efficacy, weather, effects of weather on pest populations and pesticide efficacy, and crop loss functions. The effectiveness of the control practices is compared to their cost. Submodels are linked by a core program, which estimates potential yield and crop value through the season, accumulating changes in potential yield, crop value, and management costs. While these decisions can be made by a grower or consultant, the precision and interactive economics made possible with a computer program obviously cannot be duplicated.

Integrated pest management will allow for increased sustainability in soybean production systems. This sustainability will be reflected in increased profitability and reduced losses to pests in whole production systems. Clearly, sustainable production systems must emphasize crop rotation, use of high yielding resistant cultivars, appropriate cultural practices, and loss-potential-based inputs such as fertilizer, irrigation, and pesticides for disease, insect, nematode, or weed control [11].

THE FUTURE

Soybean is affected by a great many diseases. In any given region, the complexity of the decisions needed for optimal management of these diseases increases as the number of diseases increases. Accurate decision-making aids must be developed for farmers that are based on a risk-benefit analysis for each possible decision. These decisions must cover land selection, cultivar selection, cultural practices, and chemical and biological control measures. Soybean research must accurately identify critical infection periods and/or action thresholds for pests. As an example of the present state of disease control strategies, present recommendations for control of soybean foliage, and stem and pod diseases rely on chemical treatments after the reproductive stage has been initiated.

However, recent work by Sinclair [12] has indicated that many of these diseases establish themselves during the vegetative stages, and are latent for long periods before symptoms develop in the reproductive period. Unpublished work from our group indicates that fungicide sprays applied during the vegetative period can result in large yield improvements (Figure 3). Are these yield improvements the result of latent infections, or are we reducing the severity of early infections that provide inoculum for more important disease cycles occurring later in the growing season? Perhaps there are other yield-damaging pathogens.

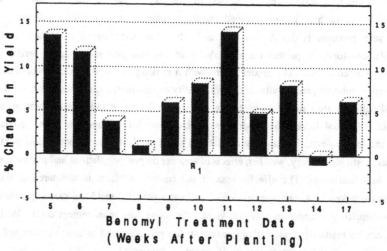

Figure 3. Yield responses of Essex cultivar soybean to single applications of benomyl (0.56 kg a.i./ha) applied at various times during the season; yield changes are in comparison to the nontreated control.

REFERENCES

1. Davis, J.M., and Sakamoto, C.M. An atlas and tables of thunderstorm and hail day probabilities in the southeastern United States. Auburn University Agric. Exp. Stn. Bull. 477, 1976, pp. 75.

2. Backman, P.A., Rodriguez-Kabana, R., Hammond, J.M., and Thurlow, D.L. Cultivar, environment, and fungicide effects on foliar disease losses in soybeans. Phytopathology, 1979, 69, 562-564.

3. Moore, W.F. Soybean Cyst Nematode, Soybean Industry Resource Committee, Extension Service, 1984, U.S. Department of Agriculture, Washington, D.C., pp. 23.

4. Jacobsen, B.J. Soybean seed quality and use of fungicide seed treatments. Ill. Seed News, 1985, 15(10), pp. 2-4.

5. Backman, P.A., Crawford, M.A., and Hammond, J.M. Comparison of meteorological and standardized timings of fungicide applications for soybean disease control. Plant Disease, 1984, 68, pp. 44-46.

6. Backman, P.A., Mack, T.P., Rodriguez-Kabana, R., and Herbert, D.A. A computerized integrated pest management model (AUSIMM) for soybeans grown in the southeastern United States. In Proc. World Soybean Research Conference IV, ed. A. Pascale, Buenos Aires, 1989, pp. 1494-1499.

7. Turner, J.T., and Backman, P.A. Factors relating to peanut yield increases after seed treatment with Bacillus subtilis. Plant Disease, 1991, 75, pp. 347-353.

8. Liu, Z., and Sinclair, J. Bacillus subtilis as a potential biological control agent for Rhizoctonia root rot of soybeans. (Abstr.) Phytopathology, 1987, 77, p. 1687.

9. Andrews, J.H. Biological control in the phyllosphere: realistic goal or false hope? Can. J. Plant Pathol., 1990, 12, pp. 300-307.

10. Kokalis-Burelle, N., Backman, P.A., Rodriguez-Kabana, R., and Ploper, L.D. Chitin as a foliar amendment to modify microbial ecology and control disease. (Abstr.) Phytopathology, 1991, 81, p. 1152.

11. Sinclair, J.B., and Backman, P.A. (eds.). Compendium of Soybean Diseases, Third Edition, 1989, APS Press, Minnesota, pp. 106.

12. Sinclair, J.B., Latent infection of soybean plants and seeds by fungi. Plant Disease, 1991, 75, 220-4.

SOYBEAN DISEASE MANAGEMENT: CHEMICAL AND BIOLOGICAL CONTROL IN TROPICAL REGIONS

G. L. HARTMAN[1] and J. B. SINCLAIR[2]

[1]Plant Pathologist, Asian Vegetable Research and Development Center,
Shanhua, Tainan, Taiwan, Republic of China, and
[2]Professor of Plant Pathology, Department of Plant Pathology,
University of Illinois at Urbana-Champaign, Urbana, IL 61801-4709, USA

ABSTRACT

Production of soybeans in the tropics has increased over the last several decades with potential for continued growth as projects aimed at increasing production and utilization are implemented at international research centers and national research programs. Diseases as well as physiological factors often are major constraints to production. The interaction of environmental conditions, host susceptibility, and the presence of virulent strains or races of pathogens all contribute to determining whether significant yield losses occur. Other factors, like cropping systems, vectors of virus diseases, and seed quality may influence the occurrence of diseases. Important diseases in the tropical environment include bacterial pustule, anthracnose, charcoal rot, frogeye leaf spot, leaf rust, red leaf blotch, rootknot nematode, and soybean mosaic virus. Some of these are broadly distributed while others are restricted in geographical distribution. Successful disease management programs begin with correctly diagnosing the disease; use of cultural techniques consisting mainly of crop rotation, host resistance, and production of high quality seeds; and the use of fungicides to control foliage and seed diseases. Research on host plant resistance has been conducted for anthracnose, bacterial pustule, frogeye leaf spot, leaf rust, and seedborne diseases. Future disease management techniques may include better diagnostic and scouting procedures to predict disease outbreaks; use and evaluation of wild perennial *Glycine* accessions for resistance to certain diseases; and more reliable recommendations for control through management and cultivar selection.

INTRODUCTION

Production of soybeans [*Glycine max* (L.) Merrill] in the tropics is much less than temperate regions. However, since the 1970s, projects aimed at increasing production and utilization in the tropics have been implemented at international research centers and national programs in many countries [1, 2, 3, 4]. Initially, except in parts of Asia where farmers have traditionally grown soybeans for centuries, the introduction of soybeans to lowland tropical areas was negligible primarily because cultivars were not adapted to growing conditions, such as tropical day lengths. The potential exists for substantial increases in production now that cultivars have been bred and selected for production in tropical environments.

Climatic conditions and seasonal differences in the tropics vary mainly due to rainfall patterns which affect the distribution and importance of diseases. Yield losses for some diseases range from relatively minor to potentially devastating. Leaf rust, caused by *Phakopsora pachyrhizi* Syd., can cause yield losses of up to 80% [5]. Yield losses caused by leaf rust have not been determined over large geographical areas other than in simulation models which have predicted a loss in yield of greater than 10% in nearly all soybean growing areas in the United States if Asian races were introduced [6]. Red leaf blotch, caused by *Dactuliochaeta glycines* (Stewart) Hartman & Sinclair, was reported to cause a 34% reduction in yield over approximately 24% of the growing area in Zambia [7]. Experimental data on losses caused by other important diseases in the tropics over large geographical regions are not available. In general, soybean production areas that have existed for centuries have more diseases than newer production areas, but often times in these newer production regions the diseases are quite severe. For example, in some subsaharan countries in Africa where soybean introduction and expansion have been relatively recent, bacterial pustule, frogeye leaf spot (*Cercospora sojina* Hara), and red leaf blotch are important diseases, but they are severe only in some regions [7, 8]. With the continued production and expansion of soybeans into tropical regions more disease limitations will be encountered as pathogens are introduced and become adapted to tropical environments. There is also a distinct possibility that pathogens not yet described on soybeans could occur as soybeans are planted in remote tropical areas.

ENVIRONMENTAL CONSIDERATIONS

Soybeans are grown throughout the year in the tropics, although production is limited in arid environments. In any given region, soybeans are usually not grown continuously, but are mostly grown for at least part of their life cycle during the rainy season. In some locations where humid and rainy conditions prevail throughout the year, production may be limited primarily because of poor seed quality which reduces stands in the field [9]. Tropical regions vary greatly in climatic conditions which often affect the distribution of pathogens. The tropical growing season is defined by rainfall, temperature, and length of growing period. Soybeans may be grown in most tropical agroecological zones, but not all pathogens occur in all zones primarily because of environmental and geographical constraints. Pathoecological zones have not been established for pathogens. However, it is well known that some pathogens, like *Macrophomena phaseolina* (Tassi) Gold. occur more frequently in semi-arid warm environments while others like *P. pachyrhizi* occur more frequently under moist conditions with somewhat cooler temperatures. Seedborne disease are often a limiting factor to continued soybean production in humid environments. Weather conditions also affect soybean maturation and disease development, and other factors such as day length influence soybean growth which in turn can affect disease development. Extended day lengths were shown to delay soybean maturity which also delayed the development of rust [10].

CROPPING SYSTEMS

In the tropics, soybeans are grown on vast tracts of land in some South American countries like Brazil, while on other continents like Africa, soybeans are grown in both large and small fields. In Asia, soybeans are frequently grown as an intercrop or as a second crop after lowland rice primarily in smaller fields. Diseases are usually more pronounced where production is more concentrated. Soybeans grown in rice-based cropping systems like those encountered in Asia generally have few root diseases. Under tropical conditions, cropping systems differ greatly and information on what diseases might predominate under a specific cropping system is lacking. Soybean production in the tropics is more vulnerable to disease because continuous cropping, alternative hosts, and vectors of virus diseases allow for a year-round inoculum supply, in contrast to the long fallow periods in temperate regions.

IMPORTANT DISEASES

In the tropics there are a number of important diseases caused by bacteria, fungi, nematodes, and viruses. Some diseases are more important on a regional or subregional basis while others occur in most if not all production areas. There are a number of diseases caused by fungi that occur in the tropics, including Alternaria leaf spot, charcoal rot, Choanephora leaf blight, downy mildew, frogeye leaf spot, pod and stem blight, Sclerotium blight, target leaf spot, and web blight. Nematode diseases, especially rootknot, and virus diseases including mungbean yellow mosaic and soybean mosaic are important. Many of the aforementioned diseases are locally important due to specific environmental conditions, alternative hosts, or a particular cropping system. Most of these diseases also occur in subtropical or temperate regions. More detail about these disease can be found in other sources [11]. Bacterial pustule, anthracnose, red leaf blotch, leaf rust, and seedborne diseases will be discussed here in more detail because of their importance in the tropics.

Pathogens that cause seedborne diseases are very important in tropical environments. Seed viability is often a constraint to the establishment of soybeans in regions that depend on seed production and storage. Along with the physiological problems associated with producing seeds at high humidities and temperatures, field-infecting pathogens like *Phomopsis* spp., *Colletotrichum* spp., and *Bacillus subtilis* (Ehrenberg) Cohn, often infect seeds in the field while species of *Alternaria*, *Aspergillus*, *Penicillium* and others infect and degrade seeds under storage conditions [9, 12, 13].

Of the bacterial diseases, bacterial pustule occurs most frequently and causes discrete brown lesions with raised pustules in the center of the lesions on the underside of the leaf. Under severe conditions plant defoliate prematurely. Yield losses have not been reported under tropical conditions, but ,in general, losses may be significant in areas where the disease is endemic, and heavy rains and high temperatures are frequent. There are other bacterial or bacterial-like diseases including machismo, bud proliferation, witches'-broom, and phyllody that also may be important in some tropical regions [11].

Anthracnose is important in all production areas. The disease is caused by several *Colletotrichum* spp. The pathogen is latent in host tissue. Fruiting structures are visible during tissue senescence or the pathogen can be detected after the tissue has been desiccated with a herbicide [14]. Under conditions of high rainfall and low light intensity, the pathogen causes veinal, stem, and pod lesions, as well as severe defoliation and death of the plant. Infected seeds are shriveled and discolored, and if planted pre- or post-emergence damping-off frequently occurs. There are few detailed reports about this disease in the tropics though it regularly occurs under humid conditions either symptomatically or asymptomatically in seed or in other plant tissue.

Downy mildew, caused by *Peronospora manschurica* (Naum.) Syd. ex Gäum, occurs worldwide and is more prevalent under moderate to cooler temperatures in the tropics. Yield losses have been reported [15], but the disease is not generally considered a threat to production. The pathogen causes pale green to light yellow spots on the upper leaf surface. These turn brown with age and under high humidity the fungus will sporulate on the lower leaf surface.

Red leaf blotch is an important disease in a restricted geographical area of subsaharan Africa. The disease has been reported to cause yield losses of up to 37% in approximately 25% of the soybean-producing area of Zambia and 10-50% losses in Zimbabwe [7]. The pathogen causes leaf, stem, and pod lesions that often coalesce to form large necrotic blotches. Diseased plants defoliate and senesce prematurely. The pathogen has been described in detail [16], but the range of disease occurrence, survival properties of the fungus, and natural hosts need further description. Because of the restricted geographical distribution of this disease, containment efforts are important to restrict its occurrence in other soybean production areas [17].

Soybean rust or leaf rust, caused by *Phakopsora pachyrhizi*, is endemic to most soybean production areas in Asia. The pathogen has a fairly wide host and geographical range [5]. *P. pachyrhizi* was first reported in the Western Hemisphere in 1976 [18]. Severity and yield losses are greater in the Eastern than the Western Hemisphere as Asian isolates have been reported to be more virulent [5]. In Asia, the disease causes yield losses of up to 80% [5]. Two distinct lesions types, TAN and RB have been described [5]. The RB lesion type has 0-2 uredia with extensive necrosis while TAN lesion type has two or more uredia without extensive necrosis. Lesions range from 2 to 5 mm in diameter. Premature defoliation is common as losses occur by a reduction in pod number, seeds per pod, and by a reduction in seed weight [19, 20, 21, 22]. The disease progresses poorly under dry conditions or when temperatures exceed 28 C. The disease is enhanced when temperatures are moderate, moisture is available as rainfall, and high humid conditions occur [23, 24].

DISEASE MANAGEMENT

Disease diagnosis is critical in a management scheme since control tactics are often based on prior knowledge about the pathogen. There are a number of publications that will aid in the accurate diagnosis of soybean diseases; the Compendium of Soybean Diseases is most complete [17]. In some tropical areas there is a lack of information and/or trained experts to determine the occurrence and severity of diseases. Sinclair [25] summarized some of the important control strategies for soybean diseases in the tropics.

Disease management practices that are used either intentionally or unintentionally include eradication, exclusion, and protection. Management of fertilizers, weeds, and water can help to reduce some diseases. There are no documented cases where eradication has been intentionally used to control a tropical soybean disease. However, cultural practices in traditional rice-based production systems in Asia have limited the occurrence of most root-infecting pathogens. Other cropping systems have not been designed to control specific diseases; however, intercropping with taller crops such as maize or sugarcane may reduce vectors of virus diseases. Most air-borne pathogens are not affected by cropping systems nearly as much as soilborne diseases. Some diseases like red leaf blotch primarily occur in a restricted geographic region whereas seedborne diseases often occur throughout soybean production areas and are especially devastating under humid rainfed conditions. To produce seed with good germination and few seedborne pathogens, planting dates and location of seed-field can be manipulated so seeds mature during drier periods. To maintain seed viability, storage should be done after reducing the seed moisture content. Good quality seed can be separated from poor seed by screening, forced air, or by flotation. Establishment problems can be reduced by mulching to help lower soil temperatures.

Examples of exclusion include limiting the movement of *D. glycines* and virulent strains of *P. pachyrhizi* that only occur in the Eastern Hemisphere to disease-free production areas, and limit the movement of seedborne pathogens into new production areas.

Biological control agents are not available commercially, but the activity of beneficial organisms under traditional cropping systems is probably very important in reducing soilborne pathogens. Efforts have been made to use biological control agents to protect seeds, but this has not been done in the tropics. There has been less effort in identifying or using hyperparasites or leaf-colonizers to control foliar diseases.

Control of diseases by protection either through breeding or fungicide applications has been actively pursued for some diseases. Breeding for disease resistance has been an active component of most tropical soybean breeding programs but the time spent on breeding in the tropics is less compared to breeding programs in temperate regions. In the last two decades, active breeding programs to evaluate and incorporate resistance to bacterial pustule, anthracnose, mungbean yellow mosaic, red leaf blotch, rust, and seedborne diseases have had priority at some research centers [1, 3, 4]. Sources of resistance have been summarized for many diseases which includes some that occur in the tropics [26]. One of the major objectives of the International Institute of Tropical Agriculture (IITA) breeding program has been to develop cultivars with improved seed longevity [1, 27]. Several screening techniques, one used to evaluate seed longevity and the other to evaluate field weathering, have been utilized to improve seed viability [28, 29]. In addition, the IITA breeding program incorporates resistance to bacterial pustule, frogeye leaf spot, and soybean mosaic virus into their advanced breeding lines [1]. At the Asian Vegetable Research and Development Center (AVRDC), the breeding program emphasizes resistance to anthracnose, bacterial pustule, downy mildew, and leaf rust. Resistance to anthracnose was reported [30], and AVRDC recently activated a screening program to identify more sources of resistance [31]. Special consideration has been given to leaf rust because of its importance in Asia. Initially, lines with rate-specific resistance were identified, but races of the pathogen complicated the selection process. Quantifying partial resistance and evaluating tolerance has been the major focus for developing high-yielding lines [3, 20].

Disease control by fungicides has been relatively recent, and it is not a widely accepted practice in the tropics, primarily because of cost and lack of recommendations in some developing countries. In some cases fungicide seed treatments such as captan, hexachlorobenzene, maneb, quintozene (PCNB), and thiram are recommended, but regulation and registration differ by country. In a few situations where production is restricted by diseases or fields are used for seed production, fungicides like benomyl, chlorothalonil, fentin hydoxide, maneb, and triadiemefon are recommended to control diseases such as anthracnose, frog-eye leaf spot, downy mildew, red leaf blotch, and rust.

There are no point systems or checklists developed for the tropical conditions like those developed for temperate regions that indicate the cost effectiveness or correct timing of fungicide application [32, 33]. Insecticides and nematicides are infrequently used in the tropics though nematodes and insect damage may be involved in predisposing the plant to other diseases.

FUTURE CONSIDERATIONS

Most soybean production occurs in temperate regions, but countries like Brazil, India, Indonesia, Nigeria, Thailand, Zambia, and Zimbabwe have at least a portion of their soybean production in tropical regions. The potential for increased production is hastened by cultivar adaptability and improved utilization technology which has increased the demand for soybeans. In conjunction with this, improvement in seed quality and disease resistance will enhance the success of tropical production. Improving seed quality may be done by breeding and by identifying locations and/or times within the year when high quality seed can be produced. In Brazil, for example seed production takes place in the off-season [2].

In some cases, disease resistance has been effective in controlling diseases. This has been true for bacterial pustule, where a source of resistance from Clemson-Non-Shatter (CNS) has been stable for many years. To alleviate potential strains of the bacterium from overcoming resistance the incorporation of other sources of resistance (gene pyramiding) is needed. For diseases which have few available commercial cultivars with high levels of resistance, sources need to be identified in the *G. max* collection or in accessions of the wild perennial *Glycine* spp. In general, the wild species form a potential reservoir of useful characteristics including disease resistance [34]. With the development of successful intraspecific crosses, new breeding lines eventually will be generated and evaluated for resistance to the major tropical pathogens.

As production increases in the tropics, more emphasis should be placed on integrated disease management strategies that will provide more information to growers on cultivar selection and optional control strategies. To aid in this process many of the national programs in various countries should begin to use current technologies such as computer-based diagnosis and they need to develop data bases on pathogen distribution and strain occurrence. The use of predictive models and forecasting systems will contribute to the process of making cost effective decisions to control diseases.

REFERENCES

1. Dashiell, K.E., Bello, L.L. and Root, W.R., Breeding soybeans for the tropics. In Soybeans for the Tropics, eds. S.R. Singh, K.O. Richie and K.E. Dashiell, John Wiley & Sons Ltd., Chichester, Great Britain, 1987, pp. 3-16.

2. Kiihl, R.A.S., Almeida, L.A. and Dall'Agnol, A., Strategies for cultivar development in the tropics. In World Soybean Research Conference III, ed. R. Shibles, Ames, Iowa, 1985, pp. 301-304.

3. Shanmugasundaram, S., The Asian Vegetable Research and Development Center's soybean program. In World Soybean Research Conference III, ed. R. Shibles, Ames, Iowa, 1985, pp. 1233-1239.

4. Singh, B.B., Breeding soybean varieties for the tropics. In Expanding the Use of Soybeans, ed. R.M. Goodman, International Soybean Program (INTSOY), Series Number 10, 1976, pp. 11-17.

5. Bromfield, K.R., Soybean Rust Monogr. 2, American Phytopathological Society, St. Paul, MN., 1984, 65 p.

6. Yang, X.B., Dowler, W.M. and Royer, M.H., Assessing the risk and potential impact of an exotic plant disease. Plant Dis., 1991, 75, 976-982.

7. Hartman, G.L., Datnoff, L.E., Levy, C., Sinclair, J.B., Cole, D.L. and Javaheri, F., Red leaf blotch of soybeans. Plant Dis., 1987, 71, 113-118.

8. Oyekan, P.O. and Naik, D.M., Fungal and bacterial diseases of soybean in the tropics. In Soybeans for the Tropics, eds. S.R. Singh, K.O. Rachie and K.E. Dashiell, John Wiley & Sons Ltd., Chichester, Great Britain, 1987, pp. 47-52.

9. Ndimande, B.N., Wien, H.C. and Kueneman, E.A., Soybean seed deterioration in the tropics. 1: Role of physiological factors and fungal pathogens. Field Crops Research, 1981, 4, 113-121.

10. Tschanz, A.T. and Tsai, B.Y., Effect of maturity on soybean rust development. Soybean Rust Newsl., 1982, 5, 38-41.

11. Sinclair, J.B. and Backman, P.A., Compendium of Soybean Diseases, The American Phytopathological Society, St. Paul, Minnesota, 1989, 106p.

12. Schiller, C.T., Ellis, M.A., Tenne, F.D. and Sinclair, J.B., Effect of Bacillus subtilis on soybean seed decay, germination and stand inhibition. Plant Disease Reporter, 1977, 61, 213-217.

13. Sinclair, J.B., Multiple fungal infections of soybean seeds in preharvest and postharvest deterioration. In Physiological-Pathological Interactions Affecting Seed Deterioration, ed. S.H. West, Crop Science Society of America, Madison, WI, 1986, pp. 65-76.

14. Cerkauskas, R.F. and Sinclair, J.B., Use of paraquat to aid detection of fungi in soybean tissues. Phytopathology, 1980, 70, 1036-1038.

15. Dunleavy, J.M., Yield reduction in soybeans caused by downy mildew. Plant Dis., 1987, 71, 1112-1114.

16. Hartman, G.L. and Sinclair, J.B., Dactuliochaeta a new genus for the fungus causing red leaf blotch of soybeans. Mycologia, 1988, 80, 696-706.

17. Sinclair, J.B., Threats to production in the tropics: Red Leaf Blotch and Leaf Rust. Plant Dis., 1989, 73, 604-606.

18. Vakili, N.G., Field observation and host range of soybean rust, *Phakopsora pachyrhizi* in Puerto Rico. In Proc. Workshop Soybean Rust West. Hemisphere, ed. N.G. Vakili, U.S. Department of Agriculture, Agriculture Research Service, Mayaguez Institute of Tropical Agriculture, Puerto Rico, 1978, pp. 4-15.

19. Chan, K.L. and Tsaur, W.L., Investigation of soybean yields lost due to rust. Annu. Rep. Dryland Food Crops Improv., 1975, **16**, 206-208.

20. Hartman, G.L., Wang, T.C. and Tschanz, A.T., Soybean rust development and the quantitative relationship between rust severity and soybean yield. Plant Disease, 1991, **75**, 596-600.

21. Ogle, H.J., Byth, D.E. and McLean, R.J., Effect of rust *(Phakopsora pachyrhizi)* on soybean yield and quality in south-eastern Queensland. Australian Journal of Agricultural Research, 1979, **30**, 883-893.

22. Yeh, C.C. and Yang, C.Y., Yield loss caused by soybean rust, *Phakopsora pachyrhizi*. Plant Protection Bull. (R.O.C.), 1975, **17**, 7-8.

23. Melching, J.S., Dowler, W.M., Koogle, D.L. and Royer, M.H., Effects of duration, frequency, and temperature of leaf wetness periods on soybean rust. Plant Dis., 1989, **73**, 117-122.

24. Tschanz, A.T., Wang, T.C. and Tsai, B.Y., Recent advances in soybean rust research. In Soybean in Tropical and Subtropical Cropping Systems, ed. S. Shanmugasundaram, Asian Vegetable Research and Development Center, Tsukuba, Japan, 1983, pp. 237-245.

25. Sinclair, J.B., Soybean disease control strategies for the tropics and subtropics. In Soybean in Tropical and Subtropical Cropping Systems, ed. S. Shanmugasundaram, Asian Vegetable Research and Development Center, Tsukuba, Japan., 1983, pp. 251-255.

26. Tisselli, O., Sinclair, J.B. and Hymowitz, T., Sources of resistance to selected fungal, bacterial, viral and nematode diseases of soybeans, INTSOY Ser. 18, College of Agriculture, University of Illinois, Urbana-Champaign, 1980, 134 p.

27. Dashiell, K.E. and Gumisiriza, G., Recent advances made in developing soybean cultivars with improved seed storability for the tropics. In World Soybean Research Conference IV, ed. A.J. Pascale, ACTAS, Buenos Aires, Argentina, 1989, pp. 2295-2302.

28. Dashiell, K.E. and Kueneman, E.A., Screening methodology for identification of soybean varieties resistant to field weathering of seed. Crop Science, 1984, **24**, 774-779.

29. Kueneman, E.A. and Wien, H.C., Improving soybean stand establishment in the tropics by varietal selection for superior seed storability: cooperation of national programs. IITA, Research briefs, 1981, **2** (2).

30. Manandhar, J.B., Hartman, G.L. and Sinclair, J.B., Soybean germ plasm evaluation for resistance to *Colletotrichum truncatum*. Plant Dis., 1986, **72**, 56-59.

31. AVRDC, Annual Report for 1991, Asian Vegetable Research and Development Center, Shanhua, Taiwan, R. O. C., 1992, (in press).

32. Jacobsen, B.J., Shurtleff, M.C., Kirby, H.W. and Melton, T.A., Condensed plant disease management guide for field crops. Univ. Ill. Coop. Ext. Serv., Circ., 1987, **1231**, 10.

33. Tekrony, D.M., Stuckey, R.E., Egli, D.B. and Tomes, L., Effectiveness of a point system for scheduling foliar fungicides in soybean seed fields. Plant Dis., 1985, **69**, 962-965.

34. Brown, A.H.D., Grant, J.E., Burdon, J.J., Grace, J.P. and Pullen, R., Collection and utilization of wild perennial *Glycine*. In World Soybean Research Conference III, ed. R. Shibles, Ames, Iowa, 1984, pp. 345-352.

NATURE AND MANAGEMENT OF FUNGAL DISEASES AFFECTING SOYBEAN STEMS, PODS, AND SEEDS

L. DANIEL PLOPER
Estación Experimental Agro-Industrial Obispo Colombres
4101 Las Talitas, Tucumán, R. Argentina

and

PAUL A. BACKMAN
Department of Plant Pathology
Auburn University, Alabama 36849, USA

ABSTRACT

Soybean diseases that affect stems, pods, and seeds can seriously limit production. Diseases in this group which are economically important include anthracnose, pod and stem blight, Phomopsis seed decay, stem canker, and Sclerotinia stem rot. An understanding of the epidemiology of these diseases, particularly the extended latent period of most of the pathogens involved, is critical in selecting management options. Most effective control will result from combining various control strategies into an integrated disease management program. Practices that should be considered are: accurate diagnosis, crop rotation, tillage, use of high quality seed, planting date, soil fertility, row spacing, weed control, irrigation management, timely harvest, cultivar selection, and chemical control. Most of these measures are equally applicable to diseases which affect stems, pods, and seeds, as well as foliage.

INTRODUCTION

Numerous fungal diseases reduce soybean [*Glycine max* (L.) Merr.] yields and quality in both temperate and tropical environments. Diseases that affect stems, pods, and seeds are among the most damaging to the crop and frequently can be limiting factors in production. With increasing production costs, disease management is a critical consideration in soybean production, especially since losses have become more serious due to changes in cropping and tillage practices and the expansion of the crop into new environments. Also, crop quality losses due to fungal infections are becoming more acute with changes in grading standards.

A characteristic shared by most of the pathogens that attack stems, pods, and seeds is their extended latent period [1]. Plants can be infected at any stage of development. However, symptoms commonly appear later in the season, usually associated with physiological changes that occur during reproductive stages. Protection of plants with fungicides early in the season has shown that these diseases can still cause significant losses, in spite of being symptomless [2]. Prolonged latent periods are probably the most important aspects of the epidemiology of these diseases and should be addressed in developing and timing control practices.

Anthracnose, pod and stem blight, Phomopsis seed decay, stem canker, and Sclerotinia stem rot are the most prevalent and economically important diseases in this group and will be reviewed in this section. Emphasis will be placed on relevant epidemiological information and suggested control strategies. A wider coverage of all these diseases, including symptomatology and detailed descriptions of causal fungi, is available in recent reviews [3,4].

Other diseases, considered primarily foliage diseases (brown spot, frogeye leaf spot, Cercospora blight and leaf spot, Rhizoctonia aerial blight, target spot, etc.), can also affect stems, pods, and/or seeds. The general control measures discussed in this chapter are equally applicable to this group of diseases.

ANTHRACNOSE

Anthracnose is a widespread disease of soybean, but it causes severe yield losses and affects the quality of the seed produced primarily in warm, humid areas. Several species of *Colletotrichum* are associated with the disease, but the most prevalent is *C. truncatum* (Schw.) Andrus & W. D. Moore (teleomorph unknown). Other important species are *C. destructivum* O'Gara [teleomorph *Glomerella glycines* (Hori) Lehman & Wolf], and *C. gloeosporioides* (Penz.) Sacc. [teleomorph *Glomerella cingulata* (Ston.) Spauld & Schrenk]. These fungi can colonize weeds as well as other agronomic crops, including perennial crops which grow on the same or adjacent fields in many tropical areas [4].

While infection can take place at any growth stage, symptoms usually are noticed only during the very early vegetative stages (pre- or post-emergence damping-off) and later in the season, as plants approach maturity (growth stages R5-R7, on the Fehr and Caviness scale). Lower branches and leaves in the canopy usually show more damage, especially under heavy disease pressure, because of their proximity to primary inoculum sources and production of secondary inoculum cycles from senescent lower leaves. Infected plants tend to senesce earlier than non-diseased plants, and consequently have reduced yields. Yield losses are more severe when the pathogen infects pods or pedicels [3].

The causal fungi survive in seeds, host debris, perennial crops, and weeds [5]. High temperatures (above 25°C) and moisture (rain, dew, or fog), especially rainy periods in late summer, favor disease development. Free water on the plant surface for 12 hr or more is required for conidia to germinate and penetrate epidermal cells [4,6].

THE DIAPORTHE / PHOMOPSIS DISEASE COMPLEX

Pod and stem blight, stem canker, and Phomopsis seed decay are important diseases in the Diaporthe/Phomopsis complex of soybean. Most of the components involved in the complex are considered to be endemic in nearly every area of soybean production in the world, and some of them are very destructive under warm, wet conditions. Losses are due to reduction in field stands, yield, and seed quality [7].

The taxonomic status and nomenclature of fungi implicated in the Diaporthe/Phomopsis complex recently were reviewed by Morgan-Jones [8]. *Diaporthe phaseolorum* (Cke. & Ell.) Sacc. and *Phomopsis phaseoli* (Desm.) Sacc. were recognized, respectively, as the valid binomials for teleomorphs and anamorphs belonging to this complex. *Phomopsis longicolla* Hobbs, a fungus associated with seed biodeterioration, was accepted as a separate entity. The previous classification of *Diaporthe phaseolorum* at variety level was considered to be unsatisfactory because of the considerable variability in morphology, physiology, and host relationships. Instead, Morgan-Jones proposed that the forma specialis concept be adopted for infraspecific designation.

Pod and stem blight, caused by *D. phaseolorum* f. sp. *sojae* (*Dps*) (syn. *D. sojae* Lehman) [anamorph: *P. phaseoli* (syn. *P. sojae* Lehman)] affects all parts of the plant, but symptoms are more conspicuous on stems, pods, and seeds. Stem infections are considered to be less important when compared to seed infection [4,7].

Stem canker is caused by two closely related but distinct organisms: *D. phaseolorum* f. sp. *caulivora* (*Dpc*) (fertile anamorph rare), which predominates in the upper midwestern United States, and *D. phaseolorum* f. sp. *meridionalis* (*Dpm*) (anamorph: *P. phaseoli* f. sp. *meridionalis*), which predominates in the southern United States. This disease can be very destructive, mainly because it kills plants from midseason to maturity, when adjacent, uninfected plants cannot compensate for the loss of infected plants. Stem canker became prevalent in the north central United States in the late 1940s and early 1950s when two highly susceptible cultivars, Hawkeye and Blackhawk, were released. Yield losses up to 50% were reported in that region [3]. Removal of the susceptible cultivars from production substantially reduced the incidence of stem canker, although it is still endemic in the region. In the southern United States, stem canker was first observed in 1973, and it became a serious threat to soybean production during the following 10 years. Losses up to 100% have been reported in highly susceptible cultivars [9]. Control of the disease in this region has been more difficult than in the northern states, mainly because resistance to cyst nematode, a more prevalent problem, is typically found in cultivars susceptible to stem canker. The disease also causes substantial yield losses in South America and Europe.

Variability among isolates of both *Dpc* and *Dpm* in pathogenicity, cultural morphology, and *in vitro* growth rates suggests the existence of races or pathotypes [9,10]. Within *Dpm*, 29 vegetative compatibility groups were identified, although one group accounted for 79% of the 297 isolates studied [11]. These groups were considered analogous to pathotypes of the pathogen, since physiologic specialization was related to vegetative compatibility. Pathogenicity may be affected in part by cytoplasmic agents such as dsRNA, which are found in some isolates of *Dpm* [8,10].

Seed decay, caused primarily by *P. longicolla* (teleomorph unknown), is the predominant problem in most countries where soybeans are grown [12,13]. Infected seeds show low germination and field emergence, and a poor appearance that leads to reductions in commercial grade. It is well known in the soybean processing industry that oil from infected seeds is of poor quality due to high levels of free fatty acids.

These fungi also can occur on other crops and weeds. At least 13 cultivated species other than soybean and 10 different weeds were reported to be colonized by *Diaporthe* and *Phomopsis* spp.

pathogenic to soybean [4,14]. The relative importance of different weed hosts as inoculum sources is unknown.

Fungi involved in this disease complex survive in seed and in crop debris. Infected seed contribute to long-distance dispersal of these pathogens and also provide a source of primary inoculum. However, in fields with a history of these diseases, infested residues are the major source of initial inoculum [15]. Pycnidia and perithecia are formed on debris in the spring, producing spores which are released in a sticky matrix and dispersed mostly by windborne and splashing water [9]. *Dps* and *P. longicolla* spores also are released from pycnidia which develop on fallen cotyledons and petioles. This secondary cycle also may occur with stem canker, although evidence for this has not been found [9]. Infections by conidia and ascospores occur over a wide temperature range, but more than 24 hr of free moisture are required for successful infections.

Although plants can be infected at any time during the growing season, symptoms of pod and stem blight and Phomopsis seed decay become visible later in the reproductive stages. Environmental conditions between R1 and crop harvest play a critical role in determining the extent of pod and seed infection. Warm weather, rainfall, and particularly high relative humidity from physiological maturity to harvest favor spread of the fungi from the pod wall to the seed. Seed infections become more severe as harvest is delayed [12]. *P. longicolla* seed infection is more prevalent in seed from lower parts of the plant, which are in close proximity to inoculum sources [13].

For stem canker, symptom occurrence and disease severity depend on the plant growth stage at the time of infection and on cultivar susceptibility. Maximum disease levels occur when susceptible plants are infected at the V3 stage, and progressively less disease develops when infection takes place from V3 to V10. No stem canker symptoms develop when susceptible plants are infected during reproductive stages or when resistant cultivars are infected at any time during the growing season. However, the pathogen can be isolated from asymptomatic tissue of resistant and susceptible cultivars throughout the season [16]. This relationship between stage of plant development and inoculum availability is believed to account for the marked year to year differences in disease severity. Stem canker severity also may be influenced by water stress, and physical and chemical properties of the soil, such as organic matter, pH, and potassium levels.

SCLEROTINIA STEM ROT

Sclerotinia stem rot, also known as white mold, occurs worldwide but causes important problems to soybean production only in temperate regions of Europe, North America, and South America, particularly Brazil and Argentina [4,17,18]. Although regarded as a minor problem in North America, incidence and severity of outbreaks are increasing in the upper midwestern United States and Ontario, Canada. These outbreaks have been associated with expansion of soybean production into areas with environmental conditions that favor disease development, changes in cultural practices, and planting in fields with a history of this pathogen in other crops. The disease is especially severe when soybean follows cabbage (*Brassica oleracea*), green and dry bean (*Phaseolus vulgaris*), lettuce (*Lactuca sativa*), lupin (*Lupinus* spp.), peanut (*Arachis hypogaea*), and sunflower (*Helianthus annuus*) [19].

The causal agent, *Sclerotinia sclerotiorum* (Lib.) de Bary [syn. *Whetzelinia sclerotiorum* (Lib.) Korf & Dumont], is one of the most nonspecific, omnivorous plant parasites. It is pathogenic to almost

400 species of plants in 64 different families. The fungus is characterized by production of a white fluffy mycelium and large (2-20 mm in diameter) black, irregularly shaped sclerotia. These resting structures are produced in and on diseased tissue and play an important role in survival and dissemination of the pathogen. With favorable conditions, a sclerotium on or near the soil surface will produce one or more, light tan to brown apothecia. These cup-shaped fruiting structures contain thousands of cylindrically-shaped, eight-spored asci [4].

Short and long range dispersal of the pathogen occurs by wind-blown ascospores, infected seed, seed contaminated with sclerotia, and movement of soil and plant debris containing sclerotia. Sclerotia are critical in the development of epidemics because of the number produced and their ability to withstand adverse conditions for extended periods of time.

Sclerotia formed on infected plants are released into the soil during harvest or following the decay of plant stems and are later redistributed within the soil profile and over the field by tillage or water movement of infested soil and debris. Survival of sclerotia in the soil is affected by various physical, chemical, and biological factors, especially activity of mycoparasites.

Primary inocula are ascospores produced after carpogenic germination of mature sclerotia located at or within 5 cm of the soil surface. A close relationship exists between environmental conditions, growth stage of the crop, inoculum production, and initiation of disease. Optimal conditions for formation of apothecia are prolonged periods of low temperatures (5-15°C) and high soil moisture (-0.25 bar) for 10-14 days [4]. These conditions usually are attained when the soybean canopy closes, which also coincides with the time when plants start flowering (R1).

Disease development is favored by extended periods of plant surface wetness and cool to moderate temperatures (12-24°C). Inoculation studies in controlled environments showed that 70-120 hr of continuous plant surface wetness at 20°C were required for disease to develop [20]. Field studies revealed that plant surface wetness lasting 40-112 hr was associated with initial disease, and that shorter wetness periods were required for development of new lesions once the epidemic started [20].

DISEASE MANAGEMENT

Several measures have been utilized to control diseases that affect stems, pods, and seeds. These include modified cultural practices, use of host resistance, and chemicals. Even though single tactics can contribute to disease reductions, they are frequently not efficacious enough to suppress losses below economic thresholds. Combinations of several control strategies in an integrated disease management program usually provide the best results. The objectives of these types of programs are to improve soybean production and reduce conditions that favor disease development. When coordinated with other crop management practices, effective disease management programs should provide economical and long lasting solutions to most soybean disease problems.

No single disease management program will be suitable for all fields. Possible strategies to be included in a program should be based on the pathogen or pathogens prevalent at each location, the potential losses, and the prevailing environmental conditions. Thus, accurate diagnosis to the level of race or forma specialis of the pathogen is a prerequisite in implementing such programs.

Control by Cultural Practices

There are a number of cultural practices that lower the incidence of these diseases, mainly by reducing inoculum, reducing the efficiency of the inoculum, or minimizing conditions that favor disease development.

Crop rotation: Most pathogens survive between growing seasons on and in soybean crop debris. Rotation with non-host crops results in declining pathogen populations as the crop debris decomposes, because most pathogens discussed here have a poor competitive saprophytic ability.

Rotations with maize (*Zea mays*), grain sorghum (*Sorghum bicolor*), cereals, and some forages are recommended to reduce inoculum of anthracnose and diseases caused by the Diaporthe/Phomopsis complex [6,13,21]. Including resistant cultivars in the rotation helps reduce stem canker damage on susceptible cultivars grown in subsequent years [16].

Crop rotation is not highly efficacious in reducing the incidence of Sclerotinia stem rot, primarily because sclerotia can survive in the soil for 3-6 years, and tillage operations ensure the presence of these resting structures at or near the soil surface. Still, this practice is considered desirable in a management program to minimize this and other diseases. Since *S. sclerotiorum* is pathogenic to many cultivated species, choosing crops for the rotation scheme should be given special consideration. Rotations with barley (*Hordeum vulgare*), buckwheat (*Fagopyrum esculentum*), and wheat (*Triticum aestivum*) have been reported to reduce incidence of the disease on soybean, whereas lupin (*Lupinus* spp.) has increased damage [17]. Lists of agronomic and vegetable crop hosts of *S. sclerotiorum* are available. In general, it is recommended to plant a non-host 1 to 2 years between soybean and other host crops [19].

Tillage practices: Tillage operations reduce levels of these diseases by favoring decomposition of soybean residue. For Sclerotinia stem rot, deep plowing of infected crop residues with subsequent shallow tillage for 3 to 5 years can reduce formation of apothecia, the main sources of primary inoculum [19]. Incidence of Sclerotinia stem rot is higher with no-till systems than with deep plowing and disking [18].

Use of high quality seed: Since all these pathogens can infect pods and seeds, the use of healthy seed is recommended to prevent their introduction into non-infested fields. Such seeds are also advantageous because they produce healthy, vigorous seedlings which are less susceptible to other soilborne pathogens and stress from environmental factors. Seed treatment fungicides are also effective in reducing primary inoculum from seed.

Planting date - Maturity of cultivar: Environmental conditions during reproductive stages have significant effects on severity of anthracnose and members of the Diaporthe/Phomopsis complex. Disease levels, especially seed infection, can be decreased if plants mature under cool, dry conditions. These conditions can be achieved by modifying the planting date or the maturity group of the cultivar used [13]. In tropical areas characterized by distinctive dry and wet seasons, planting so that maturation occurs in the dry season is suggested to reduce anthracnose severity [6].

For stem canker, late planting has been recommended in the southeastern United States to avoid initial releases of ascospores which generally occur during rainy periods in the spring. If plants do become infected, the shorter vegetative and reproductive periods induced by late planting allow less time for disease development [9].

Soil fertility and pH: Modifications of soil fertility and pH can help reduce damage caused by anthracnose. For low pH soils, liming with calcium carbonate, calcium hydroxide, or calcium sulfate was found to reduce anthracnose seedling blight [3]. Fertilization with potassium has been reported to lower anthracnose levels [22]. Soil fertility can also influence levels of seed infection. Potassium deficiency has been associated with higher levels of seed infection by *Diaporthe* and *Phomopsis* spp. [21].

Row spacing - Plant population: These practices influence the subcanopy environment. Diseases are likely to be less severe if canopy structure facilitates air circulation and leaf drying. Rows oriented in the direction of prevailing winds also may reduce humidity in the subcanopy.

For Sclerotinia stem rot, modifications of plant population and row width are suggested to promote drying out of the canopy [18]. Although yields in narrow rows (18-38 cm) are potentially higher than wide rows (75-90 cm), the canopy structure of narrow rows enhances conditions for disease development. Therefore, wide rows are recommended particularly for those fields where the disease has caused problems in previous years [19].

Weed control: In addition to increasing plant stress, weeds may contribute directly to disease severity by increasing humidity in the subcanopy, inhibiting air movement and light penetration, and interfering with chemical applications. Furthermore, since many broadleaf weeds are also hosts of the pathogens, they may favor build-up of inoculum in the field. An effective weed control is thus recommended to prevent higher disease levels.

Herbicides may affect disease incidence and the pathogens involved. Certain preplant herbicides may increase seed infection by *Phomopsis* spp. [21]. *Sclerotinia sclerotiorum* structures are affected by some herbicides. Both trifluralin and metribuzin increase germination of sclerotia and number of apothecia/sclerotium [3]. Two herbicides used in corn fields, atrazine and simazine, have a detrimental effect on sclerotia, causing abnormal carpogenic germination [19].

Irrigation management: Overhead irrigation may increase all these diseases compared to furrow irrigation. For Sclerotinia stem rot, irrigation needs to be moderate until flowering has ceased, particularly when plant canopies are dense [19].

Timely harvest: The crop should be harvested as soon as seed moisture decreases to 12-13%. Prompt harvest will reduce the incidence of seedborne organisms and improve seed germination.

Control by Plant Resistance

Use of resistant cultivars is the simplest and most efficient method of disease control. For some of these diseases, resistance is already available in adapted cultivars, whereas for others, considerable effort to identify sources of resistance and to develop resistant lines is under way.

Anthracnose: Soybean genotypes differ in their susceptibility to this disease, but high levels of resistance have not yet been found [5]. Discrepancies between results from field and greenhouse evaluations indicate that effective screening procedures are still needed to identify efficiently sources of resistance and evaluate reactions of selected lines or progeny. Soybean genotypes also differ in the degree of petiole, stem, and seed infection by *C. truncatum*.

Diaporthe/Phomopsis complex: Various evaluations have shown differential reactions of soybean genotypes to the Diaporthe/Phomopsis complex. High levels of resistance to pod and stem blight and Phomopsis seed decay have been found, but are still not available in adapted soybean cultivars. Identification and use of sources of resistance have been limited because of the strong influence that time of maturity and local environment play in final levels of seed quality. Reduced seed infection with *P. longicolla* has been reported for genotypes in maturity groups VIII, IX, and X [13]. Confusion often results because differences in disease are more frequently associated with varying maturing times during different environmental conditions than with genotypic resistance. However, differences in disease reaction have been reported for cultivars and strains of various maturity groups, even when the genotypes have matured under near-identical conditions. Several morphological and physiological characteristics of the soybean plant may account for resistance or tolerance to seed infection, including growth habit (determinate vs. indeterminate), pubescence morphology, seed coat permeability, and rate of late season maturation [12].

For stem canker, the use of genetic resistance is probably the most effective control method currently available. The replacement of susceptible varieties dramatically decreased the impact of the disease in the northern United States in the 1950s. Most cultivars now grown in that region are moderately resistant to stem canker. In the southern United States, cultivar reaction was evaluated in the field after the serious stem canker outbreaks of the late 1970s. Cultivar reactions ranged from resistant to highly susceptible [23]. Bay, Tracy-M, Braxton, Hood, and Dowling have the highest levels of resistance [9]. Most cultivars that show resistance to stem canker are susceptible to the soybean cyst nematode, which has prevented their use in certain fields. If it becomes necessary to use susceptible cultivars, other control measures such as crop rotation, seed treatment, and delayed planting should be utilized to minimize losses. Identification and use of sources of resistance to *Dpm* have been facilitated by the development of effective field and greenhouse evaluation techniques [24]. Two major dominant genes, designated Rdc_1 and Rdc_2, have been found in Tracy-M. Each gene conditions complete resistance. Lines containing individual genes are now available to identify other sources of resistance [25].

Sclerotinia stem rot: Differences in reactions of soybean cultivars to this disease have been reported [19,20]. Some of these differential reactions were due to factors which favor disease escape. Plant architecture, cultivar maturity, and lodging characteristics are all factors that influence canopy development and therefore disease severity. Open canopies, resulting from non-lodging cultivars or from particular architectural traits, reduce or prevent infection by facilitating air circulation and drying leaf and soil surfaces. Similarly, resistance found in a great number of genotypes from Maturity Groups 0-II has been attributed to their lower height [19].

In addition to disease escape or avoidance, other mechanisms of resistance are apparently involved. Field, greenhouse, and laboratory evaluations have detected important differences among soybean cultivars. Some genotypes have exhibited high and stable levels of resistance under field conditions, indicating that breeding for physiologic resistance to *S. sclerotiorum* is possible. Various methods to screen soybeans for this type of resistance have been proposed. These include screening in the field or under controlled environmental conditions [19,26]. The variable correlation values found between field and some of the proposed laboratory evaluations indicate that further research is needed in order to optimize screening techniques.

Chemical Control

Fungicide seed treatments: If seed quality is low due to fungal infection, fungicide seed treatments will improve emergence and plant stands, and reduce seedborne inoculum. These treatments are also suggested when delays in germination or emergence are expected and when seed is planted for seed production purposes. Various seed treatments are available as dusts, flowable suspensions, liquids, or wettable powders and can be applied as liquid slurries or dusts [4]. Thiram, thiabendazole, carboxin, and captan are the most widely used fungicides for treating seeds.

Foliar fungicide treatments: Fungicides applied to the foliage are recommended to increase yield and seed quality. Both systemic and protectant fungicides are available for use on soybean to control midseason stem, pod, and foliage diseases. The most widely used are the benzimidazole fungicides, such as benomyl, thiabenzadole, and thiophanate-methyl. These are systemic and can remove recently established infections. The contact fungicides chlorothalonil and fentin hydroxide (triphenyltin hydroxide) also control the mid-season disease complex. The contact fungicides do not remove established infections; instead, they protect the plant from new infections. Copper-based fungicides and the sterol biosynthesis inhibitors (demethylation inhibitors) have not been effective in controlling these diseases [2].

Fungicides applied at early pod stage (growth stage R3) can control anthracnose as well as other mid-season diseases such as brown spot and frogeye leaf spot. These treatments are particularly effective when warm, wet conditions prevail between bloom (R1-R2) and pod fill (R5). Protection with fungicides during this period provides maximal yield increases and will improve seed quality by reducing fungal infections. Yield responses are due primarily to increases in seed size and numbers per pod in the upper portion of the plant canopy [2]. A single, high rate application of benomyl or thiabendazole at R6 can be very effective in reducing seed infection and increasing seed vigor and germination, but seldom increases yield.

In many cases, responses to fungicide treatments have been erratic, indicating that a programmed approach to treatment cannot be justified, particularly when low levels of disease incidence or severity are present. However, predictive systems, developed and implemented in several regions of the United States, can help estimate the profitability of fungicide application by assessing if conditions are conducive for disease development. Some systems, such as those developed in Kentucky and Illinois, use a point scale based on field scouting, forecast and/or historical weather, cropping history, yield potential, planting date, and maturity group [4]. Auburn University in Alabama has developed a simple model for management of various mid-season diseases, including anthracnose. This model, based on observed and predicted weather, uses the number of days with measurable rain, extended periods of fog, and/or heavy dews to determine the need for fungicide application. Other methods have been developed specifically to estimate the need for fungicides to improve seed quality. A system developed in Iowa relies on field scouting and laboratory analysis of pods sampled at R6. The need for foliar fungicides is then determined by levels of pod infection by *Phomopsis* spp. [27].

For stem canker, control may be achieved with benzimidazole fungicides, except on highly susceptible cultivars. Timing of sprays is critical, as sprays should be applied during vegetative growth stages when ascospores are being actively produced and infections are occurring. Banded applications (15-20 cm) over the young plants are suggested [9].

The variability in the response to foliar fungicides also can be attributed to improper fungicide application [2]. Control of stem and pod diseases requires fungicide penetration into the plant canopy and deposition on stems, pods, and leaves, particularly those located low in the canopy where conditions favor disease development.

REFERENCES

1. Sinclair, J.B., Latent infection of soybean plants and seeds by fungi. Plant Dis., 1991, 75, 220-4.

2. Backman, P.A. and Jacobsen, B.J., Control of foliage, stem and pod diseases with fungicides. In World Soybean Research Conference IV: Proceedings, ed. A. Pascale, Buenos Aires, 1989, pp. 2091-6.

3. Athow, K.L., Fungal Diseases. In Soybeans: Improvement, Production, and Uses, 2nd ed., ed. B.E. Caldwell, American Society of Agronomy, Madison, 1987, pp. 687-727.

4. Sinclair, J.B. and Backman, P.A., eds., Compendium of Soybean Diseases, 3rd ed. American Phytopathological Society, St. Paul, 1989, 104 pp.

5. Sinclair, J.B., Anthracnose of soybeans. In Soybean Diseases of the North Central Region, eds. T.D. Wyllie and D.H. Scott, The American Phytopathological Society, St. Paul, 1988, pp. 92-5.

6. Hepperly, P.R., Soybean anthracnose. In World Soybean Research Conference III: Proceedings, ed. R. Shibles, Westview Press, Boulder, 1985, pp. 547-54.

7. Ploper, L.D., The Diaporthe/Phomopsis disease complex of soybean. In World Soybean Research Conference IV: Proceedings, ed. A. Pascale, Buenos Aires, 1989, pp. 1695-8.

8. Morgan-Jones, G., The Diaporthe/Phomopsis complex: Taxonomic considerations. In World Soybean Research Conference IV: Proceedings, ed. A. Pascale, Buenos Aires, 1989, pp. 1699-706.

9. Backman, P.A., Weaver, D.B. and Morgan-Jones, G., Soybean stem canker: an emerging disease problem. Plant Dis., 1985, 69, 641-7.

10. Kulik, M.M., Variation in pathogenicity among isolates of Diaporthe phaseolorum f. sp. caulivora. Mycologia, 1989, 81, 549-53.

11. Ploetz, R.C. and Shokes, F.M., Variability among isolates of Diaporthe phaseolorum f. sp. meridionalis in different vegetative compatibility groups. Can. J. Bot., 1989, 67, 2751-5.

12. Abney, T.S. and Ploper, L.D., Seed Diseases. In Soybean Diseases of the North Central Region, eds. T.D. Wyllie and D.H. Scott, The American Phytopathological Society, St. Paul, 1988, pp. 3-6.

13. Schmitthenner, A.F. and Kmetz, K.T., Role of Phomopsis sp. in the soybean seed rot problem. In World Soybean Research Conference II: Proceedings, ed. F.T. Corbin, Westview Press, Boulder, 1980, pp. 355-66.

14. Roy, K.W. and McLean, K.S., Host range of the Diaporthe/Phomopsis complex from soybean. In World Soybean Research Conference IV: Proceedings, ed. A. Pascale, Buenos Aires, 1989, pp. 1707-11.

15. Rupe, J., Epidemiology of the Diaporthe/Phomopsis complex. In World Soybean Research Conference IV: Proceedings, ed. A. Pascale, Buenos Aires, 1989, pp. 1712-7.

16. Smith, E.F. and Backman, P.A., Soybean stem canker: An overview. In Soybean Diseases of the North Central Region, eds. T.D. Wyllie and D.H. Scott, The American Phytopathological Society, St. Paul, 1988, pp. 47-55.

17. Martinez, C.A. and Botta, G.L., Podredumbre humeda del tallo (Sclerotinia sclerotiorum (Lib.) De Bary). In World Soybean Research Conference IV: Proceedings, ed. A. Pascale, Buenos Aires, 1989, pp. 1303-11.

18. Yorinori, J.T. and Homechin, M., Sclerotinia stem rot of soybeans, its importance and research in Brazil. In World Soybean Research Conference III: Proceedings, ed. R. Shibles, Westview Press, Boulder, 1985, pp. 582-8.

19. Grau, C.R., Sclerotinia stem rot of soybean. In Soybean Diseases of the North Central Region, eds. T.D. Wyllie and D.H. Scott, The American Phytopathological Society, St. Paul, 1988, pp. 56-66.

20. Boland, G.J. and Hall, R., Epidemiology of Sclerotinia stem rot of soybean in Ontario. Phytopathology, 1988, 78, 1241-5.

21. Sinclair, J.B., Diaporthe/Phomopsis disease complex: Control. In World Soybean Research Conference IV: Proceedings, ed. A. Pascale, Buenos Aires, 1989, pp. 1718-23.

22. Sij, J.W., Turner, F.T. and Whitney, N.G., Suppression of anthracnose and Phomopsis seed rot on soybean with potassium fertilizer and benomyl. Agron. J., 1985, 77, 639-42.

23. Weaver, D.B., Cosper, B.H., Backman, P.A. and Crawford, M.A., Cultivar resistance to field infestations of soybean stem canker. Plant Dis., 1984, 68, 877-9.

24. Keeling, B.L., Measurement of soybean resistance to stem canker caused by Diaporthe phaseolorum var. caulivora. Plant Dis., 1988, 72, 217-20.

25. Kilen, T.C. and Hartwig, E.E., Identification of single genes controlling resistance to stem canker in soybean. Crop Sci., 1987, 27, 863-4.

26. Nelson, B.D., Helms, T.C. and Olson, M.A., Comparison of laboratory and field evaluations of resistance in soybean to Sclerotinia sclerotiorum. Plant Dis., 1991, 75, 662-5.

27. McGee, D.C., Prediction of Phomopsis seed decay by measuring soybean pod infection. Plant Dis., 1986, 70, 329-33.

MANAGEMENT OF FOLIAR FUNGAL DISEASES
IN SOYBEAN IN BRAZIL

JOSE TADASHI YORINORI
Centro Nacional de Pesquisa de Soja
EMBRAPA-CNPSo
Caixa postal 1061, 86001, Londrina, PR, Brasil

ABSTRACT

Eleven foliar fungal diseases have been found on soybeans [*Glycine max* (L.) Merrill] in Brazil. Of these, frogeye leaf spot (*Cercospora sojina*), brown spot (*Septoria glycines*) and Cercospora leaf blight (*Cercospora kikuchii*) are of major concern. Downy mildew (*Peronospora manshurica*), target spot (*Corynespora cassiicola*) and soybean rust (*Phakopsora pachyrhizi*) have occasional outbreaks. Of all the foliar diseases, frogeye leaf spot has caused the greatest impact on yield (up to 100% loss) but it is currently under control by use of resistant varieties. Brown spot and Cercospora leaf blight are the most important late season diseases. They may cause more than 20% yield loss (ca. US$ 1 billion worth) annually, in disease favorable years. The possibilities of reducing losses due to late season diseases through varietal resistance or tolerance and fungicide sprays are demonstrated but these practices are not used routinely. Chemical seed treatment and integrated crop management for reducing the inoculum potential in the field are emphasized.

INTRODUCTION

Soybeans [*Glycine max* (L.) Merrill] in Brazil, are affected by more than 40 diseases caused by fungi, bacteria, nematodes and viruses. The importance of each disease varies from year to year and from region to region, depending on the climatic condition of each growing season.

The following foliar fungal diseases have been identified: 1. Alternaria leaf spot (*Alternaria* sp.), 2. Ascochyta leaf spot (*Ascochyta* sp.), 3. brown spot (*Septoria glycines* Hemmi), 4. downy mildew [*Peronospora manshurica* (Naoum) Syd.], 5. frogeye leaf spot (*Cercospora sojina* Hara), 6. Myrothecium leaf spot (*Myrothecium roridum* Tode ex. Sacc.), 7. Phyllosticta leaf spot (*Phyllosticta* sp.), 8. powdery mildew (*Microsphaera diffusa* Cke. Pk.), 9. Cercospora leaf blight [*Cercospora kikuchii* (Mats. & Tomoy.) Gardner], 10. Rust (*Phakopsora pachyrhizi* H.& P. Syd.), 11. target spot [*Corynespora cassiicola* (Berk. & Kurt) Wei] (1, 2, 3). The most widespread and important diseases are frogeye leaf spot, brown spot and Cercospora leaf blight. Downy mildew and target spot are commonly found in all soybean producing regions but seldom reach epidemic proportions. The soybean rust was first identified in Brazil in 1979 (1) and has since shown occasional outbreaks in the high plains (800-1100m altitudes) of Central Brazil. The other diseases occur sporadically or in localized outbreaks within a field.

Following is an account of the development, research and management of the most important foliar fungal diseases of soybeans in Brazil.

Frogeye leaf spot (*C. sojina*)

The disease was first found in Brazil during the 1970/71 growing season, in a field for seed increase of variety Bragg introduced from the United States (4). In the following five years (1971-75), frogeye leaf spot caused severe losses of susceptible varieties in the southernmost states of Parana, Santa Catarina and Rio Grande do Sul, often reaching 100% loss. By 1975, the area of susceptible varieties was reduced from more than 80% in some counties to a few hectares.

With the increase in the area of resistant varieties and with several years of unfavorable weather, frogeye leaf spot almost disappeared. As the disease became less conspicuous, many farmers went back to growing susceptible varieties, and after a few years, large areas of soybeans in southern Brazil again became vulnerable to the disease.

With the expansion of soybean production to the warmer and rainy savanna regions of West-Central and North Brazil, several new varieties were released without previous tests for reaction to frogeye leaf spot. In the 1987/88 growing season, an equivalent to 200,000 hectares of susceptible varieties EMGOPA 301 and Doko were devastated. An estimated half a million metric tons of soybeans (ca. US$ 11 million worth) were lost in the states of Goias, Mato Grosso and Mato Grosso do Sul. As a consequence, the susceptible varieties were replaced mostly by the resistant "FT-Cristalina", creating a new potentially dangerous situation where one variety is now grown in ca. 60% (2.5 million hectares) of more than four million hectares in the West-Central region.

The pathogen *C. sojina* has been quite variable in Brazil. From 1982 to 1989, 22 races of the fungus were identified (5, 6). Most races occur on traditionally susceptible varieties but new virulent races have developed. In 1988, race Cs-15 (7) broke the resistance in cultivar Santa Rosa, after more than 20 years of production. This variety was frequently used in breeding for resistance to frogeye leaf spot. Before the development of race Cs-15, 85 (69%) of 123 commercial varieties grown in Brazil were resistant to the occurring races. After the race Cs-15, 55 (45%) of the varieties remained resistant. Currently, all soybean breeding programs in Brazil are joining efforts to release only resistant varieties, but many farmers still insist on growing susceptible soybeans.

In order to cope with a possible new outbreak or breakdown of resistance by a new race, chemical control of frogeye leaf spot was investigated. Fungicides for seed treatment and field sprays were evaluated.

Seed treatment: Naturally infected seed samples of varieties Bragg (49.4% seeds infected) and Uniao (33.6% infected) were used in a blotter test with fungicides captan (150g a.i./100kg of seed), carboxin (20g), thiabendazole (20g) and thiram (140g). The fungicides thiabendazole and thiram were more efficient, reducing the number of seeds with *C. sojina* to 0.2% and 0.6%, respectively, on variety Bragg, and to 0.4% and 1.6%, respectively, on variety Uniao. The fungicides captan and carboxin reduced the number of seeds with *C. sojina* to 7.0% and 18.2%, respectively, on variety Bragg, and to 18.6% and 16.2%, respectively, on variety Uniao (8).

Currently, the following fungicides are recommended for general soybean seed treatment: 1. captan (150g a.i/100kg seed), 2. carboxin+thiram (75g+75g), 3. thiabendazole (20g), 4. thiabendazole+thiram (17g+73g), 5. thiram (210g), and 6. tolcoflos-methyl+captan (60g+120g)(9).

Assessment of losses due to frogeye leaf spot: Field experiments with artificial inoculations and fungicide treatments were carried out to measure: a. the effect of frogeye leaf spot upon yield; b. the curative effect of systemic fungicide, and c. the effect of late season diseases (brown spot and Cercospora leaf blight) occurring simultaneously with frogeye leaf spot. Six varieties (Bossier, Bragg, BR-5, BR-6 (Nova Bragg), IAS-5, and Uniao) were studied. Variety BR-6 (Nova Bragg) was resistant to all known races of *C. sojina*, until race Cs-15 developed; varieties Bossier were regarded as susceptible but more tolerant than varieties Bragg, BR-5 and Uniao, which were higly susceptible.

The experiment consisted of the following treatments: I. unsprayed and uninoculated control; II. two inoculations with *C. sojina* (ca. 15,000 conidia/ml), the first at early flowering stage (R1)(10); III. two inoculations (same as II.) + two fungicide sprays [benomyl (0.25 kg a.i) + mancozeb (1.6 kg a.i.)/300 liters of water/ha], seven days after each inoculation, and IV. fungicide sprayed every 15 days, the first at seven days after the first inoculation in treatments II. and III.(total of four sprays). The plots measured 3m x 6m and had six rows spaced 0.5m apart. At maturity, two central rows were harvested for yield.

The results presented in Table 1, show that *C. sojina* caused considerable yield reduction of susceptible varieties Bragg (I-II=22.6%), BR-5 (I-II=23.5%) and Uniao (I-II=15.2%). Varieties Bossier (I-II=8.9%) and IAS-5 (I-II=2.9%) were less affected. Variety BR-6 (Nova Bragg) was not affected by *C. sojina* but was affected by the late season diseases.

The curative effect of benomyl is shown in the comparisons between treatments I-III and III-II. In I-III, the small yield differences indicate that the infection by both inoculations was halted by the fungicide sprays made seven days after each inoculation. This is also supported by the comparison between treatments III-II, where the inoculated but unsprayed plots (II) had considerable reduction in yield of varieties Bragg (21.2%), BR-5 (21.8%) and Uniao (13.0%). Comparisons between treatments IV and I (control), shows the effect of naturally occurring late season diseases, particularly the brown spot which was more severe. The comparison between the treatments IV and II, shows the extent of reduction in yield caused by *C. sojina* infection and the late season diseases. The latter results are veryclose to the combined effects on yield measured by I-II, and by IV-I.

The results of this experiment showed that it is possible to control frogeye leaf spot with resistant varieties and by chemical sprays with systemic, curative fungicides or by preventive treatment of susceptible varieties . The total yield loss measured had the combined effects of frogeye leaf spot, and the naturally occurring brown spot and Cercospora leaf blight.

Brown spot (*S. glycines*) and Cercospora leaf blight (*C. kikuchii*).

Both diseases are widely disseminated in all soybean producing regions and are particularly serious in the warmer and rainy savanna region of West-Central Brazil. Their effects are most visible when the soybeans

TABLE 1

Effect of *Cercospora sojina* (frogeye leaf spot) upon yield of soybeans subjected to artificial inoculations and fungicide sprays with benomyl (0.25 kg a.i.) + mancozeb (1.6 kg a.i./300 liters water/ha). Average of two years (1982/83 - 1983/84) experiment. EMBRAPA-CNPSo, Londrina, PR. (Yorinori, unpublished).

Treat-ment	Bossier		Bragg		BR-5		BR-6		IAS-5		Uniao	
	kg/ ha	%	kg/ ha	%	kg/ ha	%	kg/ ha	%	kg/ ha	%	kg/ ha	%
I [1]	2380 [2] a		2601 [2] ab		2805 [2] ab		2597 [2] ab		2771 [2] ab		2954 [2] a	
II	2169 a		2012 b		2146 c		2416 b		2689 b		2504 b	
III	2120 a		2553 ab		2743 b		2700 ab		3024 a		2875 ab	
IV	2593 a		3106 a		2986 a		2972 a		3037 a		3208 a	
C.V.(%)	13.64		15.78		4.38		9.05		6.62		8.96	
I-II	211 [3]	8.9 [4]	589	22.5	659	23.5	180	6.9	82	2.9	450	15.2
I-III	260	10.9	48	1.8	62	2.2	-130	-2.2	-253	-9.1	79	2.7
III-II	-49	-2.3	541	21.2	597	21.8	283	10.5	335	11.0	375	13.0
IV-I	213	8.2	505	16.3	181	6.1	375	12.6	265	8.7	254	7.9
IV-II	424	16.3	1094	35.2	840	28.1	555	18.7	347	11.4	704	21.9
IV-III	473	18.2	553	17.8	243	8.1	272	9.1	52	1.7	333	10.4

Cultivar and treatment effect upon yield

[1]. I = Uninoculated and unsprayed control; II = inoculated twice (ca. 15,000 conidia/ml)(at R1/R2 stage and 15 days later); III = same as in II + fungicide sprayed 7 days after each inoculation; and IV = fungicide sprayed every 15 days, til maturity, the first at 7 days after first inoculation in II and III.

[2]. Average yield (kg/ha). Numbers followed by the same letters do not differ at 5% level by Duncan's test.

[3]. Average yield difference between treatments (kg/ha).

[4]. Average yield difference between treatments (%).

reach the full pod stage and begin to senesce (early R7). The natural yellowing of the senescing leaves is quickly replaced by a light-brown (brownspot) or dark-brown to reddish-brown (Cercospora leaf blight) discoloration and defoliation occurs before the pods are fully mature. Generally, brown spot and Cercospora leaf spot occur simultaneously and distinction between them is difficult, thus, they are referred to as "late season disease complex". These diseases are one reason why soybean yield in Brazil is very low (1,800kg/ha), when the potential yield may be as high as 4,000kg/ha. On favorable years, annual losses may amount to more than 20% or ca. US$1 billion (11).

Reliable sources of resistance to brown spot has not been found (12) but under field conditions it is possible to distinguish genotypes that are more or less susceptible. Most commercial varieties are susceptible to both diseases, but some seem to react differently. One example is variety Davis which is highly susceptible to brown spot but has good resistance to Cercospora leaf blight (13)

The assessment of yield losses, the screening for tolerance and the control of late season diseases have been studied using fungicide sprays.

Fungicide sprays and yield loss assessment: The experiments for the fungicide trials and yield loss assessments were carried out at Londrina, in replicated, randomized complete block design.

The studies showed that some fungicides (carbendazim, fentin acetate, fentin hydroxide and thiabendazole were effective against the late season diseases and promoted yield increase by more than 10%. The higher yield was due to the greater seed weight as a consequence of longer vegetative period (by as many as 19 days of delayed harvest) promoted by the disease control (Table 2).

The late season infection and its effect upon yield was determined by experiments where cumulative sprays, starting at early flowering stage (R1), were compared with reduced number of sprays. In each subsequent treatment, the first spray was delayed by 15 days. The fungicides benomyl (0.25kg a.i.) and fentin hydroxide (0.20kg a.i./300 liters of water/ha) were used. The results shown in Table 3, demonstrate that two to six sprayings before full pod stage had the same effect as that applied at early full pod stage (early R6), thus, indicating that earlier applications were ineffective or unnecessary. Even one spray between late full pod (late R6) and early maturing stage (R7) resulted in some yield increase.

Screening for tolerance: Screening for tolerance to late season diseases was attemped by comparing sprayed and unsprayed plots of commercial varieties arranged according to their maturity groups. Each variety had one 10m row replicated four times in a completely randomized split-plot design. A similar procedure was used in screening for tolerance to soybean rust in Taiwan (14). Half (5m) of each 10m-row was sprayed with a mixture of benomyl (0.25 kg a.i.) + mancozeb (1.6 kg a.i.)/300 liters of water/ha. Early varieties were sprayed four times and later varieties were sprayed five to six times, starting at late flowering (R3) to pods fully extended (R4). The unsprayed portion was regarded as the control plot and was subjected to natural infection by late season diseases. Treated and untreated rows were harvested as they reached the stage of harvest maturity. Comparisons of yield differences (kg and %/ha) were made among varieties that had the unsprayed plots harvested close or on the same date. The varieties with least differences in yield between treated and untreated plots were considered more tolerant to late season diseases.

The results shown in Table 4, seems to indicate that there are possibilities for screening for tolerance to brown spot and/or Cercospora leaf blight by the method used. The varieties with least yield differences also had more normal and simultaneous yellowing of the leaves, pods and stems. On varieties with greater yield differences the leaves were shed before pods had fully matured. Differences were also noticeable in the date of harvest and seed weight between treated and untreated rows.

Currently, studies are under way for selection of more tolerant genotypes and integrated crop management involving crop rotation/succession, seed treatment, and tillage practices are recommended. Fungicides are not routinely used, except for seed treatment, but they can be effective and economically feasible, especially in the rainy savanna region of West-

TABLE 2

Effect of fungicides on yield (kg/ha), defoliation (%), days to harvest maturity and on 100 seed weight of variety Davis, during 1982/83 growing season). EMBRAPA-CNPSo, Londrina. (Yorinori, unpublished).

Fungicide [1]	Yield [2]		Defoliation [3]	Days to [4]	100 seed weight [5]	
(kg a.i./ha)	kg/ha	%	%	maturity	g	%
Carbendazim 75 WP (0.25)	3093 a	18.0	51.0	19	15.8	13.7
Fentin hydroxide 40F (0.20)	3041 a	16.0	46.0	19	16.1	15.8
Fentin acetate 20 WP (0.30)	3004 a	14.6	45.0	19	15.7	12.9
Thiabendazole 40 F (0.40)	2979 a	13.7	65.0	13	15.0	7.9
Benomyl 50 WP (0.50)	2859 ab	9.1	30.0	19	15.5	11.5
Mepronil 75 WP (2.25)	2757 ab	5.2	97.6	0	14.2	2.0
Control	2620 b		99.0		13.9	
C.V. (%)	7.62					

[1]. Sprayed four times between R3 and R6/R7 (300 liters of water/ha).
[2]. Yield (kg/ha): based on four 5m² plots/treatment; (%) difference between fungicide treated and untreated control. Numbers followed by the same letters do not differ at 5% level by Duncan's test.
[3]. Defoliation (%): based on visual assessment when control plots had reached near harvest maturity (late R7/early R8).
[4]. Days to maturity: difference in days to harvest (delayed harvest) between sprayed and control plots.
[5]. 100 seed weight: average weight (g) of 5 x 100 seeds and % difference between treated and control.

Central Brazil. Their efficient use is much dependent upon the right timing of application (R5.4 to early R6 growth stages) and the weather condition of the growing season.

Rust (*P. pachyrhizi*)

Rust is one of the most destructive diseases of soybeans in the tropical and sub-tropical countries of Asia and Oceania (14, 15, 16, 17). It was first identified in Brazil in 1979 (1) and has since occasionally caused severe defoliation to soybeans in the high plains (800 to 1.100m altitudes) and where temperatures are mild (15 to 26C), in Central Brazil. In 1990/91 season, rust was found causing severe defoliation in commercial fields of variety FT-Seriema in Central Brazil.

TABLE 3
Effect of cumulative sprays of benomyl (0.50kg a.i./ha) and fentin
hydroxide (0.20kg a.i./ha) on variety Davis, sown on Dec. 4., 1983, at
Londrina. EMBRAPA-CNPSo, Londrina, PR. (Yorinori, unpublished).

№ of sprays	Growth stage at first spray	Benomyl		Fentin hydroxide	
		kg/ha	Effect (%)	kg/ha	Effect (%)
6 [1]	R1 [2]	3131 [3] a	26.4 [4]	3080 [3] abc	14.6* [4]
5	R3	2676 bc	8.0**	2837 bc	5.5*
4	R4/R5.1	2856 abc	15.3	3266 a	21.5
3	R5.4	2781 abc	12.3	3297 a	22.6
2	R6	2956 ab	19.3	3317 a	23.4
1	R6/R7	2748 abc	11.0	3175 ab	18.0
0	(Control)	2477 c		2689 c	
C.V. (%)		8.44		8.11	

[1]/ № of sprays in each treatment.
[2]/ Growth stage at first spray, and followed every 15 days.
[3]/ Average yield (kg/ha), based on four 5m² plots. Numbers followed by the
same letters do not differ at 5% level by Duncan's test.
[4]/ Percent yield differences between sprayed and control plots.
 * Phytotoxicity due to cumulative sprays.
 ** Affected by localized low fertility soil.

   P. pachyrhizi isolates from different regions of Eastern and Western
Hemisphere, were separated into four races, but no pathogenic variability
seems to occur in the rust fungus population of the Western Hemisphere
(15). Sources of resistance to the most virulent races have been rare (14,
15, 18). Although no genotypes were found immune to the rust fungus, many
more reliable sources of resistance are available for the milder race occurring in Brazil (19).
   Where rust is a limiting factor for soybean production, control of
the disease with fungicides is considered as most reliable and economically feasible (17).

Downy mildew (*P. manshurica*)

The disease is widely distributed in Brazil but has rarely caused serious
damage. It is most common when a prolonged period of wet, overcasting and
cool weather follows during the early vegetative to flowering stages. Most
varieties grown in Brazil have good resistance to downy mildew.

Target spot (*C. cassiicola*).

The disease is present in all soybean producing regions in Brazil and has
occasionally caused serious damage on susceptible varieties. Losses of
12-32% have been reported in the United States (20).

TABLE 4

Effect of late season diseases (brown spot and Cercospora leaf blight) on yield and days to harvest of commercial soybeans, measured by fungicide sprays with benomyl (0.25kg a.i.) + mancozeb (1.6kg a.i.)/300 l water/ha at Londrina, PR. EMBRAPA-CNPSo, Londrina, 1986. (Yorinori, 1986, unpublished).

Variety[1]	Yield[2]		Effect of treatment[3]		Days to maturity[4]		
	kg/ha A	kg/ha B	kg/ha A-B	%	A	B	A-B
OC.5= Piquiri	1775 a	1209 b	566	31.9	114	107	7
Parana	2168 a	1416 b	752	34.7	114	107	7
OC.3= Primavera	2924 a	1851 b	1073	36.7	119	111	8
IAS-5	2720 a	2131 b	589	21.6	119	114	5
OC.4= Iguacu	2477 a	1904 b	573	23.1	129	114	15
FT-7 (Taroba)	3588 a	2674 b	914	25.5	122	114	8
FT-9 (Inae)	2769 a	2002 b	767	27.7	122	114	8
Davis	3466 a	2135 b	1331	38.4	122	114	8
Sertaneja	3190 a	2287 b	903	28.3	122	119	3
FT-2	2878 a	2518 a	360	12.5	134	122	12
BR-6 (Nova Bragg)	3476 a	2754 b	722	20.8	134	122	12
Bragg	3359 a	2582 b	777	23.1	134	122	12
OC.2= Iapo	3130 a	2391 b	739	23.6	134	122	12
BR-13 (Maravilha)	3356 a	2274 b	1081	38.2	134	122	12
FT-6 (Veneza)	3071 a	2988 a	83	2.7	134	125	9
FT-3	3446 a	2892 b	554	16.1	134	130	4
FT-10 (Princesa)	4055 a	4452 a	-397	-9.8	137	134	3
Bossier	3290 a	2981 a	308	9.4	140	134	6
FT-4	3200 a	2392 b	808	25.2	145	137	8
FT-5 (Formosa)	4763 a	4625 a	138	2.9	145	145	0
FT-8 (Araucaria)	3889 a	3429 a	460	11.8	152	145	7
Paranagoiana	3431 a	3240 a	191	5.6	160	152	8
Cristalina	3904 a	3390 b	514	13.2	160	152	8
Average	3231	2631	600	20.1			
C.V.(%)	5.67						

[1]. Varieties recommended for 1991/92 crop season in the state of Parana.

[2]. Yield of fungicide treated (A) and utreated (B) plots. Numbers followed by the same letters in each variety do not differ at 5% by F test (F= 3.9888).

[3]. Effect of treatment: yield differences between treated (A) and untreated (B) plots (kg/ha and % of difference).

[4]. Days to maturity: days from sowing (Nov. 22, 1985) to harvest in A and B, and difference in days to harvest between treated and untreated plots (A-B).

The fungus *C. cassiicola* has frequently been found associated with root rot and premature death of plants under no-tillage farming in Southern and West-Central Brazil. Assessment of yield losses made at pre- harvest and based on disease incidence and on differences in seed yield between healthy and infected plants, showed yield reductions of 6,2% and 11.5% in varieties Davis and FT-4, respectively. Variety Davis was grown where the previous crops had been oat, soybean, oat, corn and wheat; FT-4 was grown following oat, soybean, wheat, soybean and oat, thus, after two successive crops of soybeans (21)

Leaf symptoms and root rot frequently are not observed in the same field but isolates from root were capable of causing typical target spots under artificial inoculations (22; Yorinori, unpublished).

Control of target spot in the temperate region have been achieved with resistant varieties (23). No studies were found on screening for resistance to target spot in tropical soybeans but field observations seem to indicate that most commercial varieties have high levels of field resistance. The fungus is most adapted to warm and humid conditions and has a potential for causing severe damage in susceptible varieties in the rainy savanna region of West-Central Brazil.

REFERENCES

1. Deslandes, J.A. Ferrugem da soja e de outras leguminosas causadas por *Phakopsora pachyrhizi* no Estado de Minas Gerais ("Rust of soybeans and other legumes caused by *Phakopsora pachyrhizi* in the State of Minas Gerais"). Fitopatologia Brasileira, 1979, 4, 337-339. (English Summary).

2. Sinclair, J.B., and Backman, P.S. (eds.). Compendium of Soybean Diseases (3rd. ed.), American Phytopathological Society, St. Paul, 1989, 106 p.

3. Yorinori, J.T. Doencas da soja no Brasil ("Soybean diseases in Brazil"). In Soja no Brasil Central (3a. ed.), ed. Fundacao Cargill, Campinas, 1986, pp. 301-364. (In Portuguese).

4. Yorinori, J.T. Doencas ("Diseases"). In Soja no Parana, Ministerio da Agricultura, Instituto de Pesquisa Agropecuaria Meridional, Curitiba, PR, 1971, Circ. 9, pp. 13-16. (In Portuguese).

5. Yorinori, J.T. Frogeye leaf spot of soybean (*Cercospora sojina* Hara). In Proceedings, World Soybean Research Conference, 4, Buenos Aires, 1989, Asociacion Argentina de la Soja, Buenos Aires, 1989. Vol. III. pp. 1275-1283.

6. Yorinori, J.T. Identificacao de racas de *Cercospora sojina* Hara e distribuicao geografica no Brasil ("identification of races of *Cercospora sojina* Hara and geographical distribution in Brazil"). In Resumos, S eminario Nacional de Pesquisa de Soja no Brasil, 5, Campo Grande, 1989, EMBRAPA-CNPSo, Londrina, 1989. p. 31. (In Portuguese).

7. Yorinori, J.T., Garcia, A., Kiihl, R.A.S., and Hirooka, T. Nova raca de *Cercospora sojina* Hara, patogenica ao gene de resistencia da culti var Santa Rosa ("New race of *Cercospora sojina* Hara, pathogenic to the

resistant gene in cultivar Santa Rosa"). In Resumos, Seminario Nacio
nal de Pesquisa de Soja, 5, Campo Grande, 1989, EMBRAPA-CNPSo,
1989. p. 31. (In Portuguese).

8. Yorinori, J.T. Tratamento de sementes de soja para controle da disse
minacao de *Cercospora sojina* Hara (mancha "olho-de-ra")["Soybean seed
treatment for the control of the dissemination of *Cercospora sojina*
Hara (Frogeye leaf spot)]. In Seminario Nacional de Pesquisa de Soja,
3, Campinas, EMBRAPA-CNPSo, Londrina, 1984, pp. 33. (Abstract). (In
Portuguese).

9. Henning, A.A., Krzyzanowski, F.C., Franca Neto, J.B. and Yorinori,J.T.
Tratamento de sementes de soja com fungicidas ("Soybean seed treatment
with fungicides"), EMBRAPA-CNPSo, Londrina, 1991, Comunicado Tecnico,
49. 4 p. (In Portuguese).

10. Ritchie, S., Hanway, J.J., and Thompson, H.E. How a soybean Plant De
velops, Iowa State University of Science and Technology, Coop.
Ext. Serv., Ames, Special Report 53, 1982. 20 p.

11. Yorinori, J.T. Soybean diseases and yield loss assessment in Brazil,
In Abstracts, International Congress of Plant Pathology, 5, Kyoto,
1988, p. 288.

12. Lim, S.M. Brown spot severity and yield reduction in soybean. Phytopathology, 1980, 70, 974-977.

13. Walters, H.J. Purple seed stain and Cercospora leaf blight. In Procee
dings, World Soybean Research Conference, 3, Aimes, 1984, ed. R. Shibles, Westview Press, Boulder, 1985, pp. 503-506.

14. Tschanz, A.T., and Shanmugasundaram, S. Soybean rust. In Proceedings,
World Soybean Research Conference, 3, Aimes, 1984, ed. R. Shibles,
Westview Press, Boulder, 1985, pp. 562-567.

15. Bromfield, K.R. Soybean rust. American Phytopathological Society, St.
Paul, 1984, Monograph 11. 65 p.

16. Tschanz, A.T. Rust. In Compendium of Soybean Diseases (3rd. ed.), ed.
J.B. Sinclair and P.A. Backman, American Phytopatological Society, St.
Paul, 1989, 106 p.

17. Yeh, C.C. Soybean rust. In Proceedings, World Soybean Research Confe
rence, 4, Buenos Aires, 1989, Asociacion Argentina de la Soja, 1989,
Vol. III, pp. 1269-1274.

18. Hinson, K. and Hartwig, E.E. Soybean Production in the Tropics. FAO,
Plant Production and Protection Paper, 4, AGPS/MISC/35. FAO, Rome.
1977. 92 p.

19. Yorinori, J.T. Reacao de cultivares comerciais de soja a ferrugem, em
avaliacao a campo ("Reaction of commercial soybean cultivars to rust
under field condition"), Resultados de Pesquisa de Soja,1988/89,
EMBRAPA-CNPSo, Londrina, 1990. pp. 174-175. (In Portuguese).

20. Hartwig, E.E. Effect of target spot on yield of soybeans. Plant Dis.
Reptr., 1959, 43, 504-505.

21. Yorinori, J.T. Podridao radicular da soja nas areas de plantio direto da Colonia Castrolanda, Castro, PR ("Root rot of soybeans under no-tillage at Castrolanda, Castro, PR"). In Resultados de Pesquisa de Soja 1986/87, EMBRAPA-CNPSo, Londrina, 1988, pp. 207-209. (In Portuguese).

22. Spencer, J.A. and Walters, H.J. Variation in certain isolates of *Corynespora cassiicola*. Phytopathology, 1989, 59, 58-60.

23. Snow, J.P. and Berggren, G.T., Jr. Target spot. In Compendium of Soybean Diseases (3rd. ed.), ed. J.B. Sinclair and P.A. Backman, American Phytopathological Society, St. Paul, 1989, 106 p.

NATURE AND MANAGEMENT OF SOILBORNE FUNGAL DISEASES OF SOYBEAN

JOHN C. RUPE
Department of Plant Pathology
University of Arkansas
Fayetteville, AR 72701 USA

ABSTRACT

Soilborne fungal diseases of soybeans fall into two catagories: seedling diseases and mid- to late-season diseases. Environmental conditions and cultivar susceptibility play critical roles in disease development, particullaly soil temperature and moisture. Other factors that may be important are fertility, herbicide use, and other pathogens, especially nematodes. Control of these diseases may involve: altering the environment through planting date, tillage, drainage, or soil fertilization; reducing the pathogen population by crop rotation or tillage; or protecting the plant through resistance or by fungicides. Optimum control usually involves the integrated use of several control measures.

INTRODUCTION

A wide range of soilborne fungal diseases affect soybean from planting through harvest. These diseases may cause little or no yield loss or may devastate the crop. Disease severity and subsequent yield loss depends on the relationships between inoculum density, environment, host susceptibility, and interactions with other microorganisms. The complexity of these relationships and the difficulties in manipulating the soil environment can put severe constraints on management practices. Additionally, variability in the pathogen and economic considerations further limit management options available to the grower.

Soybean diseases caused by soilborne fungi fall into two broad categories: seedling diseases and mid- to late-season diseases. While numerous soilborne fungi cause diseases in soybeans, only those that are of major concern to soybean production will be covered in this paper. These diseases include: seedling diseases, Phytophthora root rot, sudden death syndrome, southern blight, brown stem rot, and charcoal rot.

SEEDLING DISEASES

Seedling diseases usually are associated with conditions, particularly high soil moisture, that delay seed germination or seedling emergence and establishment (1). These diseases can cause significant reductions in stands, but not always reductions in yield, because crop growth can compensate for lost plants early in the growing season and because growers commonly overplant (1).

The most common soilborne fungal pathogens of seedling soybeans are Pythium spp., Fusarium spp., and Rhizoctonia solani. Phytophthora root rot also causes seedling disease, but will be covered separately. There are several species of Pythium that attack soybeans and all cause similar symptoms. Pythium spp. commonly cause a pre- or post-emergence damping-off, infecting either the hypocotyl or the roots (2). Symptoms begin as water soaked lesions which later turn dark brown as the tissue collapses. Small infected roots may decay and break off. The cortex of larger roots may slough off leaving the woody central core. Plants can recover from root infections by producing new lateral roots.

The Pythium spp. that attack soybeans can be divided into two groups based on temperature (2). Diseases caused by P. ultimum Trow, P. irregulare Buis., and P. debaryanum Hesse are favored by soil temperatures of 10-15 C and cause little disease at 22 C. These fungi cause problems early in the season when the soils are cool. Diseases caused by P. aphanadermatum (Edson) Fitz. and P. myriotylum Drechs. are favored by soil temperatures of 25 to 36 C and cause problems in double cropped soybeans which are planted into warm soils. Infection by both groups of Pythium spp. is favored by saturated soil conditions that aid zoospore dissemination and stress plants by reducing the oxygen content of the soil (3,4). As in other crops, the optimum time for infection by these Pythium spp. is during the initial stages of seed germination (2). Plants become more resistant with age.

A number of Fusarium spp. can be isolated from soybean and a few cause seedling diseases. F. oxysporum Schlecht. causes a pre- and post-emergence damping-off of emerging seedlings and a root rot of established seedlings (5). This fungus is associated with stunted plants (6) and yield loss (7). Infection is favored by cool (14-23 C) wet conditions (5). It causes destruction of the lower tap root and lateral roots . This results in slow emergence and stunted, weak seedlings. Plants may be predisposed by the soybean cyst (Heterodera glycines Ichinohe), root-knot (Meloidogyne incognita (Kofoid & White) Chitwood), or sting (Belonolaimus longicaudatus Rau) nematodes or other fungal pathogens such as Rhizoctonia solani or Pythium ultimum (5,8). Other strains of F. oxysporum can cause a vascular wilt of soybeans and other species of Fusarium also cause seedling disease (5).

Rhizoctonia solani Kuhn (teleomorph Thanatephorus cucumeris (Frank) Donk) causes a pre- and post-emergence damping-off under warm, wet conditions (9). Sunken, reddish lesions form on the stem and roots and can girdle the stem killing the plant. Cankers may form on older plants and, under very humid conditions, aerial hyphae may colonize above ground parts of the plant causing aerial blight. Most of the isolates of R. solani that attack soybean belong to anastomosis group AG-4, but a few may belong to other anastomosis groups. Of significance in rice-growing areas is AG-1 which attacks both soybeans and rice, causing sheath blight in rice. R. solani survives as sclerotia and as mycelium in plant debris.

There are many methods of seedling disease control. Most involve optimizing conditions for the emergence and establishment of the plant. Planting date is very important. Delaying planting until the soil temperature favors rapid emergence and establishment greatly reduces losses due to most Pythium spp. and Fusarium spp. but may favor infection by P. aphanidermatum, P. myriotylum, and R. solani (2,5,9). Improving drainage reduces seedling diseases directly by reducing the spread of zoospores of Pythium and indirectly by not subjecting the plant to low oxygen stress (3,4). Reduced tillage practices, especially under cool planting conditions, may increase seedling disease problems because of lower soil temperature, higher soil moisture, and higher soil bulk density compared to conventional tillage. The use of high quality vigorous seed is important because they germinate and emerge faster than low quality seed and have lower levels of seed exudates that stimulate germination of pathogen propagules (1). Damage by P. ultimum is more severe in seed lots with high levels of seed infection by Phomopsis longicola. Cultivation can reduce the damage caused by seedling diseases. Pre-emergence cultivation may be useful if soil crusting will prevent or delay seedling emergence. Cultivation of surviving diseased seedlings and maintaining optimum soil moisture helps these seedlings develop adventitious roots (2). Chemical seed treatments can be effective if the proper fungicide is used. Metalaxyl and etridiazole used on seed or in-furrow controls Pythium spp. and Phytophthora root rot. Carboxin, thiram, captan, and quintozene control R. solani and Fusarium spp. (10).

PHYTOPHTHORA ROOT ROT

An important disease of both seedlings and older plants is Phytophthora root rot (PRR) caused by Phytophthora megasperma var. glycinea Kuan & Erwin (PMG). Yield losses to this disease average 5% in the USA, but as high as 100% have occurred in individual fields (11). The disease occurs throughout most of the world, but has not been reported in China or in Central or South America.

PMG causes a pre- or post-emergence damping-off indistinguishable from symptoms caused by Pythium spp. (12). Unlike Pythium spp., PMG can continue to kill susceptible plants throughout the growing season. Symptoms in older plants vary with the tolerance of the plant. Susceptible plants develop brown or black lesions that girdle the stem beginning below the soil line and can progress as high as 10 nodes up the plant (11). In susceptible cultivars, the root systems are destroyed and wilting, chlorosis, and death soon follow. Leaves and petioles remaining attached to the dead plant. More tolerant cultivars may have a lesion on one side of the stem that never girdles the stem and very little root rot. These cultivars are stunted and yield less than resistant cultivars, but are not killed. Infected plants may not exhibit symptoms, but still yield less than uninfected plants.

Development of PRR is very dependent on environmental conditions. Infection is favored by flooding within a week after planting (11). Like Pythium root rot, saturated soil conditions cause low oxygen stress to the plant and aid in dissemination of zoospores and seed exudates. Because of this dependence on soil moisture, PRR generally is confined to low spots in a field and is more prevalent on clay soils than on coarse-textured soils. Optimum temperature for disease ranges from 25-28 C, but the pathogen is active at temperatures as low as 15 C (13). Soil compaction, reduced tillage, and high fertility also increase disease intensity (11). It is not clear how long PMG can survive in a field, but disease increases with mono-cropping and can be reduced by rotations with corn or sorghum.

Besides the influence of the environment, Phytophthora root rot development is very dependent on host resistance. Cultivars may have qualitative resistance or varying levels of tolerance. Qualitative resistance is conferred by a single dominant gene which provides resistance to several races of PMG (11,12). There are 12 such genes at 6 loci. In the north central region of the United States, use of qualitative resistance has led to the development of new races with virulence to these genes. Currently there are more than 25 known races of the pathogen. In the southern United States resistance has not been overcome by new races. Some breeding programs are pyramiding two genes for resistance into cultivars providing resistance to all known races of the pathogen (11). Tolerance (also known as field resistance or rate-reducing resistance) is conferred by a number of genes and is not race-specific. Generally, tolerant plants are susceptible as seedlings, but are resistant after the first trifoliate leaf stage. Yield losses occur with tolerant cultivars but are much less than with susceptible cultivars. Tolerant cultivars do not put selection pressure on the pathogen to develop new races.

Control of PRR is accomplished primarily by the use of resistant cultivars, however these cultivars can be compromised by the development of new races. An alternative control strategy uses tolerant cultivars as part of an integrated control program including drainage, conventional tillage, crop rotation, proper herbicide use, and seed treatment. Certain herbicides, particularly trifluralin, 2,4-DB and glyphosate, increase phytophthora root rot (14). The most effective fungicide against PMG is metalaxyl which can be applied as a seed treatment or an in-furrow treatment. Since metalaxyl is translocated acropetally, the in-furrow treatment is better than seed treatment since more of the roots are protected, but in-furrow treatments are prohibitively expensive. Work is currently under way to develop methods for assessing the risk of PRR in a field with the use of antibodies and a baiting technique (11).

SUDDEN DEATH SYNDROME

Sudden death syndrome of soybean (SDS) is a disease affecting soybeans with high yield potential occurring during reproductive development (15). The disease causes a root rot, but is characterized by the foliar symptoms. These symptoms begin as interveinal chlorotic spots that enlarge into chlorotic streaks. Chlorotic areas may become necrotic and affect all of the interveinal leaf tissue, the veins remaining green. Leaflets with severe symptoms abscise, but the petioles remain attached to the stem. A vascular discoloration extends up the stem, but the pith remains white. First observed in Arkansas in 1971, SDS occurred sporadically until 1982 when a severe outbreak of the disease occurred. In 1984, another severe outbreak occurred in Arkansas and the disease was reported in several midwestern states. SDS is now reported in seven states: Arkansas, Kentucky, Illinois, Indiana, Mississippi, Missouri, and Tennessee. Yield losses through pod and seed abortion can be as high as 80 to 100% in severely affected areas.

SDS is caused by a strain of the fungus Fusarium solani (Mart.) Appel & Wollenw. emend. Snyd. & Hans. (16,17). This strain produces abundant macroconidia, few microconidia, and has a distinctive blue color in culture. Other strains of F. solani collected from soybean roots do not produce the typical SDS foliar symptoms (16). The soybean cyst nematode (SCN), Heterodera glycines, has been associated with SDS and is commonly found in fields with SDS. While SCN is not necessary for SDS development, SCN may hasten and intensify disease development (16) and may predispose some, but not all, cultivars to the disease (18).

Disease development is strongly related to environmental conditions and the growth of the plant. Symptoms appear during mid-reproductive development in the midwest, but may appear several weeks before flowering in the south (18). Significant increases in disease development in the south occur after flowering and are related to the number of days after planting (unpublished data).

Along with plant age, the environment plays an important role in disease development. SDS occurs primarily in soybeans growing under optimum conditions and is closely associated with optimum soil moisture. In Arkansas, disease distribution is closely related to irrigation with the greatest disease occurring in areas of the field receiving the most water. Little or no disease occurs in non-irrigated areas (unpublished data). Temperature is also important. Severe outbreaks of SDS have occurred in Arkansas after periods of cool weather in August. Preliminary studies indicate that both disease and in vitro growth of the fungus are optimum between 25 and 30 C with little activity occurring at 35C (unpublished data).

Soil populations of F. solani are correlated with disease incidence (unpublished data). Populations vary during the year with the highest occurring during the winter and the lowest at planting. It is not clear what effect other crops have on the soil populations of F. solani but severe disease outbreaks have occurred in fields previously planted in corn, cotton, sorghum, rice, and soybeans (15).

The main control for SDS is the use of resistant cultivars. There are clear differences in cultivar responses to SDS ranging from almost immune to highly susceptible (18). The nature of this resistance has not been determined, but may involve resistance to a nonhost-specific toxin produced by the fungus. This toxin has been used to determine cultivar susceptibility by a cotyledon inoculation with the culture filtrate (19). Results of the cotyledon test corresponded with field results. While strong differences in cultivar reaction occur, these reactions may vary from location to location. Some of these differences may be due to variability in the pathogen. A preliminary report found 12 isolates of the pathogen fit into 5 groups based on the reaction of 10 differential cultivars (20). In another study, the reaction of some cultivars appeared to be dependent on their susceptibility to the race of SCN present in the field (18).

SOUTHERN BLIGHT

Southern blight caused by Sclerotium rolfsii Sacc. attacks a wide range of hosts including soybeans in the tropics and in warmer temperate soybean growing areas (21). Generally, southern blight appears as isolated patches of a few affected plants. Leaves turn brown but cling to the stem. The most characteristic sign under humid conditions is a white mat of hyphae at the base of the stem and eventually the formation of sclerotia. Mature sclerotia have a tan to brown rind, a white interior and measure 1-2 mm in diameter. These sclerotia serve as a means of overwintering and disseminating the pathogen.

Southern blight is favored by temperatures between 25 and 35 C, high soil moisture and a dense canopy, but drought may precede severe disease outbreak (21). The fungus is highly aerobic so infection occurs at or just below the soil surface and is favored by sandy soils. Burying or disking the sclerotia may lower infection.

Control of southern blight includes avoiding heavily infested areas, deep plowing, rotation with less susceptible crops, such as corn or sorghum, and planting tolerant cultivars (21).

BROWN STEM ROT

Brown stem rot (BSR), caused by Phialophora gregata (Allington and Chamberl.) W. Gams., is an important soybean disease in the midwest United States and in Canada. Often going undetected, yield losses due to this disease range as high as 44%, depending on the environment, cultivar, and strain of the fungus present (22). The fungus infects through the roots but symptoms appear during pod set as a reddish brown discoloration of the vascular elements and the pith (23). This discoloration begins in the roots or crown of the plant and can extend up the plant. The length of the stem discoloration depends on the susceptibility of the cultivar. Later in the season, the lower stem turns a dull brown and a sudden blighting of the interveinal leaf tissue killing the leaves. Severely blighted leaves abscise and plants may mature 20 to 30 days earlier than normal.

BSR development is favored by moisture stress and moderate temperatures, especially if preceded by moist conditions. Disease development occurs from 15 to 27 C with no development at 32 C (23). This temperature range corresponds to spore germination (opt. 21-25 C, very slow at 30 C), sporulation (optimum 19-25 C, none above 25 C), and toxin production (optimum 19-25 C, none at 30 C) (24). The role of moisture stress is less clear. Generally, BSR is associated with moisture stress in July and August, but in Wisconsin BSR is favored by irrigation (25).

Crop rotation is an effective control strategy for BSR. Rotations with corn or sorghum for 2 or 3 years effectively control the disease (23). Rotations with other legumes, such as alfalfa and clover, are not effective. However rotation with a resistant cultivar is as effective as corn or sorghum (26), because the biomass of P. gregata is much less in a resistant than in a susceptible cultivar (27).

The development of resistant cultivars has greatly changed our understanding of BSR. While resistance was first reported in 1968 in PI 84.946-2, it was not until 1978 that a BSR-resistant cultivar was released (22). Lack of a greenhouse screening method, inconsistent field results, and reduced yield potential of the resistant lines delayed development of BSR-resistant cultivars.

Greenhouse screening methods were improved when it was determined that there are two types of the fungus. One type produces the stem browning but no foliar symptoms and the other type produces both symptoms (24). The development of foliar symptoms in the greenhouse is highly correlated to resistance in the field (27). The foliar symptoms are caused by toxins produced by the fungus. These toxins are not produced by the strain of the fungus which only causes stem browning. There are five related toxins produced by P. gregata (24). Gregatin A produces stem browning and foliar symptoms. In the field, both the extent of stem browning and foliar symptoms are used to evaluate cultivar resistance. One problem with resistant cultivars is that they yield less than susceptible cultivars in the absence of BSR (22). However, in the presence of BSR, resistant cultivars can yield 30 to 34% more than susceptible cultivars. Current breeding programs are attempting to incorporate BSR resistance into high yielding cultivars.

CHARCOAL ROT

The most important disease of non-irrigated soybeans is charcoal rot caused by
Macrophomina phaseolina (Tassi) Goid. Favored by hot, dry environmental
conditions, this pathogen attacks over 500 plant species world wide (28). Yield losses
in Missouri average 5% per year with losses of 30 to 50% in some fields (29).
Dissemination and overwintering of the fungus is by as microsclerotia in the soil or
debris. Survival of the fungus is much longer in dry soils than wet soils. The fungus
infects roots all season and grows intercellularly as long as the plant is not under
moisture stress (29). This intercellular growth causes no apparent damage to the
plant. However, under moisture stress, particularly during reproductive development,
the fungus begins to grow intracellularly and forms microsclerotia that plug the
vascular tissue. Diseased plants mature early. The leaves die, but remain attached to
the stem. Stems become discolored turning silver to gray and the roots turning gray
to black with microsclerotia visible beneath the epidermis. In the tropics, seeds can
become infected leading to a reddish brown lesion at the soil line on the resulting
seedlings. Infected seedlings may die under hot, dry conditions or recover under
cool, moist conditions, but harbor a latent infection (27).

Charcoal rot is strongly influenced by the environment. Temperature optimum
for growth is from 28 to 35 C (29). Disease development has a similar temperature
range but requires moisture stress as well. Levels of moisture stress have not been
related to disease development but the root colonization is reduced by irrigation (30).
In a study of 112 fields in Missouri, increases in soil populations of M. phaseolina
were related to increases in soil pH and potassium and decreases in phosphorus,
rainfall, and clay content of the soil (29). Some herbicides (chloramben and 2,4-DB)
can increase root colonization others (glyphosate and vernolate) have no effect. Soil
populations of the fungus can be reduced to relatively low levels by rotation with
corn, grain sorghum, or, especially, cotton. However, under favorable conditions,
even an 80% reduction in the soil population of M. phaseolina may have no effect on
disease development (31).

Control measures for charcoal rot are directed at relieving plant stress and
reducing soil populations. These measures include rotation with corn, sorghum, or
cotton for up to 2 years, reduced seeding rates, narrower rows to speed canopy
closure, and irrigation (28). There is no known resistance to this pathogen.

CONCLUSIONS

There are a number of measures available to growers to control soilborne fungal
diseases of soybean. Some of these measures rely on changes in cultural practices
such as rotation, tillage, drainage, soil fertility, planting date, seed quality, or
herbicide use. Others depend on cultivar resistance or tolerance or, with seedling
diseases, the use of seed or soil fungicide treatments. The effectiveness and use of
these control practices, used either singly or in combination, is dependent on the
environment, the pathogen, and cost of the control practice. Changes in the
environment may enhance disease, obviating the effect of tolerant cultivars and many
cultural control practices, or suppress disease, obscuring yield improvement by use of
the control practices. The pathogen may have races limiting growers to those
cultivars that are resistant. Cultivars currently resistant to the pathogen may be
compromised by the development of new races. Pathogens may also interact with
other organisms, especially nematodes, complicating control. Finally, cost determines

what, if any, control measures are used. The cost of control may outweigh the perceived benefits if the disease occurs sporadically and use of the control measure results in lower yields or is difficult to fit into other farm operations. In the future, scientists must develop methods for predicting the risk from soilborne diseases and develop more effective and practical control measures.

REFERENCES

1. Ferriss, R. S., Seedling establishment-An epidemiological perspective. P. 7-13. In Soybean Diseases of the North Central Region. eds. T. D. Wyllie and D. H. Scott. APS Press, St. Paul. pp. 149, 1988.

2. Anonymous, Pythium rot. p. 43-44. In Compendium of Soybean Diseases, 3rd edition. eds. J. B. Sinclair and P. A. Backman, APS Press, St. Paul, pp. 106, 1989.

3. Schlub, R. L., and Schmitthenner, A. F., Effects of soybean seed coat cracks on seed exudation and seed quality in soil infested with Pythium ultimum. Phytopathology, 1978, 68:1186-1191.

4. Stanghellini, M. E., and Hancock, J. G., Radial extent of the bean spermasphere and its relation to the behavior of Pythium ultimum. Phytopathology, 1971, 61:165-168.

5. Datnoff, L. E., Fusarium blight or wilt, root rot, and pod and collar rot. p. 33-35. In Compendium of Soybean Diseases, 3rd edition. eds. J. B. Sinclair and P. A. Backman. APS Press, St. Paul, pp. 196, 1989.

6. Anderson, T. R., Olechowski, H., and Welacky, T., Incidence of rhizoctonia and fusarium root rot of soybean in southwest Ontrario. Can. Plant Dis. Surv., 1988, 68:143-145.

7. Leath, S., and Carroll, R. B., Use of ridge regression to predict yield reduction by Fusarium sp. in selected soybean cultivars. Can. J. Plant Pathol., 1985, 57:58-66.

8. Griffin, G. J., Importance of Pythium ultimum in a disease syndrome of cv. Essex soybean. Can. J. Plant Pathol., 1990, 12:135-140.

9. Liu, Z., Rhizoctonia diseases. p. 45-47. In Compendium of Soybean Diseases, 3rd edition. eds. J. B. Sinclair and P. A. Backman, APS Press, St. Paul pp.106, 1989.

10. Mitchell, J. K., Holland, B. T., and Kirkpatrick, T. L., Plant Disease Control Handbook, MP154. Cooperative Extension Service, University of Arkansas, Little Rock, pp. 148, 1991.

11. Schmitthenner, A. F., Phytophthora root rot: Detection, ecology, and control. In World Soybean Research Conference IV. ed. A. J. Pascale. p 1284-1289, 1989.

12. Schmitthenner, A. F., Phytophthora rot. p. 25-38. In Compendium of Soybean Diseases, 3rd edition. eds. J. B. Sinclair and P. A. Backman, APS Press, St. Paul, pp. 106, 1989.

13. Athow, K. L, Phytophthora root rot of soybean. In World Soybean Research Conference III: Proceedings. ed. R. Shibles. p. 575-581, .1984

14. Schmitthenner, A. F., Phytophthora rot of soybean. p. 71-80. In Soybean Diseases of the North Central Region. eds. T. D. Wyllie and D. H. Scott, APS Press, St. Paul, pp. 149, 1988.

15. Rupe, J. C., Hirrel, M. C., and Hershman, D. E., Sudden death syndrome. p. 84-85. In Compendium of Soybean Diseases, 3rd edition. eds. J. B. Sinclair and P. A. Backman, APS Press, St. Paul pp. 106, 1989.

16. Roy, K. W., Lawrence, G. W., McClean, K. S., and Killebrew, J. F., Sudden death syndrome of soybean: Fusarium solani as incitant and relation of Heterodera glycines to disease severity. Phytopathology, 1989 79:191-197.

17. Rupe, J. C., Frequency and pathogenicity of Fusarium solani recovered from soybeans with sudden death syndrome. Plant Dis., 1989, 73:581-584.

18. Rupe, J. C., Gbur, E. E., and Marx, D. M., Cultivar response to sudden death syndrome of soybean. Plant Dis., 1991, 75:47-50.

19. Lim, S. M., Song, H. S., and Gray, L. E., Phytotoxicity of culture filtrates from Fusarium solani isolated from soybean. Phytopathology, 1990, 80:1044 (abstr.).

20. Lim, S. M., and Jin, H., Pathogenic variability in Fusarium solani isolated from soybeans with sudden death syndrome symptoms. Phytopathology, 1991, 81:1236 (abst.).

21. Gazaway, W. S. and Hagan, A. K., Sclerotium blight. p. 48-49. In Compendium of Soybean Diseases, 3rd edition. eds. J. B. Sinclair and P. A. Backman, APS Press, St. Paul, pp. 106, 1989.

22. Tachibana, H., Use and management of resistance for control of brown stem rot of soybeans. p. 102-105. In Soybean Diseases of the North Central Region, eds. T. D. Wyllie, and D. H. Scott, APS Press, St. Paul, pp. 149, 1988.

23. Gray, L. E., Brown stem rot. In Compendium of Soybean Diseases, 3rd edition, p. 29-30. ed. J. B. Sinclair and P. A. Backman, APS Press, St. Paul, pp. 106, 1989.

24. Gray, L. E., Brown stem rot of soybeans. In World Soybean Research Conference III: Proceedings. ed. R. Shibles. p. 598-601, 1984.

25. Lockwood, J. L., Late season diseases: Summary and discussion. In Soybean Diseases of the North Central Region. eds. T. D. Wyllie and D. H. Scott, APS Press, St. Paul, pp. 149, 1988.

26. Tachibana, H., "Prescription plant medicine" for control of brown stem rot of soybeans. In World Soybean Research Conference IV, ed. A. J. Pascale, p. 1296-1302, 1989.

27. Mengistu, A., Grau, C. R., and Gritton, E. T., Comparison of soybean genotypes for resistance to and agronomic performance in the presence of brown stem rot. Plant Disease, 1986, 70:1095-1098.

28. Wyllie, T. D., Charcoal rot. p. 30-33. In Compendium of Soybean Diseases, 3rd edition. eds. J. B. Sinclair and P. A. Backman, APS Press, St. Paul, pp. 106, 1989.

29. Wyllie, T. D., Charcoal Rot of soybeans-Current status. p. 106-113. In Soybean Diseases of the North Central Region. eds. T. D. Wyllie and D. H. Scott, St. Paul, APS Press, pp. 149, 1988.

30. Kendig, S. R., and Rupe, J. C., Relationship of irrigation regime to populations of Macrophomina phaseolina microsclerotia in root tissue and yield of soybean. Phytopathology, 1989. 79:1179-1180. (abstr.).

31. Pearson, C. A., S, Schwenk, F. W., Crowe, F. J., and Kelley, K., Colonization of soybean roots by Macrophomina phaseolina. Plant Disease, 1984, 68:1086-1088.

PRESENT SITUATION OF SOYBEAN DISEASE MANAGEMENT IN JAPAN

Kazufumi Nishi
Department of Plant Protection,
National Agriculture Research Center
3-1-1, Kan-non dai, Tsukuba City, Ibaraki 305, Japan

ABSTRACT

Presently in Japan, soybean disease management depends mainly on the application of fungicides. However, recently research on the development of methods of biological control has become very active in addition to the use of resistant varieties and improvement of cultural practices. Recent research highlights on biological control are as follows: Control of Verticillium disease by using non pathogenic Fusarium oxysporum, control of soybean mosaic disease by using the attenuated virus, and control of soybean cyst nematode by using parasitic fungi. In addition, a new method, soil sterilization will be introduced in this report.

INTRODUCTION

Soybean cultivation has been practiced in Japan for a long time. Historically, soybean was described in "Taiho-ritsuryo" a completion of old Japanese laws in 701, in which soybean was listed as a taxable crop.

There are two types of theories regarding the origin of soybean in Japan, soybean was introduced to Japan from China through Korea in the 2000s or 200s B. C. and soybean originated in Japan. The Japanese people appreciate soybean very much and have devised many kinds of foods based on soybean such as soy sauce, tofu, fermented soybean and miso soup. Therefore, soybean has been one of the most important materials in Japan since older times.

SOYBEAN CULTIVATION IN JAPAN

In Japan, soybean is cultivated throughout the country. The major producing centers are located in the Hokkaido, Tohoku, Kanto, Chubu or Kyushu districts (Fig. 1). The area with soybean in Japan which exceeded 400,000 ha. at the beginning of the 20th century, gradually decreased. Before World War II, the cultivated area was 320,000 ~ 350,000 ha. During the War, it decreased rapidly, and finally it reached 223,100 ha. in 1947. After the War, conversely the area increased rapidly (429,900 ha. in 1954), then decreased again to 79,300 ha. in 1977. In 1978, due to the implementation of a new policy, the so-called reorganization of paddy field utilization, soybean cultivation was actively promoted. The cultivated area gradually increased and presently it covers 120,000 ~ 150,000 ha.(Fig. 2). Current production of mature grains of soybean is about 270,000 ton. Since the amount of domestic consumption is nearly 5 million ton, the self-sufficiency ratio is only about 5 %. Judging from the amount of soybean consumption for fresh use, the self-sufficiency ratio ranges from 30 to 35 %. Though the yield had fluctuated arround 1.1 ~ 1.2 ton/ha previously, it increased and amounts now to about 1.7 ton per ha.

There are three types of soybean cultivation. The first is where the final product consists of mature grains of soybean. About 15 % of production is for fresh use such as soy sauce, tofu and fermented soybean, and almost 80 % is for processing for soybean oil.

The second type of soybean cultivation is aimed of the production of green vegetable soybean which has been used as a relish for the consumption of alcoholic beverage long time. The area for a cultivated with soybean for this purpose is about 14,000 ha. but recently consumption has gradually increased (Fig. 2).

The third type of cultivation of soybean is as a green manure or forage crop. The cultivated area for this purpose which covered 150,000 ~ 200,000 ha. before World War II, is presently negligible (Fig. 2).

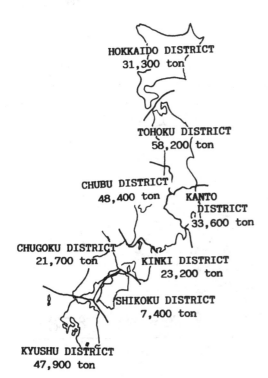

HOKKAIDO DISTRICT
31,300 ton

TOHOKU DISTRICT
58,200 ton

CHUBU DISTRICT
48,400 ton

KANTO DISTRICT
33,600 ton

CHUGOKU DISTRICT
21,700 ton

KINKI DISTRICT
23,200 ton

SHIKOKU DISTRICT
7,400 ton

KYUSHU DISTRICT
47,900 ton

Figure 1. Distribution of areas cultivated with soybean in Japan and amount of production in 1989.

IMPORTANT DISEASES

In Japan, soybean diseases are 7 viral diseases, one mycoplasmalike

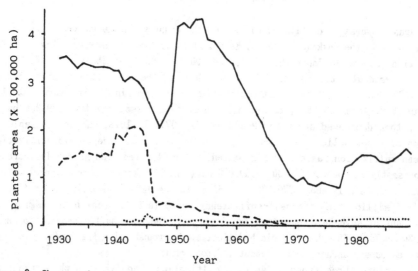

Figure 2. Changes in the area cultivated with soybean in Japan (——— : for mature grains, ········: for green vegetable, ————: for green manure or forage crop).

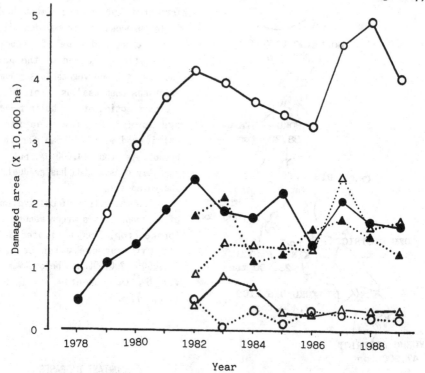

Figure 3. Changes in area of soybean damaged by major diseases in Japan (O——O : Downy mildew, ●——● : purple stain, ▲······▲ : viral diseases, △······△ : bacterial pustule, △——△ : soilborne diseases, O······O : Sclerotinia stem rot).

organism (MLO) disease, two bacterial diseases, 33 fungal diseases and three nematode diseases. Among them, about 10 are major diseases. Changes in the area damaged
by these diseases is summarized in Figure 3.

Recently, soilborne fungus diseases have become increasingly important following the adoption of the rotational system of paddy fields promoted by the Japanese
government since 1967. After a 10 year research period, 13 kinds of soilborne diseases including 4 new ones in Japan were identified. Their rough distribution and an
outline of the measures implemented for control were summarized by Nishi and Takahashi (1). Among these soilborne diseases, Sclerotium blight caused by Corticium
rolfsii, red crown rot caused by Calonectria crotalariae, Phytophthora rot caused by
Phytophthora megasperma var. sojae are very common and cause severe damage to
soybean. The distribution of Verticillium disease caused by Verticillium dahliae is
limited but the damage is considerable.

Control measures for major diseases are summarized in Table 1. Application of
agricultural chemicals is presently one of the most effective measures for controlling these diseases, although other strategies can also be applied.

BREEDING OF RESISTANT VARIETIES

Use of varieties resistant to diseases still remains an important control measure in
Japan. The most advanced study deals with the breeding of varieties resistant to
soybean mosaic virus, and a new variety "Dewamusume" was developed in 1977 (2). In
1982, "Suzuyutaka" which is resistant to the same virus was developed (3). This
variety is high-yielding and shows a good adaptability under various cultural conditions. Both varieties are also resistant to soybean stunt virus. A variety resistant to soybean dwarf virus, "Tsuru-kogane" was released in 1984(4). "Toyosuzu",
"Okushirome", "Nambushirome" were developed as recommended varieties resistant to
soybean cyst nematode. Research on varieties with higher resistance to the same
nematode was initiated in 1960s and "Suzu-hime" was developed in 1980 (5). Similar
studies have been pursued also for the breeding of varieties resistant to other
diseases.

RECENT RESEARCH HIGHLIGHTS

Recently research on soybean disease management has covered biological control. At
the same time use of resistant varieties and improvement of cultural practices etc.
are being promoted. Here, some recent research highlights will be introduced.

Biological control
Suwa and Hayashi reported that the pre-treatment of soybean seedlings with a
nonpathogenic strain of Fusarium oxysporum isolated from the conductive tissues
of plants with Verticillium disease resulted in a considerable decrease in the
occurrence of Verticillium disease caused by Verticillium dahliae (6). After the

TABLE 1
Major diseases and practical control measures implemented in Japan.

Name of disease	Region with damage[*]	Control measures
Purple stain (Cercospora kikuchii)	Nationwide	Use of healthy seeds. Drainage of field. Seed disinfection with chemicals such as thiram + benomyl, thiram + thiophanate methyl. Chemical application of benomyl, thiophanate methyl, copper sulfate.
Mosaic (Soybean mosaic virus)	Nationwide	Use of healthy seeds. Control of aphids. Use of resistant varieties.
Dwarf (Soybean dwarf virus)	Nothern region	Use of resistant varieties. Control of foxglove aphid.
Sclerotinia rot (Sclerotinia sclerotiorum)	Hokkaido	Avoidance of continuous cropping. Avoidance of manure application at high level. Chemical application of thiophanate methyl, iprodione, procymidone, vinclozolin.
Sclerotium blight (Corticium rolfsii)	Southwestern region	Crop rotation with gramineous plants Plowing to replace surface soil with subsoil. Removal of diseased plants. Soil fumigation with chloropicrin.
Phytophthora rot (Phytophthora megasperma var. sojae	Northern region	Avoidance of continuous cropping. Use of resistant varieties. Drainage of field.
Red crown rot (Calonectria crotalariae)	Tohoku, Kanto, Chyugoku, Hokuruku, Shikoku	Avoidance of continuous cropping. Drainage of field. Use of resistant varieties.
Anthracnose (Colletotrichum truncatum)	Tohoku, Kanto	Use of healthy seeds. Chemical application of copper sulfate.
Downy mildew (Peronospora manshurica)	Nationwide	Use of healthy seeds. Drainage of field.
Cyst nematode disease (Heterodera glycines)	Nationwide	Avoidance of continuous cropping. Use of resistant varieties. Soil fumigation with chemicals such as 1,3-dichloropropene, chloropicrin.

* : Name of district is shown in Figure 1.

roots of young seedlings (10 days after sowing) were dipped into a solution containing nonpathogenic F. oxysporum ($10^7 \sim 10^9$ propagules/mℓ) for 1~6 hour, the treated plants were transplanted to the field. The method is also effective for controlling Verticillium diseases of other crops such as tomato and sweet pepper (6).

Kosaka and Fukunishi observed that pre-treatment with attenuated soybean mosaic virus markedly decreased the occurrence of mosaic disease (7). The attenuated virus was isolated from soybean plants grown at a low temperature (15 ℃) for 14 days. Similar studies are being carried out by Honkura et al. (8).

Akasaka reported that Paeciromises sp. isolated from eggs of soybean cyst nematodes frequently displayed a biological control activity on the nematode (9). These studies are in progress and it is anticipated that they will provide successful control of the nematode.

Soil sterilization

A new soil sterilization method in which hot water (about 80 ℃) is injected into fields infested with soilborne pathogens was reported by Nishi et al. (10) who stated that the occurrence of red crown rot of soybean decreased considerably (Table 2). Details of the method are as follows. Hot water is prepared by a portable boiler and is sprinkled through a heat proof vinyl hose on the surface of the field covered with a heat-resistant vinyl film until the soil temperature (at a depth of 20 cm from the soil surface) reached 55℃. This method is also applicable to the control

TABLE 2

Effect of "Soil sterilization method" on the control of red crown rot of soybean caused by Calonectria crotalariae.

Date of treatment	Treatment	Disease severity[*]	Yield (g/row)
July, 22, 1990	Soil sterilization	16.31	843.4
	No treatment	41.80	550.2
May, 11, 1991	Soil sterilization	9.60	958.2
	No treatment	35.10	923.7
May, 30, 1991	Soil sterilization	7.08	1275.1
	No treatment	37.21	889.1
June, 25, 1991	Soil sterilization	18.46	1415.4
	No treatment	27.83	1100.4

[*] : Disease severity=$\dfrac{n_1+2n_2+3n_3+4n_4+5n_5}{5(n_0+n_1+n_2+n_3+n_4+n_5)} \times 100$, where n_0

~ n_5 indicate the number of plants showing damage rate from 0 (healthy) to 5 (severe disease and plant died), respectively. In n_0, n_1, n_2, n_3, n_4, n_5 of the formula, each numeral indicates the disease score of plants, and n indicates the number of diseased plants for the respective score.

of other soilborne diseases and nematodes such as Fusarium wilt of spinach caused by Fusarium oxysporum f. sp. spinaciae and sweet potato root knot nematode disease caused by Meloidogyne incognita (11).

REFERENCES

1. Nishi, K. and Takahashi, H., An introduction of soilborne diseases in soybean (in Japanese). Natl Agric. Res. Center, Tsukuba, 1990, pp. 1-32.

2. Ishikawa, M., Matsumoto, S., Nagasawa, T., Hashimoto, K., Koyama, T., Kokubun, K., Murakami, S., Nakamura, S., Miyahara, T., Matsumoto, S., Konno, Z., Iizuka, N., Takahashi, K. and Yunoki, T., A new soybean variety "Dewamusume" (in Japanese with English summary). Bull. Tohoku Natl. Agric. Exp. Stn., 1979 , 59, 71-86.

3. Hashimoto, K., Nagasawa, T., Kokubun, K., Murakami, M., Koyama,T., Nakamura, S., Matsumoto, S., Matsumoto, S. and Sasaki, K., A new soybean variety "Suzuyutaka" (in Japanese with English summary). Bull. Tohoku Natl. Agric. Exp. Stn.,1984, 70, 1-38.

4. Banba, H., Tanimura, Y., Matsukawa, I., Ushirogi, T., Mori,Y. and Chiba, I., A new soybean variety "Tsuru-kogane" (in Japanese with English summary). Bull. Hokkaido Pref. Agric. Exp. Stn., 1985, 52, 53-64.

5. Sunada, K., Sakai, S., Gotoh, K., Sanbuichi, T., Tsuchiya, T., and Kamiya, M., A new soybean variety "Suzu-hime" (in Japanese with English summary). Bull. Hokkaido Pref. Agric.Exp. Stn., 1981, 45, 89-100.

6. Suwa, S. and Hayashi, N., Biological control of Verticillium diseases by the application of nonpathogenic Fusarium oxysporum (Abst. in Japanese). Ann. Phyto-path. Soc. Japan, 1989, 55, 506.

7. Kosaka, Y., and Fukunishi, T., Control of soybean mosaic disease with an attenu-ated soybean mosaic virus. Proceeding of International Seminar : Biological control of plant diseases and virus vectors (FFTC book series No. 42), Food and Fertilizer Technology Center for the ASPAC Region, Taipei, 1991, pp. 103-110.

8. Honkura, R., Tsuji, H., Tomioka, K., Nakamura, S., Hashiba, T. and Ehara, Y., Biological control of soybean mosaic by the application of attenuated soybean mosaic virus (Abst. in Japanese). Ann. Phytopath. Soc. Japan, 1991, 57, 461.

9. Akasaka, Y., Control of soybean cyst nematode, Heterodera glycines by natural enemy. 1. Parasitics of fungi isolated from eggs of H. glycines to H. glycines (in Japanese with English summary). Ann. Rept. Plant Prot. North Japan, 1989, 40, 149-151.

10. Nishi, K., Kuniyasu, K. and Takahashi, H., Effect of soil sterilization with hot water injection on soybean root necrosis caused by <u>Calonectria</u> <u>crotalariae</u> (inJ apanese with English summary). <u>Rept.</u> <u>Tottori</u> <u>Mycol.</u> <u>Inst.</u> 1990, **28**, 293-305.

11. Kuniyasu, K., Nishi, K., Momota, Y. and Takeshita, S., Soil sterilization by hot water injection (in Japanese). <u>Shokubutsu-boeki</u>, 1991, **45**, 247-251.

DISEASE MANAGEMENT IN SOYBEAN: USE OF CULTURAL TECHNIQUES AND GENETIC RESISTANCE

DAVID B. WEAVER
Department of Agronomy and Soils

and

RODRIGO RODRIGUEZ-KABANA
Department of Plant Pathology

Alabama Agricultural Experiment Station
Auburn University, AL 36849 USA

ABSTRACT

The soybean is a good host for many fungal, bacterial, viral, and nematode pathogens. Environmental and economic concerns often dictate that cultural practices and genetic resistance are the only disease management options open to a grower. Primary cultural options include crop rotation and burial of crop residue by tillage. Crop rotation is very effective in reducing yield loss to diseases and nematodes, but is not always economically feasible. One of the most stable, economical, and environmentally safe disease and nematode management strategies is the use of genetic resistance. Genetically resistant cultivars are available for a variety of diseases and nematodes, including Phytophthora rot, brown stem rot, stem canker, frogeye leaf spot, soybean cyst nematodes, and root-knot nematodes. Vertical resistance is widely used for a number of diseases and is a major objective in a number of breeding programs. Horizontal resistance and tolerance are used to a lesser extent.

INTRODUCTION

The soybean (*Glycine max* (L.) Merr.) is a good host for many fungal, bacterial, viral, and nematode pathogens [1]. They can be windborne, endemic to the soil, or transmitted by crop debris, seed, insect vectors, or contaminated machinery [1]. Soybean production, world-wide and especially in tropical and subtropical areas, such as the southeastern United States, takes place in an environment where several pathogens typically are present in the same field. Management of these polyspecific mixtures of nematodes and other pathogens is complex, requiring careful cultivar selection and other management methods. However, economic restraints can limit the choice of control strategies in soybean. The soybean is a comparatively low-value crop, with an average worth of less than $500 (U. S.) per hectare. Environmental issues also affect production, as many inexpensive, effective pesticides, such as EDB (ethylene dibromide) and DBCP (1,2-dibromo-3-chloropropane), are no longer available to the grower. Yield responses to applications of these halogenated hydrocarbons permitted profitable soybean production even in fields with severe nematode problems [2]. These fumigants were removed because of adverse toxicological and ecological problems associated with their manufacture and use [3]. Current nematicides available for use in soybean are very few and dosages required are too expensive and impractical [4] and thus they are not presently an option for the management of nematode problems in soybean. Thus economic and environmental issues limit disease control strategies in soybean to low-cost or no-cost environmentally safe production practices such as crop rotation, burial of crop debris by tillage, and cultivar selection (genetic resistance). These three management practices form the basis for sound disease control strategies in soybean.

CULTURAL TECHNIQUES

Crop rotation for control of soilborne diseases and nematodes and tillage for control of pathogens that survive on crop debris are the most important cultural techniques. Some of the best examples of successful implementation of these techniques can be found in the southeastern U. S. where disease problems are generally more severe and varied than in the midwestern U. S. Discussion of the effects of cultural techniques on all soybean diseases is not within the scope of this paper. Examples of their effectiveness and some considerations will follow with more discussion on the use and effectiveness of genetic resistance. Other management methods (no-till, reduced tillage, delayed planting) will not be considered in this paper because of the paucity of information available with regards to their effects on nematodes and most soilborne fungal pathogens.

Crop rotation

The root-knot nematode (*Meloidogyne* spp. Goeldi) (RKN) and the soybean cyst nematode (*Heterodera glycines* Ichinohe) (SCN) are the principal phytonematodes that affect soybean. Both are highly responsive to crop rotation [5,6], particularly when both species occur together in the same field [7,8]. The types of rotations depend as much on economics as on their efficacy for suppression of problems caused by nematodes and other pathogens. Rotations with graminaceous crops are usually very effective in reducing problems caused by root-knot and cyst nematodes. Rotations of soybean with corn (*Zea mays* L.) or sorghum (*Sorghum bicolor* (L.) Moench) are very effective for the management of nematode problems in fields infested with root-knot or cyst nematodes, as well as in fields with mixtures of these species [5,7]. When corn can be produced under irrigation or in situations where yields exceed 6.3 Mg hectare^{-1}, the soybean-corn rotation is a profitable management tool. Corn yields in the midwestern U. S. are often over 6.3 Mg hectare^{-1}, so the corn-soybean cropping system is very popular there. Sorghum can be produced for grain or for forage. Grain production is economically feasible only where the grain can be marketed for poultry feed or more rarely for oil extraction. Soybean-sorghum rotations where sorghum is used as a forage crop are limited to farms or producers connected with cattle operations. Sorghum has the advantage over corn because it is more drought tolerant. Sorghum-soybean rotations are generally more effective than soybean-corn rotations for management of root-knot and cyst nematodes [6,7]. When implementing rotations of soybean with corn or sorghum, it should be understood that both corn and sorghum can be hosts to the peanut RKN (*M. arenaria* (Neal) Chitwood) and cotton RKN (*M. incognita* (Kofoid and White) Chitwood). Since both crops have a range of susceptibilities to these nematode species, the choice of corn or sorghum cultivar to be used in the rotation is important [9].

Rotations of soybean with bahiagrass (*Paspalum notatum* Flugge) also can be very successful in combatting problems caused by mixed infestations of root-knot and cyst nematodes [8]. In contrast with sorghum or corn, bahiagrass is not a host for the economically important RKN species or a host for SCN. It is a vigorous plant with a root system capable of penetrating through the hardpans that develop from traffic by agricultural machinery in some soils. Soybean-bahiagrass rotations are most suitable to deal with severe polyspecific infestations of root-knot and cyst nematodes. Soybean yields in such fields are improved by as much as 80-90% when the bahiagrass rotation is implemented [8]. The degree of yield response to the rotation depends on soybean cultivar; however, this dependency is not as important the first year after bahiagrass as it is for later years. The first year after bahiagrass, nematode susceptible soybean cultivars with high yield potential performed as well as or better than resistant (or tolerant) cultivars. In the second and third years after bahiagrass, soybean cultivars resistant to root-knot and cyst nematode demonstrated a clear advantage in yield over high-yielding susceptible cultivars [10]. For both RKN and SCN, the greatest yield benefit from the soybean-bahiagrass rotation and other successful rotations is realized during the first year in soybean. Yield responses to the rotation crop decline quickly so that by the third or fourth year the effect of the rotation crop on yield is negligible [6,10]. Population dynamics of RKN follow exponential models [11]. Thus RKN can regenerate populations to severe infestation levels even when very few specimens survive adverse conditions posed by the absence of suitable hosts for one or more years.

Available work clearly shows that disease and nematode problems in soybean can be managed by rotations. However, the use of rotations is limited by the economic constraints faced by each producer. These constraints are many and as varied as there are producers, so that a rotation suitable for a producer with cattle operations, for example, is not economically feasible for those without such operations. Thus new profitable rotation crops are needed so producers can deal with nematode and other soilborne problems in soybean. In subtropical and temperate regions, tropical plant species can be imported and used as rotation crops with soybean. Some of these plants, e.g., castorbean (*Ricinus communis* L.), sesame (*Sesamum indicum* L.), and velvetbean (*Mucuna deeringiana* L.), are nonhost for many soybean nematodes and produce root exudates that are nematicidal or nematostatic. Rotations of soybean with other legumes can be effective for the management of nematode problems. Soybean following one year of hairy indigo (*Indigofora hirsuta* L.) or american jointvetch (*Aeschynomene americana* L.) resulted in increased soybean yields in a field infested with *M. arenaria* and *H. glycines* [12]. Hairy indigo and american jointvetch are warm weather legumes which produce in excess of 10 Mg hectare^{-1} of green manure and can be used as cattle feed. These rotations would be particularly useful if they were combined with corn or other N requiring crops to take advantage of the positive effect of the legumes on soil fertility.

Tillage

Burial of crop debris by tillage has long been recognized as an effective disease management method for pathogens that survive on crop debris. This practice can be especially important in situations where adequate genetic resistance to a disease is unavailable, or economic restraints preclude the use of rotations or chemical applications. When southern stem canker, caused by *Diaporthe phaseolorum* (Cke. & Ell.) Sacc. f. sp. *meridionalis*, became a problem in the southeastern U. S. in the 1980s, few highly resistant cultivars were available. It was possible to grow many moderately resistant and moderately susceptible cultivars with the aid of tillage for burial of diseased crop debris [13]. Another disease that can be managed by tillage is pod and stem blight, caused by a complex of *Diaporthe phaseolorum* var. *sojae* (Lehman) Wehm. and *Phomopsis longicolla* Hobbs. Genetic resistance to this complex is available, but has not been incorporated into cultivars. The primary source of inoculum for this disease complex is crop debris [14], suggesting either tillage or crop rotation as good disease management practices. Effects of tillage on nematodes are not well documented; however, there is evidence that reduced tillage may suppress the buildup of SCN [15].

GENETIC RESISTANCE

In addition to economic and environmental issues, genetic resistance for disease management in soybean has other advantages. Economic concerns go beyond just the added grower production costs of chemical applications. The cost of supporting public and private breeding programs is extremely low compared to the potential increase in profit that can be realized from the utilization of genetic resistance. The cost of a genetically resistant cultivar is usually no greater than the cost of susceptible cultivar seed, essentially resulting in a "no cost" control option. The grower management expertise required is usually no greater than being able to identify the disease problem and selecting the proper cultivar. Certain types of genetic resistance in soybean are very stable and have been used for years with no loss of effectiveness.

Types of Genetic Resistance

Types of genetic resistance used in soybean disease management strategies range from single-gene (vertical, qualitative) resistance, to polygene (horizontal, quantitative) resistance, to tolerance. Single-gene resistance is by far the most important and has been used to manage a wide array of economically important soybean diseases. In many cases, single-gene resistance may be the only practical management option available to a grower. Horizontal resistance and tolerance are used to a much lesser extent.

One of the best examples of race-specific, single-gene resistance is the series of alleles and genes controlling resistance to Phytophthora rot, caused by *Phytophthora megasperma* Drechs. f. sp.

glycinea Kuan & Erwin [1]. At least 25 physiologic races of the pathogen have been identified [1]. A survey of soybean cultivars developed and released by public institutions in the U. S. since 1980 reveals that 69 out of 134 cultivars have at least one major gene for resistance to Phytophthora rot, most at the rps_1 locus. Seven cultivars have two genes. Cultivars that do not have specific resistance genes were either released for production in the southeastern U. S. (where Phytophthora rot is not a problem) or are specialty cultivars for which genetic resistance to disease is not a breeding objective. The gene Rps_1^k [16] is currently the most popular resistance gene used in programs where *Phytophthora* resistance is a major objective. Rps_1^k confers resistance to almost all important physiologic races of *Phytophthora*. Although single genes for resistance to *Phytophthora* have been identified at six loci, with multiple alleles at the rps_1 and rps_3 loci, the only genes present in commercially available cultivars are Rps_3 [17], Rps_6 [18], and genes of the allelomorphic series at the rps_1 locus. Still other genes have been identified and are in the process of being described in the literature. Cultivars with two genes for *Phytophthora* resistance have the combination of Rps_3 with either Rps_1^b or Rps_1^c. One cultivar, Archer, has the combination of Rps_1^k and Rps_6. Pyramiding resistance genes is not done often for *Phytophthora* for three reasons: (1) the most widely-used genes are part of an allelomorphic series at the rps_1 locus and thus cannot be combined into a single homozygous genotype, (2) certain genes at the rps_1 locus give resistance to many of the important physiologic races, lessening the possible advantages of pyramiding, and (3) plant breeders fear the development of "super races" of the pathogen. The backcross method is by far the preferred method of introducing these genes into adapted cultivars. Resistance genes usually are backcrossed from the source, which may be a plant introduction or obsolete cultivar, into a widely-adapted cultivar. The new backcross-derived cultivar is then used as a source for backcrossing the gene into other cultivars, or is entered into biparental crosses to generate populations of homozygous lines from which resistant progenies are selected.

Single-gene resistance has been especially important in management of SCN. "Single-gene" may be a misnomer in this case, because up to 10 gene pairs with varying levels of dominance may be required to confer SCN resistance to more than one SCN race. Also, resistance or susceptibility is based on an arbitrary level of nematode reproduction (10%) relative to a susceptible check. Thus a line with 9% nematode reproduction relative to the check is considered resistant, while a line with 11% nematode reproduction relative to the check is considered susceptible. The recessive genes rhg_1, rhg_2, rhg_3, [19] and the dominant gene Rhg_4 [20] must all be present for resistance to the economically important race 3 of SCN. Since these genes were first identified, 16 races of SCN have been described [21], indicating a large degree of genetic diversity in the pathogen. Several resistance sources have been identified, including PI 437654 which has resistance to all economically important races [22]. Practical experience indicates that SCN populations change (new races develop) in response to the continuous culture of a resistant cultivar. When genes conferring resistance to the new races are identified, these genes are routinely pyramided into existing resistant cultivars, usually with a modified backcross procedure. Races 3, 4, 5, and 14 are the most economically important in the U. S. Only one cultivar, Hartwig, has been developed that has resistance from PI 437654, and is resistant to all economically important races.

Of course, the development of new resistance-breaking physiologic races is a problem common to almost any situation where single-gene, race-specific resistance is used. However, some single genes in soybean have been used to control certain diseases for years without a breakdown in resistance. A single gene used to control frogeye leaf spot (caused by *Cercospora sojina* Hara) in some soybean cultivars, Rcs_3 (sometimes referred to as the Davis gene, after the cultivar in which it was first identified) [23], has been shown to be effective against all known races of frogeye leaf spot, including the many isolates described in Brazil [24]. This gene gives a phenotype similar to horizontal-type resistance genes, in that a small degree of disease development often occurs when plants are artificially inoculated (sometimes referred to as a "flecking" response) [23]. This characteristic is not associated with other described genes for frogeye leaf spot resistance, which are very race-specific. Nevertheless, Rcs_3 currently is found in only a limited number of U. S. cultivars. Other genes for resistance to frogeye leafspot exist and occur with high frequency in adapted germplasm, particularly in Maturity Groups 5 and higher. Of 110 public and private U. S. cultivars in Maturity Groups 4, 5, 6, 7, and 8, 68% were rated resistant when artificially inoculated with race 5 of *C. sojina* [25]. Many have genes for which inheritance and relationships to the described genes for resistance to frogeye leaf spot have not been determined. Some of these cultivars, although resistant to race 5, have been observed to be highly susceptible in the field, indicating that other races are important.

Another example of single-gene resistance that has been very stable is the gene *rpx* [26], which confers resistance to bacterial pustule, caused by *Xanthomonas campestris* pv. *glycines* (Nakano) Dye. This gene currently is present in almost all southern U. S. cultivars (maturity groups 5 and higher) and many northern cultivars, and has been used for over 30 years without a breakdown in resistance. Other diseases (and their respective resistance genes) that are primarily controlled by single genes include soybean rust, caused by *Phakopsora pachyrhizi* Syd., (Rpp_1, Rpp_2, and Rpp_3, either of which confers resistance) [27]; bacterial blight, caused by *Pseudomonas syringae* pv. *glycinea* (Coerper) Young, Dye & Wilkie (Rpg_1) [28]; brown stem rot, caused by *Phialophora gregata* (Allington and Chamberlain) W. Gams, (Rbs_1) [29]; and southern stem canker, (Rdc_1 and Rdc_2, either of which confers resistance) [30].

Horizontal resistance has been used to a lesser extent for disease management in soybean. The terms "field resistance", "rate-reducing resistance", "field tolerance", or "tolerance" have been used to describe some soybean cultivars that do not have specific genes for resistance to hypocotyl inoculation with Phytophthora rot, but suffer less yield loss than other susceptible cultivars when planted in infested fields. However, this type of resistance is not as effective in reducing losses to *Phytophthora* as race-specific resistance [31]. Heritability estimates for field resistance to *Phytophthora* are high, ranging from 79 to 89% on an entry-mean basis when evaluated under controlled conditions in the greenhouse [32].

Horizontal resistance is the only form of resistance to the major species of rootknot nematode (*Meloidogyne* spp. Goeldi). Recent evidence indicates a fairly small number of genes confer resistance [33], but inheritance studies are difficult to conduct because of the large influence of the environment. Although horizontal, resistance to RKN is species-specific. Apparently, a different set of genes is needed to confer resistance to *M. incognita* (Kofoid and White) Chitwood than is needed for resistance to *M. arenaria* (Neal) Chitwood [34]. Cultivars have been developed, however, that have resistance to these two species plus *M. javanica* (Treub) Chitwood by combining different resistance sources. One plant introduction, PI 417125, has been found to be resistant to all three species, but whether resistance is controlled by the same genes for all three species is not known [35].

Field resistance also may be important in controlling southern stem canker. Although single genes have been found that confer resistance [30], a wide range of disease reaction phenotypes have been observed [36], indicating the presence of modifying genes.

Tolerance has been found to be significant in soybean only for nematode pests. In this case, a tolerant genotype is susceptible, allowing high nematode reproduction, but suffers less yield loss or suppression than intolerant genotypes. Tolerance to SCN [37] and Columbia lance nematode (*Hoplolaimus columbus* Sher) [38] has been reported, but only for the characterization of current cultivars for relative tolerance or intolerance. Tolerance also has been suggested as at least a partial mechanism for RKN resistance [11]. Although tolerance to nematodes has been found to exist in soybean cultivars, it is not generally a breeding objective.

Management of genetic resistance in the field

Genetic resistance, particularly individual vertical resistances, cannot be depended upon to manage a disease indefinitely. Generally, some combination of genetic resistance and other management practices, such as rotation with a nonhost crop, will give the best results. Although some studies have shown that long-term use of resistant cultivars for SCN management will lead to the development of resistance-breaking biotypes [6], other studies have not. Hartwig *et al.* [39] found that no treatment, including nematicides, rotation of resistant and susceptible cultivars, and rotation with a grain sorghum, was superior to the continuous culture of race 3 resistant Centennial over a period of 10 years. In practical terms, sound management systems combine the use of cultivars resistant or tolerant to nematodes and rotations that suppress nematode and other soilborne problems.

221

REFERENCES

1. Sinclair, J.B. and Backman, P.A. (eds.), Compendium of Soybean Diseases. APS Press, St. Paul, Minn., 1989.

2. Weaver, D.B., Rodríguez-Kábana, R. and Carden, E.L., Multiple-species nematode resistance in soybean: effect of genotype and fumigation on yield and nematode numbers. Crop Sci., 1988, 28, pp. 293-8.

3. Johnson, A. W. and Feldmesser, J., Nematicides - a historical review. In Vistas on Nematology, eds. J.A. Veech and D.W. Dickson, Society of Nematologists, Inc., Hyattsville, Md., 1987, pp. 448-54.

4. Rodríguez-Kábana, R., Robertson, D.G., King, P.S. and Weaver, C.F., Evaluation of nematicides for control of root-knot and cyst nematodes on a tolerant soybean cultivar. Nematrópica, 1987, 17, pp. 61-70.

5. Kinloch, R.A., Soybean and maize cropping models for management of *Meloidogyne incognita* in the Coastal Plain. J. Nematol., 1986, 18, pp. 451-8.

6. Slack, D. A., Riggs, R. D., and Hamblen, M. L., Nematode control in soybeans: Rotation and population dynamics of soybean cyst and other nematodes. Ark. Agr. Exp. Stn. Rep. Ser., 1981, 263, pp. 1-36.

7. Rodríguez-Kábana, R., Weaver, D.B., Robertson, D.G., King, P.S. and Carden, E.L., Sorghum in rotation with soybean for the management of cyst and root-knot nematodes. Nematrópica, 1990, 20, pp. 111-9.

8. Rodríguez-Kábana, R., Weaver, D.B., Garcia, R., Robertson, D.G. and Carden, E.L., Bahiagrass for the management of root-knot and cyst nematodes in soybean. Nematrópica, 1989, 19, 185-93.

9. Windham, G.L. and Williams, W.P., Resistance of maize inbreds to *Meloidogyne incognita* and *M. arenaria*. Plant Dis., 1988, 72, pp. 67-9.

10. Rodríguez-Kábana, R., Weaver, D.B., Robertson, D.G., Carden, E.L. and Pegues, M.L., Additional studies on the use of bahiagrass for the management of root-knot and cyst nematodes in soybean. Nematrópica, 1991, 21, (in press).

11. Rodríguez-Kábana, R. and Weaver, D.B., Soybean cultivars and development of populations of *Meloidogyne incognita* in soil. Nematrópica, 1984, 14, pp. 46-56.

12. Rodríguez-Kábana, R., Weaver, D.B., Robertson, D.G., Young, R.W. and Carden, E.L., Rotations of soybean with two tropical legumes for the management of nematode problems. Nematrópica, 1990, 20, pp. 101-10.

13. Rothrock, C. S., Phillips, D. V., and Hobbs, T. W., Effects of cultivar, tillage, and cropping system on infection of soybean by *Diaporthe phaseolorum* var. *caulivora* and southern stem canker symptom development. Phytopathology, 1988, 78, pp. 266-70.

14. Garzonio, D. M., McGee, D. C., Comparison of seeds and crop residues as sources of inoculum for pod and stem blight of soybeans. Plant Dis., 1983, **67**, pp. 1374-6.

15. Edwards, J. H., Thurlow, D. L., and Eason, J. T., Influence of tillage and crop rotation on yields of corn, soybean, and wheat. Agron. J., 1988, **80**, pp. 76-80.

16. Bernard, R.L. and Creemans, C.R., An allele at the rps_1 locus from the variety 'Kingwa'. Soybean Genet. Newsl., 1981, **8**, pp. 40-2.

17. Mueller, E.H., Athow, K.L. and Laviolette, F.A., Inheritance of resistance to four physiologic races of *Phytophthora megasperma* var. *sojae*. Phytopathology, 1978, **68**, pp. 1318-22.

18. Athow, K.L. and Laviolette, F.A., Rps_6, a major gene for resistance to *Phytophthora megasperma* f. sp. *glycinea* in soybean. Phytopathology, 1982, **72**, pp. 1564-67.

19. Caldwell, B.E., Brim, C.A. and Ross, J.P., Inheritance of resistance of soybeans to cyst nematode, *Heterodera glycines*. Agron. J., 1960, **52**, pp. 635-6.

20. Matson, A.L. and Williams, L.F., Evidence of a fourth gene for resistance to the soybean cyst nematode. Crop Sci., 1965, **5**, p. 477.

21. Riggs, R. D. and Schmitt, D. P., Complete characterization of the race scheme for *Heterodera glycines*. J. of Nematol., 1988, **20**, pp. 392-5.

22. Anand, S.C., Gallo, K.M., Baker, I.A. and Hartwig, E.E., Soybean plant introductions with resistance to races 4 or 5 of soybean cyst nematode. Crop Sci., 1988, **28**, pp. 563-64.

23. Boerma, H.R. and Phillips, D.V., Genetic implications of the susceptibility of Kent soybeans to *Cercospora sojina*. Phytopathology, 1983, **74**, pp. 1666-8.

24. Yorinori, J.T., Frogeye leafspot of soybean. In Proceedings, World Soybean Research Conf. IV, ed. A. J. Pascale, 1989, pp. 1275-83.

25. Pace, P.F., Ploper, L.D. and Weaver, D.B., Evaluation of soybean cultivars for reaction to frogeye leaf spot race 5. Biol. and Cult. Tests for Cont. of Plant Dis., 1992, **9**, (in press).

26. Hartwig, E.E. and Lehman, S.G., Inheritance of resistance to the bacterial pustule disease in soybeans. Agron. J., 1951, **43**, pp. 226-9.

27. Hartwig, E.E. and Bromfield, K.R., Relationships among three genes conferring specific resistance to rust in soybeans. Crop Sci., 1983, **23**, pp. 237-9.

28. Mukherjee, D., Lambert, J.W., Cooper, R.L. and Kennedy, B.W., Inheritance of resistance to bacterial blight (*Pseudomonas glycinea* Coerper) in soybeans (*Glycine max* L.). Crop Sci., 1966, **6**, pp. 324-6.

29. Willmot, D.B. and Nickell, C.D., Genetic analysis of brown stem rot resistance in soybean. Crop Sci., 1989, **29**, pp. 672-4.

30. Kilen, T.C. and Hartwig, E.E., Identification of single genes controlling resistance to stem canker in soybean. Crop Sci., 1987, 27, pp. 863-4.

31. Anderson, T.R., Plant losses and yield responses to monoculture of soybean cultivars susceptible, tolerant, and resistant to *Phytophthora megasperma* f. sp. *glycinea*. Plant Dis., 1986, 70, pp. 468-71.

32. Walker, A.K. and Schmitthenner, A.F., Heritability of tolerance to Phytophthora rot in soybean. Crop Sci., 1984, 24, pp. 490-1.

33. Luzzi, B.M., Boerma, H.R. and Hussey, R.S., Inheritance of resistance to the southern root-knot nematode in soybean. Agron. Abstr., 1990, p. 99.

34. Weaver, D.B., Breeding cultivars resistant to root-knot nematodes. In Proceedings, World Soybean Research Conf. IV, ed. A. J. Pascale, Buenos Aires, Argentina, 1989, pp. 1161-6.

35. Luzzi, B.M., Boerma, H.R. and Hussey, R.S., Resistance to three species of root-knot nematode in soybean. Crop Sci., 1987, 27, pp. 258-62.

36. Weaver, D.B., Cosper, B.H., Backman, P.A. and Crawford, M.A., Cultivar resistance to field infestations of soybean stem canker. Plant Dis., 1984, 68, pp. 877-9.

37. Boerma, H.R. and Hussey, R.S., Tolerance to *Heterodera glycines* in soybean. J. Nematol., 1984, 16, pp. 289-96.

38. Nyczepir, A.P. and Lewis, S.A., Relative tolerance of selected soybean cultivars to *Hoplolaimus columbus* and possible effects of soil temperature. J. Nematol., 1979, 11, pp. 27-31.

39. Hartwig, E.E., Young, L.D. and Buehring, N., Effects of monocropping resistant and susceptible soybean cultivars on cyst nematode infested soil. Crop Sci., 1987, 27, pp. 576-79.

DISEASE MANAGEMENT IN SOYBEAN: USE OF CHEMICAL CONTROL AND APPLICATION TECHNOLOGY

J. P. Snow and G. T. Berggren, Department of Plant Pathology and Crop Physiology, Louisiana Agricultural Experiment Station, Louisiana State University Agricultural Center, Baton Rouge, Louisiana 70803, USA.

ABSTRACT

Although soybean is parasitized by fungal, viral and bacterial pathogens, only those diseases caused by fungi are controlled by chemical means to any large degree. The economic importance of the disease, the value of the crop in a given year, the price and effectiveness of the fungicide, and the comparative effectiveness of other control measures such as cultural practices combine to determine whether fungicides will be used to control a particular disease. The need for fungicides is determined partially by geographical location since many of the more destructive fungal diseases are prevalent in areas of the world where rainfall is high and temperatures are warm during the growing season. In addition, the type of disease will determine, to a limited extent, whether or not a fungicide will be effective in controlling a disease. Foliar diseases are, in general, easier to control with fungicides than stem and root diseases, because of ease of delivering the fungicide to the appropriate location. Chemical control of soybean diseases includes control of seedling diseases with seed treatments or soil-applied chemicals, control of root and stem diseases and control of foliar and pod diseases. Technological advances involved in delivering the chemicals and improvement of predictive systems have combined to make chemical control of soybean diseases a viable, effective part of disease management programs.

INTRODUCTION

Soybeans grown throughout the world are attacked by a wide array of diseases including those that affect the seed, seedling, root, stem, foliage, flower and pod. In the southern United States of America, a 10.4% yield loss from all diseases was reported in 1990 (1).

For more than 200 years, seedling diseases have been controlled by coating seed with chemicals (2). Copper sulfate treatments were developed in the early 1800's followed by formaldehyde and copper carbonate in the late 1800's and early 1900's. Organic mercury treatments became available in 1913, but were discontinued in the 1970's due to concerns about wildlife and environmental contamination. Until the introduction of the first systemic fungicides, the oxathiins, in 1966 (3), treatments were effective only against pathogens located in or on superficial tissues. The systemic chemicals offered control of pathogens located deep in seed tissues for the first time. The benzimidazole (4,5,6,7) fungicides were introduced in the late 1960's providing effective control of many pod, leaf and stem diseases. Before these fungicides became widely available, control of pod, leaf and stem diseases relied on the use of cultural practices and the availability of resistant cultivars. With the discovery of the acylalanine fungicides, control of Phytophthora and Pythium became possible (5,8).

Although chemicals such as benzimidazoles are very effective, research has shown that repeated use of these chemicals can lead to a decrease in efficacy due to development of pathogen resistance (9). Fungicides which are characterized by single-site activity develop pathogen resistance more rapidly than those with multiple-site activity. However, the multiple-site compounds may offer decreased uptake and more rapid metabolic detoxification.

A thorough understanding of the epidemiology of each disease is essential to the efficient and effective use of chemicals (4). Timing of fungicide application to take advantage of pathogen vulnerability or to prevent sporulation or infection can reduce the number of applications needed to control a disease. Computer-based systems designed to predict disease loss based on environmental factors such as rain events, aid in deriving cost/benefit ratios for chemical use (10).

Seed, Seedling and Root Diseases

Soybean seed and young plants are attacked by a wide array of seedborne and soilborne pathogens that cause seed decay or pre-and post-emergence diseases (11). Poor stands and the production of plants with low vigor and poor performance in the field may result from infection by these organisms especially during cold, wet planting seasons. Losses from these diseases may vary with soil type as well as weather. Under proper environmental conditions, these pathogens may also cause root and stem problems on mature plants. While seedling diseases may vary dramatically in local areas from one year to the next, over an eight-year period (1983-1990) in the southern United States of America, losses varied from a low of 0.8% in 1986 (12) to a high of 1.1% in 1983 (13).

Soybean seedling diseases caused by species of Rhizoctonia, Fusarium and Phomopsis have been effectively controlled (11,14,15) with fungicides applied directly to the seed as dusts, liquids or thick water suspensions. Seed treatment fungicides that can be applied to the seed include thiram, captan, pentachloronitrobenzene, carboxin and etridiazole. Most of the seed treatments act as protectants, however, carboxin will act systemically providing fungicidal activity to deeper seed tissues and to the young seedling. In some cases, these chemicals are applied as liquids or granules to the soil around the seed to provide control over a longer period of time than seed treatments. However, the chemicals applied directly to the seed are easier to apply and offer the advantage of lower cost. Control of Pythium and Phytophthora requires the use of metalaxyl applied as a seed or soil treatment. Seed treatment chemicals also may be used to control diseases other than seedling diseases. For example, downy mildew can be reduced with seed treatment fungicides (11).

The most common method used in the application of fungicides to soybean seed is by hopper-box application at planting. Materials are applied directly to the seed in the planter box in either a dry powder or liquid formulation. The use of hopper-box applied liquids greatly increased in the 1980's as formulations of these materials were developed which were not only effective, but were also relatively easy for the producer to apply. In general, the acceptance of a practice such as hopper-box treatment at planting, is related to the ease of application and cost. The use of hopper-box applied dry powders is relatively simple and inexpensive, but the effectiveness of the material in coating and adhering to the seed surface is limited. Oil or aqueous-based liquid seed treatments are much more effective, while they are proportionally more expensive to use, and slightly more time consuming to apply.

In certain areas of the United States of America, soybean seed treated with a selected fungicide can be obtained for planting. This treatment is performed by commercial seed producers or contracted handlers. The advantages of pretreated seed include less time spent in the planting process, as hopper-box treatment is eliminated, and uniform coverage of seed with the protectant. The major disadvantages of this method include the cost of commercial application of the fungicide and the non-marketability of pretreated seed for use other than planting.

While application of granular fungicides to the soil in the area where the seed is to be planted has been successful in the control of seedborne diseases in many crops, particularly cotton, this technique has not been practiced on a wide scale in soybean production. The lack of effective registered materials and unfavorable economics have limited the acceptance of this practice.

Leaf, Pod and Stem Diseases

The severity of leaf, pod and stem diseases may vary considerably from year to year. For example, the average loss from stem canker in the southern United States of America was 0.03% in 1985 (16) and 2.0% in 1989 (17). Pod and stem blight losses in the same area averaged 0.5% in 1990 (1) and 2.5% in 1986 (12). Although foliar, pod and stem diseases were known for many years, their importance as yield-limiting factors in soybean production was reported in 1975 (18). In these early studies using benomyl, triphenyltin hydroxide and thiabendazole, yield increases were attributed primarily to an increase in seed weight rather than seed number. Plants were inoculated with Corynespora cassiicola (target spot), Diaporthe phaseolorum var. sojae (pod and stem blight) and Cercospora sojina (frogeye leaf spot) or with each of the fungi individually. Although yields were not decreased by inoculating with individual fungi, yields of plants receiving combination inoculations were significantly decreased. Yield reduction due to the brown spot disease was demonstrated in 1980 (19). In these studies, yield increases from benomyl treatment were attributed to an increase in seed weight. Effect of stem canker on soybean yields was reported in 1986 (20).

Protectant fungicides offer a rather limited effect on invading pathogens by coating the leaf surface and preventing the organism from germinating and/or infecting the plant. Complete coverage with these materials is important since any unprotected leaf area is subject to infection. Removal of the protectants by dew or rainfall may require additional applications to achieve the same degree of continuous control as systemic fungicides. Once infection has occurred, the protectant fungicides will no longer be effective in retarding the progress of the pathogen. On the other hand, the systemic fungicides become distributed throughout the plant. Although a pathogen may penetrate the external plant barriers, the systemic fungicide will limit further disease development. Unfortunately, however, systemic fungicides are also more expensive than protectants.

The use of foliarly-applied fungicides for the control of leaf, pod and stem diseases reached a peak (in percent of crop treated) in the United States of America in the early 1980's. The fungicides most commonly applied include benomyl, thiabendazole, thiophanate-methyl and chlorothalonil (21). The economics of fungicide application has been of major concern in the past decade, as soybean prices have been depressed, causing producers to reduce inputs into the crop whenever possible. Research in the area of fungicide application indicates that timing of application, rate of material, and environmental factors are of critical importance in the control of soybean diseases (18,22,23,24). Application of fungicides from the R_2-R_6 growth stages (25) has been shown to be most effective in limiting yield losses (11). Other factors critical to control of soybean foliar, pod and stem diseases include the rate of application, the site or sites of application and the volume of carrier. Application in commercial production of soybean is commonly by the use of fixed wing aircraft. While aerial application of fungicides delivers only 60-80% of the spray to the upper half of the canopy, it is generally the favored technique due to the speed of treating large areas versus the use of ground-driven equipment. Additionally, aerial application is less destructive to the dense canopy of the soybean crop.

Ground-driven equipment can be more effective in smaller, irregular-shaped fields having row widths greater than 90 cm (7).

The most important aspect of fungicide application to the crop is deposition of the material. Deposition is dependent on several factors including carrier volume, droplet size, and physical drift. Systemic fungicides require ± 20 droplets per square centimeter, and protective (or contact) fungicides ± 70 droplets per square centimeter of plant surface (7). Physical drift is greatest when precautions are not taken to regulate droplet size (26). Jacobsen et al. (27) concluded, "to minimize drift losses, it is important to minimize the percentage of spray volume contained in droplets less than or equal to 100μm". Additionally, the authors suggested that a hollow-cone nozzle with a volume mean diameter of 350μm will have only about 2% of the output volume in droplets less than 100μm. Subsequently, 98% of the spray will be deposited in the target area. Aerial application of fungicides to soybean is generally recommended at the rate of 46.8 1/ha (18,28). Control of diseases by ground-applied fungicides is more effectively achieved with a higher volume of carrier (187 1/ha (29).

Fungicide applications for the control of several foliar, pod and stem diseases can be effectively achieved by either ground or aerial application of fungicides. The most critical factor of either method is deposition of the material. Control of seed and seedborne diseases is most effectively obtained through the use of either hopper-box (liquid or dust) seed treatment or seed pretreated with fungicide by the seed supplier.

REFERENCES

(1) Sciumbato, G.L., Southern United States soybean disease loss esimates for 1990. Proc. Southern Soybean Disease Workers, 1991, 18, 32-8.
(2) Roberts, D.A. and Boothroyd, C.W. Fundamentals of Plant Pathology. W.H. Freedman and Co., New York, 1984, 402 pp.
(3) Sharville, E.C., Chemical Control of Plant Diseases, Prestige Press, Fort Worth, 1969, 340 pp.
(4) Backman, P.A. and Jacobsen, B.J. Control of foliage, stem and pod diseases with fungicides. In Pascale, A.J., ed. Proc. World Soybean Research Conf. IV, Buenos Aires, Argentina, 1989, pp. 2091-96.
(5) Edgington, L.V., Martin, R.A., Bruin, G.C., and Parsons, I.M. Systemic fungicides: a perspective after 10 years. Plant Disease, 1980, 64, 19-23.
(6) Nesmith, W.C., Changes in fungicide use patterns. Plant Disease, 68, 834-35.
(7) Jacobsen, B.J., and Backman, P.A.. Application considerations. In Pascale, A.J., ed. Proc. World Soybean Research Conf. IV, Buenos Aires, Argentina, 1989. pp. 1368-72.
(8) Schmitthenner, A.F., Phytophthora root rot: detection ecology, and control. In Pascale, A.J., ed. Proc. World Soybean Research Conf. IV, Buenos Aires, Argentina, 1989, pp. 1396-61.

(9)   Staub, T. and Sozz, D.   Pesticide resistance: a continuing challenge. Plant Disease, 1984, 68, 1026-31.

(10) Backman, P.A., Crawford, M.A. and Hammond, J.M. Comparison of meterological and standardized timings of fungicide applications for soybean disease control. Plant Disease, 1984, 68, 44-6.

(11) Sinclair, J.B. and Backman, P.A. Eds. Compendium of Soybean Diseases, Third Edition, APS Press, St. Paul, MN. 1989, 106 pp.

(12) Mulrooney, R., Southern United States soybean disease loss estimates for 1986. Proc. Southern Soybean Disease Workers, 1987, 14, 39-43.

(13) Mulrooney, R., Southern United States soybean disease loss esimates for 1983. Proc. Southern Disease Workers, 1984, 11, 2-9.

(14) Colyer, P.D., ed. Soybean Disease Atlas. Shreveport Associated Printing Professionals, 1984, 43 pp.

(15) Scott, D.H., Soybean disease management in the North Central region of the United States. In Pascale, A.J., ed. Proc. World Soybean Research Conf. IV, Buenos Aires, Argentina, 1989, pp. 1284-89.

(16) Mulrooney, R., Southern United States soybean disease loss esimates for 1985. Proc. Southern Soybean Disease Workers, 1986, 13, 9-13.

(17) Sciumbato, G.L., Southern United States soybean disease loss esimates for 1989. Proc. Southern Soybean Disease Workers, 1990, 17, 11-14.

(18) Horn, N.L., Lee, F.N. and Carver, R.B.   Effects of fungicides and pathogens on yields of soybeans. Plant Disease Reptr., 1975, 59, 724-28.

(19) Lim, S.M., Brown spot severity and yield reduction in soybean. Phytopathology, 1980, 70, 974-77.

(20) Harville, B.G., Berggren, G.T., Snow, J.P., and Whitman, H.K. Yield reductions caused by stem canker in soybeans. Crop Science, 1986, 26, 613-16.

(21) Berggren, G.T., McGawley, E.C., Marshall, J.G., Pace, M.E., Gershey, J.S., Horn, N.L., Snow, J.P., Freedman, J.A., Winchell, K.A., and Joye, G.F. Foliar fungicides for soybean production in Louisiana. Louisiana Agriculture, 1985, 28, 21-2.

(22) Backman, P.A., Rodriguez-Kabana, R., Hammond, J.M., and Thurlow, D.L. Cultivar, environment, and fungicide effects of foliar disease losses in soybean. Phytopathology, 1979, 69, 562-64.

(23) Berggren, G.T., McGawley, E.C., Pace, M.E., Gersehy, J.S., and Joye, G.F. Strategies for controlling soybean aerial blight in Louisiana, abstr., Phytopathology, 1984, 74, 872.

(24) Jacobsen, B.J., Effect of foliar fungicides on soybean yield and seed quality. Proc. Ninth Soybean Seed Research Conf., pp. 49-55.

(25) Fehr, W.R., Caveness, C.E., Burmond, D.T., and Pennington, J.S. Stages of development descriptions for soybeans, Glycine max (L) Merrill Crop Science, 1971, 11, 929-31.

(26) Akesson, N.B. and Yates, W.E. The use of aircraft in agriculture. FAO Development. Paper #94. FAO, Rome, Italy, 1974, 217 pp.

(27) Jacobsen, B.J., Calibration and use of aircraft for applying fungicides. In Hickey, D.H., ed. Methods of Evaluating Pesticides for Control of Plant Pathogens, APS Press, St. Paul, MN, 1986, pp. 39-44.

(28) Whitam, H.K., Hollier, C.A. and Overstreet, C. Louisiana Plant Disease Control Guide. Louisiana Cooperative Extension Service, 1991, 180 pp.

(29) Horn, N.L., Carver, R.B. and Fort, T.M. Annual report to the
Soybean Promotion Board. Dept. of Plant Pathology, Louisiana
Agricultural Experiment Station, 1976, 25pp.

FUNGICIDES AS PART OF AN INTEGRATED SYSTEM IN THE MANAGEMENT OF SOYBEAN DISEASES

H. V. MORTON
CIBA-GEIGY Corporation, Post Office Box 18300, Greensboro, NC 27419

ABSTRACT

Disease control in soybeans is achieved primarily through cultural practices and genetic resistance. Fungicides are used as seed-treatments, in-furrow at planting, or as foliar treatments aimed at specific diseases to increase plant stands, yields, and quality of the harvested beans. The opportunities for integrating fungicides as a means of controlling soybean diseases is discussed.

INTRODUCTION

To understand the role fungicides play as a part of the management of soybean disease, it is important to recognize that a specific soybean pathogen may be very destructive one season and difficult or impossible to find the next season. More than 100 pathogens are known to affect soybeans, of these about 35 are economically important [1]. The current soybean acreage planted annually in the U.S. is approximately 60 million, placing soybeans as the third largest agronomic crop behind corn and wheat.

Selection of soybean cultivars genetically resistant to disease is regarded as the best means of crop protection; generally, this method is the most economical and efficient. Yet annually, the estimated percent loss of soybean yields has ranged between 10 and 33 since 1974 in the southern states [2]. In 1987 Ploper [3] estimated the losses in Indiana to be 14.9% (fungi 9.8%, bacteria 0.1%, viruses 0.2%, and nematodes 4.8%). The importance of these losses are recognized by the American soybean farmers. In a recent survey of the American Soybean Association's 34,000 members, those responding placed the two most important research needs to increase yields as breeding for disease control and weed control.

In order to address the role of fungicides to improve disease control, this paper will be divided into three parts, viz: seed treatments, soil pathogens, and foliar pathogens.

SEED TREATMENTS

Plant disease control in soybeans is achieved almost entirely through cultural practices and genetic resistance. The most commonly used fungicides have been seed treatments aimed at seed-borne diseases, particularly when planting in cool wet soils or when seed quality is low because of seed-borne disease(s). Generally, seed lots with greater than 80% germination will not benefit from seed treatments when planted at recommended seeding rates [4].

At least 26 fungal, 7 bacterial, and 7 viral diseases in soybeans are seed-borne. The most consistently cited fungi isolated from soybean seeds are Diaporthe/Phomopsis spp., Cercospora kikuchii, Alternaria spp., and Fusarium spp. [3], Wall et. al. [5] showed that in Indiana a fungicide seed treatment improved emergence of seed when more than a 15% Phomopsis infestation was present. Carboxin-thiram mixtures and Captan are registered as seed treatments, with the former being the product of choice (see Table I).

TABLE 1
Seed Treatments for Control of Soybean Seed-Borne Diseases

		Per Cent Disease*		
Treatment	Rate Per cwt.	Phomopsis	Fusarium	Alternaria
Captan 400	2 fl. oz.	34.	17.	8
Vitavax 200	4	3	21	3
Untreated Check		53	17	11

*Average across three soybean varieties.

Cultural practices which can reduce the incidence of seed-borne fungi include:
1) Crop rotation, 2) Planting date, 3) Tillage, and 4) Cultivar selection.

Split or single applications of foliar fungicides have also been effective in reducing, but not eliminating, the incidence of seed-borne pathogens. Foliar fungicides have been used on a limited basis because they must be applied before disease symptoms are visible and before it is known if economically significant levels of seed infection will occur. The incidence of seed-borne pathogens is dependent on the proximity of inoculum and field environment during seed maturation.

Selection of soybean cultivars genetically resistant to these diseases could be one of the best means of crop protection; generally this method is the most economical and efficient. Identification of sources of resistance has been limited because of the strong influence that the time of maturity and particularly environmental conditions between the yellow pod and mature pod stages have on seed quality [6].

Warm temperatures and high humidity during these final stages of seed maturation are known to enhance disease development, thus, cultivars that mature early in the season generally have a higher incidence of seed-borne fungi than do cultivars maturing later, (when temperatures and humidity are lower).

SOIL PATHOGENS

The National Academy of Sciences completed a pest-control assessment of present and alternative technologies in 1975 [7]. Volume II of this assessment addressed corn and soybeans, and included broad range recommendations. I would like to address two of these which CIBA-GEIGY has diligently pursued.

This study team, therefore, recommends that an increased public commitment be made to monitor pest populations in the field and to develop effective surveillance systems for detecting potential pest build-ups, new pests, and new races of current pests.

CIBA-GEIGY established an agreement with Agri-Diagnostics in 1986 to develop an immunologically based Phytophthora detection kit. One of the original goals of this project was to develop monoclonal antibodies to P. megasperma which could be used to assay infected plant tissues and infested soils. While the plant tissue assay was readily accomplished, it took a great deal more research to develop a soil assay. The major obstacle to this assay was oospore collection and processing; oospores had to be pulverized for the antibody to react using ELIZA (See Table 2).

TABLE 2
Comparison of a Phytophthora on Site Diagnostic Kit and Bioassay
for P. megasperma f. sp. glycinea

Assay	Advantages	Disadvantages
ELIZA Kit	Good prediction of a particular sample, within a relatively short amount of time.	· Due to pathogen distribution, difficult to get an accurate field assessment.
		· Not species, or race specific.
Bioassay	Effective for detection.	· Not quantitative.
		· Slow (2 weeks).
		· Interference from other pathogens.

After four years of research, our conclusion was that due to the limitations in soil sampling it was difficult to obtain a representative measure of Phytophthora from within a field. This, along with the fact that wet weather immediately following planting was equally important in developing disease (even from low levels of inoculum), led to the conclusion that the kit alone could not gauge the risk of disease development within a particular field.

The study team recommends that strategies for back-up systems of pest control in the corn/soybean sector be developed. This includes:

· Public and private attention to building inventories of chemicals that differ from those in common use.

· Legal and administrative procedures to move quickly against pest outbreaks with currently restricted products if necessary, and

· Comprehensive and diverse inventories of plant genetic materials from which new resistance can be developed as quickly as possible.

The NAS Committee reported [7] soybean varieties grown on most of the U.S. acreage trace to only six introductions from Manchuria, these six introductions are similar in type and are probably somewhat related. Clearly the U.S. has a rather narrow germplasm base represented in our commercial varieties, presenting a great risk, particularly when the main form of disease resistance is genetic resistance.

Of the 35 economically important soybean diseases, two of these, unless managed, can cause significant losses. Interestingly, both are soil-borne pathogens viz: Phytophthora megasperma f. sp. glycinea (Pmg.) and Heterodera glycines (cyst nematode). Table 3 shows the potential importance of soil-borne pathogens in causing losses [7].

Initially, the directions for use of metalaxyl called for a broadcast application at planting with rates up to 1.25 lbs. a.i./A to provide season long control of Pmg. Further research showed the performance of metalaxyl could be optimized by applying the product in-furrow at planting at low rates, (0.3-1.1 fl. oz. Ridomil 2E/linear foot of row)* along with planting cultivars tolerant to Pmg. The value of metalaxyl is it protects the seedlings prior to the onset of the tolerance mechanism, while also providing the crop the opportunity for a good vigorous start. Metalaxyl seed treatments last 2-3 weeks, while the in-furrow treatments afford 6 weeks protection. Genetic tolerance then takes over in mid-season and any root tip damage caused by Pmg at this stage is of no economic concern.

Each year in Ohio a series of performance trials are conducted with soybean varieties. From several years of research Schmitthenner [8] has designed the following set of recommendations for the use of metalaxyl (See Table 4).

*1 fl. oz. contains 7.1 gr. a.i.

TABLE 3
Infectious Soybean Diseases in the Corn Belt - Present and Potential
Importance in Causing Crop Losses

Seed and Soil Borne Diseases	Casual Agent	Potential Losses Should Present Control Methods Fail or Not Be Used			
		Local Areas	In Sector	Local Areas	In Sector
Seed Decay	Pythium and other fungi	L	VL	L	VL
Seedling Blight	Pythium and other fungi	L	VL	L	VL
Phytophthora Rot	Phytophthora megasperma f. sp. glycinea	L	VL	H	L
Rhizoctonia Root and Stem Rot	Rhizoctonia solani	L	VL	L	L
Cyst Nematode	Heterodera glycines	M	VL	H	M
Purple Seed Stain	Cercospora kikuchii	VL	VL	VL	VL
Storage Rot	Aspergillus and Penicillium spp.	L	VL	L	L

VL = Very Low (0-1%)
L = Low (2-10%)
M = Medium (11-35%)
H = High (36-50%)

NOTE: These estimates are based upon empirical data where available, and otherwise, on the general knowledge and judgment of several scientists in the field.

The integration of several single genes for resistance to Pmg into commercial soybean cultivars has led to the identification of 30 races of the pathogen. An effective strategy for preventing proliferation of Pmg races would be to grow highly tolerant cultivars and apply metalaxyl. Alternatively, breeders are left with pyramiding 2 of the 14 genes for Phytophthora resistance together as Athow did in the varieties Miami and Keller. The need for giving this further consideration becomes more relevant since the isolation of races that can defeat the Rps-1K gene for Phytophthora resistance .

When soybean plants are invaded by pathogens, the plants respond by producing a phytoalexin called glyceollin. This compound has been implicated in the growth retardation of Pmg in resistant plants [9, 10]. Metalaxyl treated susceptible soybeans produced concentrations of glyceollin, when inoculated with Pmg, which approach the level found in inoculated resistant plants. These results suggest that metalaxyl interacts with host resistance phenomena associated with Pmg.

One of the new uses of metalaxyl being researched in the Mississippi Delta is to grow Group 4 soybeans; these genotypes require 30 days less time in the field than the Group 5 varieties normally grown in this area. The key to the strategy is to apply metalaxyl to the soil and plant early (April 15 in place of May 15). Metalaxyl is needed to control several Pythium spp. which cause severe seed decay at this early planting date. (See Table 5). The Group 5 soybeans normally set and fill during the months of August and September, and are harvested in October. The problem is the months of August and September normally coincide with a drought. Thus, the same, if not better yields, are obtained with Group 4 cultivars as compared to Group 5, as they are harvested by mid to late August.

TABLE 4

Soybean Cultivar Reaction to Phytophthora megasperma f. sp. glyinea
and the Recommended Need for Metalaxyl

Category	Percentage of Cultivars (1989)	Phytophthora Tolerance Rating*	Metalaxyl Use Recommendation
Resistant	11.8	1	None.
High Tolerance	17.3	1.1-4	APRON seed or RIDOMIL soil.
Moderate Tolerance	29.8	4.1-5	RIDOMIL soil.
Low Tolerance	26.3	5.1-6	RIDOMIL soil.
No Tolerance	14.9	6.1-10	RIDOMIL will prolong the life of the crop - but not necessarily increase the yield.

*1 = no root rot and very vigorous
 2 = no root rot better than average vigor
 3 = no root rot and average vigor
 4 = no root rot slight-stunting
 5 = 10% dead plants slight-stunting
 6 = 20% dead plants moderate stunting
 7 = 50% dead plants with moderate-severe stunting
 8 = >50% mortality severe stunting
 9 = all plants dead before flowering
10 = all plants dead soon after emergence

TABLE 5

Stands and Yield of Group 4 Soybeans (Tennessee 486) Planted in
Mississippi on May 1, 1990

Treatment	Rate of Ridomil 2E Per Acre	Stand (5/10/90)	Vigor (6/15/90)	Yield (9/22/90)
Untreated	--	44	78	40
In-Furrow	0.25 pts. prod.	100	96	42
7" Band at Planting	1.5 pts. prod.	94	94	46
		(per 10')	(0-100)	(bu/ac)

The average soybean yield in Mississippi in 1990 was 21.0 bu/ac.

Another project - CIBA-GEIGY undertook was to develop a new applicator to apply the in-furrow treatments of metalaxyl, utilizing a drop tube assembly (See Figures 1 and 2). These efforts were initiated to ensure correct application of low rate metalaxyl as a service to the grower, thereby fulfilling two goals - (1) product stewardship and (2) an important facet of integrated control.

FIGURE 1. In-Furrow Liquid Applicator

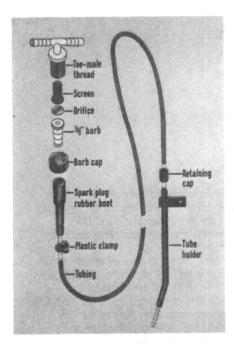

Installation Instructions Ridomil Kits

1. Attach tube holder on back side of double disc opener.
2. Starting on each end of drill or corn planter, fasten an elbow-male thread by barb, with nylon straps, followed by tee-male thread by barbs to the center of the implement.
3. Cut 1/2" hose to fit between nozzle body fittings and fasten in place.
4. When you have placed the orifice holders on front or rear frame, place one screen and an orifice in each nozzle fitting, placing a cap and hose barb on top.
5. Place 1/4" tubing into tube holder leaving 1" to 2" out bottom of the tube holder.
6. Place red retaining cap over hose and tighten on top of the tube holder to hold hose in place.
7. Attach 1/4" line by placing spark plug boot over line and pushing 1/4" line inside 3/8" hose barb, and spark plug boot to outside of hose barb to hold in place.
8. Having completed the above, connect the supply line from pump to center of implement on 1/2" line.
9. There is a flow chart available to determine the correct pump pressure.
10. Do calibration of unit before applying Ridomil.

FIGURE 2. Mounting of In-Furrow Liquid Applicator

Evidence of the importance of soil-borne diseases in contributing to yield loss is clear from the survey shown in Table 6 [11].

Table 6
Disease Survey of Soybean Root and Stem Rots in Ohio in 1986
(100 Fields in 46 Counties)

	No. of Fields With Disease
Disease	
Phytophthora	42
Rhizoctonia	25
Stem Canker	19
Sclerotinia	7
No Disease	24
Severe Disease	22

The importance of soil-borne diseases in soybeans will increase due to the recently discovered inverse correlation between sudden death syndrome (SDS) caused by Fusarium solani and resistance to Pmg (Pmg resistance cultivars are more susceptible to SDS). Clearly, the agro-ecosystem demands integration of several practices (including metalaxyl) to achieve an acceptable level of soybean disease management.

It should be noted that several herbicides have been shown to have an effect on soil pathogens, something that should be considered in developing IPM programs. Specifically on soybeans, it was shown by Duncan and Paxton [12] that trifluralin increased Phytophthora root rot from slight losses, up to 80% mortality, depending upon the inoculum density and soil condition. Today, trifluralin is used on approximately 30% of the U.S. soybean acreage and the related compound pendimethalin, on close to 12% of the acres. Dinitroanaline herbicides are used on more than 40% of the U.S. soybean acreage and should be a concern in soil-borne disease management as researchers have shown that trifluralin inhibits root development under adverse growing conditions [13]. Duncan and Paxton [12] noted an increase in oospore production resulting from exposing Pmg to trifluralin, thus postulating increased inoculum potential as a possible explanation for their findings.

FOLIAR FUNGICIDES

Four foliar fungicides are registered on soybeans in the U.S.. (See Table 7).

TABLE 7
Foliar Fungicides Registered on Soybeans in the U.S.

Trade Name	(Common Name)	Rate
Benlate 50WP	Benomyl	½ - 1 lb./A
Bravo 500	Chlorothalonil	2 - 3½ pts./A
Mertect 340-F	Thiabendazole	6 - 10 fl. oz./A
Topsin M 45G	Thiophanate-Methyl	10-20 fl. oz./A

Each of these may be applied twice per season, usually at late flowering to early pod set (R-3) and then 14-21 days later (R-5). Whether the investment for a foliar fungicide will be economical depends on the price of soybeans and the disease pressure in a given field. However, the goal of the treatment is to limit disease loss and to improve seed quality from the treated crop.

TABLE 8
Infectious Soybean Diseases in the Corn Belt - Present and Potential Importance in
Causing Crop Losses

Stem and Foliar Diseases	Casual Agent	Potential Losses Should Present Control Methods Fail or Not Be Used			
		Local Areas	In Sector	Local Areas	In Sector
Phyllosticta Leaf Spot	Phyllosticta sojicola	VL	VL	VL	VL
Brown Spot	Septoria glycines	L	VL	L	L
Bacterial Blight	Pseudomonas glycinea	L	VL	L	L
Bacterial Pustule	Xanthomonas phaseoli var. sojensie	L	VL	L	VL
Wildfire	Pseudomonas tabaci	VL	VL	VL	VL
Soybean Mosaic	SM virus	M	VL	M	L
Bud Blight	TRS virus	L	VL	L	VL
Downy Mildew	Peronospora manshurica	L	VL	L	VL
Frogeye Leaf Spot	Cercospora sojini	L	VL	L	VL
Stem Canker	Diaporthe phaseolorum var. caulivora	L	VL	M	L
Pod and Stem Blight	D. phaseolorum var. caulivora	M	L	M	L
Alternaria Leaf Spot	Alternaria spp.	VL	VL	VL	VL
Brown Stem Rot	Cephalosporium gregatum	M	L	M	M
Charcoat Rot	Macrophomina phaseoli	L	VL	L	VL

VL = Very Low (0-1%)
L = Low (2-10%)
M = Medium (11-35%)

NOTE: These estimates are based upon empirical data where available, and otherwise, on the general knowledge and judgment of several scientists in the field.

Over the years, more than a dozen states have developed a point system to determine if the use of foliar applied fungicides is likely to give an economic return. The key feature of these systems is the attempt to prevent disease from reaching high levels, rather than curing disease that is already present. These systems take into account factors that influence the development of diseases and typically weigh the importance of each factor by assigning points for specific conditions for each of the checklist items. Generally, the systems from the southern states are designed to increase yields, whereas those for the Midwest focus more on maintaining high seed quality. Foliar diseases of soybeans do not generate a high

level of potential loss, and only Brown stem rot attains the moderate rating. (Table 8).
Today the major foliar diseases of soybeans include:
· Anthracnose (Colletotrichum dematium, Glomerella glycines).
· Pod and Stem Blight (Digsorthe phaseolorum var. sojae).
· Cercospora leaf blight (Cercospora spp.).
Of the current fungicides used on soybeans, approximately 60% are used foliarly,
Table 9. The quantity of the products used has remained relatively constant over the past
five years. The total expenditures of fungicides on soybeans, which is the third largest ag-
ronomic crop in the U.S., represents only 2% of the total U.S. fungicide market.

TABLE 9

Estimated Market Share of the Major Fungicides Used on Soybeans in the U.S.

Product	Market Share	
	1985	1990
Foliar		
Benomyl	50%	50%
Thiophanate-methyl	10	10
Soil		
Metalaxyl	10	30
Seed		
Captan	10	2
Others	20	8
Total Expenditure	$12 million	$10 million
Price of Soybeans	$5.05/bu	$5.75 bu

REFERENCES

1. Sinclair, J. B. (ed.) Compendium of Soybean Diseases, 2nd ed. American Phytopathological Society. St. Paul, MN, 104 pp., 1982.

2. Sciumbato, G. Disease loss estimates from sixteen Southern states. Proc. Southern Soybean Disease Workers, 1991. Vol. XVIII, 32-28. (In Press) 1991.

3. Ploper, L. D. Influence of Soybean Genotype on Rate of Seed Maturation and Its Impact on Seed-Borne Fungi. Ph.D. Thesis. Purdue University, West Lafayette, IN. 182 pp, 1987.

4. Schmitthenner, A. F. and Kmetz, K. T. Role of Phomopsis sp. in the Soybean Seed Rot Problem. Pages 355-366 in: World Soybean Research Conference II: Proceedings. F. T. Corbin, ed. Westview Press, Boulder, CO. 897 pp, 1980.

5. Wall, M. T., McGee, D. C., and Burris, J. S. Emergence and Yield of Fungicide-Treated Soybean Seed Differing in Quality. Agronomy Journal, 1983 Vol. 75, 969-973.

6. Wilcox, J. R., Abney, T. S., and Frankenberger, E. M. Relationship Between Seed-Borne Soybean Fungi and Altered Photoperiod. Phytopathology, 1985, 75:797-800.

7. Houck, J. P. (Chairman), Corn/Soybean Study Team of National Academy of Sciences, 1975. Vol. II. Corn/Soybean Pest Control, pp. 169.

8. Schmitthenner, A. .F. Experiments on Fungicide and Cultural Control of Phytophthora Root Rot on Soybean in Ohio, 1987. Plant Path. Mimeo Series #73, OARDC/OSU, Wooster, OH, 1983.

9. Yoshi Kawa, M. Yamauchi, U., and Masago, H. Glyccollin: Its Role in Restricting Fungal Growth in Resistant Soybean Hypocotyles Infected with Phytophthora Megasperma var. Sojae. Physiol. Plant Pathol., 1983, 12:73-82.

10. Ward, E.W.B., Lazarovitz, G., Stossel, P., Barrie, S. D., and Unwin, C. H. Glyceollin Production Associated with Control of Phytophthora Rot of Soybeans by the Systemic Fungicide, Metalaxyl. Phytopathology, 1980, 70:738-740.

11. Schmitthenner (1986) Report on Ohio Soybean Disease Survey - Personal Communications.

12. Duncan, D. R. and Paxton, J. D. Trifluralin Enhancement of Phytophthora Root Rot of Soybean Plant Disease, 1981, Vol. 65 #5, 435-436.

13. Klingman, G. C. and Ashton, F. M. (ed.), Weed Science Principles and Practices. Published by John Wiley and Sons, Inc. 431 pp, 1975.

DETECTION AND DIAGNOSIS OF SOYBEAN DISEASES FOR IMPROVED MANAGEMENT

DENIS C. MCGEE
Seed Science Center and Department of Plant Pathology
Iowa State University
Ames, Iowa 50011, U.S.A.

ABSTRACT

In recent years a considerable amount of information has been obtained on prediction or forecasting of soybean diseases. These systems vary in methodology and the extent to which they have been developed and applied. They all, however, have a common objective of improving an aspect of management of the crop or product. This paper groups the systems into four classifications: general disease prediction; disease prediction for individual diseases; disease prediction for seed diseases before harvest; and detection of diseases that affect seed quality after harvest. It describes the epidemiological basis of the systems, how they are used, and their application in management of soybean diseases.

INTRODUCTION

Detection and diagnosis is the first step in addressing a plant disease problem. Most often, the information is used simply to suggest general control practices to avoid the problem in future years. Detection and diagnosis, however, can be an important part of disease management, particularly in relation to prediction or forecasting.

Effective disease prediction or forecasting systems require comprehensive data on the crop and its development throughout the season, the pathogen, and on interactive effects of the environment on disease development (1). The classical late blight of potato and apple scab systems were developed empirically by defining "infection periods" based on the occurrence of certain weather events. More recently, regression analysis has been employed to integrate the various epidemiological factors that influence a disease. These systems vary from a simple linear regression between one parameter and the disease, as in the relationship between soilborne inoculum of _Plasmodiophora brassicae_ and clubroot disease of cabbage, to multiple regression using several factors that influence the disease, as in the epidemimetic model EPIMAY for southern corn leaf blight of corn caused by _Helminthosporium maydis_ (1).

In recent years, a considerable amount of information has been obtained on prediction of soybean diseases. These systems vary in methodology and the extent to which they have been developed and applied. They all, however, have a common objective of improving an aspect of management of the crop or product. This paper groups the systems into four categories, and describes the epidemiological basis for each system, how they are used, and their application in management of soybean diseases.

GENERAL DISEASE PREDICTION SYSTEMS

Soybeans are susceptible to several foliar and stem diseases caused by fungi, including anthracnose (_Colletotrichum_ _truncatum_), pod and stem blight (_Phomopsis_ and _Diaporthe_ spp.), brown spot (_Septoria_ _glycines_), and frogeye leafspot (_Cercospora_ _sojina_). These can be controlled by foliar fungicides. Although yield increases up to 20% have been attributed to this control, responses generally were erratic when fungicide applications were originally introduced in the 1970s (2). Predictive systems were, therefore, developed in an attempt to define conditions under which fungicide application could be justified. These systems do not measure the diseases or the pathogens, but utilize a group of agronomic, climatic and economic criteria to define the conditions under which fungicide application may improve crop yield. They range from point systems (3) to a computerized Integrated Pest Management Model (AUSIMM) (4). Further comments on these systems are not made, because they are discussed a greater length in other papers in this publication.

PREDICTION SYSTEMS FOR PARTICULAR DISEASES

Aspects of disease prediction systems also have been investigated for particular soybean diseases.

Brown Stem Rot
This disease, caused by _Phialophora_ _gregatum_, is a significant economic problem in North America. Although resistant cultivars showed yield increases of 17-34% in infested soils, no yield advantage could be detected in the absence of significant disease pressure. It, therefore, was proposed that these cultivars be used only in fields where more than 75% of plants were infected in a recent year. This can be determined for individual soybean fields by examining stubble for internal browning of cut ends after harvest (5).

Charcoal Rot
Charcoal rot, caused by _Macrophomina_ _phaseolina_, tends to be a problem in warm, dry soybean production areas throughout the world (6). Annual losses recently were estimated at 5% statewide in Missouri, with some growers experiencing 30-50% loss (7). The main inoculum source is small, black microsclerotia in soil and crop debris of hosts. Populations of microsclerotia have been directly related to severity of charcoal rot, which, in turn, was inversely related to soybean yields (8). Factors that favor survival and germination of microsclerotia include the number of years of planting soybeans, dry soils (9,10), high C:N ratios of amendments (9), low bulk density (10), and oxygen concentrations greater

than 16% (7). Computer models that account for these factors are being developed to manage the disease by predicting changes in these populations (7).

Soybean Rust

This is a major disease caused by Phakopsora pachyrhizi in the tropical and subtropical areas of the eastern hemisphere. The potential for the disease to survive and spread in the United States has been examined extensively. As no aecial host is known (11), the pathogen seems to be totally dependent on uredospores to initiate new infections (12). Epidemiological data showed that germination of rust uredospores and penetration into leaves is favored by warm temperatures (18-26 C) and long periods of leaf wetness (6h at 20-25 C, and 10h at 18-26 C) (12). These conditions do exist throughout much of the U.S. soybean production areas, therefore, uredosporic infection could occur in the growing season. Although uredospores would not survive winter to begin new infections, they could persist on perennial hosts in southern states (12), as they do on native legumes in Australia (13). A computer model projected potential losses of $7.2 billion per year should the pathogen be introduced into the continental United States (14).

Red Crown Rot

This is a serious disease of peanuts that was reported on soybeans in the United States in 1973 (15), and is now found in several southern states. Using statistical models, the percentage yield loss for susceptible cultivars was calculated based on percentage of plants on which perithecia of the causal organism, Cylindrocladium crotalariae, could be detected. An incidence of 100% of plants infected, for example, predicted a yield loss of 50% (16). Soilborne microsclerotia are the only known inoculum source, and have been measured at densities up to 98/g soil in severely infested fields in Louisiana (17). A multiple regression model using microsclerotial density, plot elevation and soil moisture within plots in individual fields was tested to explain movement of the pathogen to flood waters. The model, however, had a poor fit (16).

Bean Pod Mottle

Bean pod mottle virus (BPMV), was first discovered in southern United States in the late 1950s (18,19), and is now widespread across soybean producing states. The virus can be transmitted to soybeans from Desmodium paniculatum, a perennial host for BPMV (20,21) by the bean leaf beetle (Ceratoma trifurcata), the main vector for the virus (22). ELISA and Ouchterlony gel diffusion assays can detect the pathogen in bean leaf beetles and leaf tissues (23,24). A good correlation between the degree of BPMV infection of leaves and bean leaf beetles within the same rows in a soybean field suggested the possibility of using beetle assays to predict plant infection and the need for control by insecticide sprays (25).

PREDICTION SYSTEMS FOR SEED DISEASES

Several diseases have adverse effects on the appearance, germination and vigor of soybean seeds. Because of the demand for high quality seeds throughout the world, considerable research effort has been expended on control of seed diseases.

Phomopsis Seed Decay

This disease is caused by a complex of fungi, including _Phomopsis longicolla_ and _Diaporthe phaseolorum_ var. _sojae_, the causes of pod and stem blight, and _Diaporthe phaseolorum_ var. _caulivora_, the cause of soybean stem canker. Seed germination can be reduced by as much as 90% in severely infected crops (26). Effective control can be achieved by applications of benzimidazole fungicides at growth stage R6 (27). Severity of infection, however, can vary greatly from year to year (28). Predictive methods, therefore, were developed in Kentucky and Iowa to identify fields that should be sprayed.

The Kentucky method (29) is a point system that uses four criteria, cropping history, cultivar selection, planting date, and rainfall, all of which have been shown to be related to increased severity of Phomopsis seed decay. Seed infection was greater in fields with a rotational history of continuous soybeans than corn-soybeans (30). Early-maturing cultivars or early planted crops are at a high risk of developing Phomopis seed decay because seed matures at a time when warm wet weather, favorable for seed infection, is most likely to occur (31,32). It also has been shown that rainfall between flowering and physiological maturity favors seed infection (32,33).

The Iowa method uses the level of pod infection at growth stage R6 to predict the risk of severe seed infection. Epidemiological studies had shown that pods are the primary pathway for infection of the seed (28,34). The feasibility of this predictive method was established by showing that, although pod infection can occur at any time from flowering onwards, extensive seed infection will not occur until growth stage R7 (28,35) and then, only under precise conditions of humidity and temperature (31,36). Laboratory and field experiments showed that three, four, and five days at 95-100% relative humidity at temperatures averaging 25, 20 and 15 C, respectively, were needed to cause substantial seed infection in infected pods (31). It also was necessary to show that inoculum reaching pods after R7 would not result in significant seed infection (28). In practice, the method requires sampling pods in the field at growth stage R6. These are immersed in 1.3% solution of sodium hypochlorite for one min, followed by five sec in 1:4 solution of the herbicide Basagran (the sodium salt of bentazon). Pods are incubated on damp blotters in clear plastic boxes for 7 days at room temperature under continuous light and then examined under a stereoscopic microscope for typical pycnidia of _Phomopsis longicolla_. If more than 50% of pods are infected fungicide sprays are recommended, and if less than 25% pods are infected sprays are not recommended (37). This method and the Kentucky method have been validated in commercial fields (28,38).

The Kentucky method has the advantage of providing immediate information at the time fungicides should be applied. It relies entirely on indirect estimates of disease severity, however, and cannot account for circumstances that might favor disease development. Because inoculum with the potential to infect seeds is measured directly, this is not a problem with the Iowa method. The Iowa method, however, requires equipment and materials, although these are inexpensive and readily obtainable. A more serious disadvantage of the method is that it requires a 7 day incubation period, when there is only a 14 day period over which sprays can be applied. In both systems there are ranges of values that give indefinite recommendations on the need to spray, but this can be overcome by combining the two methods (39,40). Neither method can account for the effect on seed infection of weather that occurs between R7 and harvest, but they do establish a sufficient degree of the risk of severe seed disease occurring to be of practical value (40).

Purple Seed Stain

This disease, caused by Cercospora kikuchii, does not adversely affect yields or other germination of the seed (41), but the discoloration symptom affects marketability of the seed, whether it is used for planting or processing. In Japan, epidemiological data indicated that severity of seed infection could be predicted by high airborne spore counts of the pathogen in August. Spores were trapped on glass slides installed horizontally at a height of 20 cm in the plant canopy. Thiophanate-methyl, benomyl and Bordeaux mixture are the fungicides that may be used in this disease forecasting program (42).

Downy Mildew

Although widespread throughout world, this disease, caused by Peronospora manshurica, rarely causes significant economic losses. Oospore-encrusted seeds have little adverse effect on germination, but may cause problems to seed producers regarding visible seed quality. A disease forecasting system has been developed in Russia, based on soil waterlogging and temperatures during leaf wetness (43).

Storage Fungi

Aspergillus and Penicillium spp. cause deterioration of soybeans stored at seed moisture contents in the range 13-20%, resulting in reduced seed germination (44) and downgrading of grain for reasons of heat damage and mustiness (45). These storage fungi have been implicated in increase in free fatty acids in deteriorated soybeans (46). Measurements of total or individual fatty acids have not, however, been reliable as a means of predicting storability of soybeans (44,47).

DETECTION OF DISEASES THAT AFFECT SEED QUALITY

Options are available to manage soybean seed diseases after harvest. A necessary step in implementing these actions is to test for seedborne pathogens. Some pathogens can be recognized by characteristic symptoms on the seeds. Phomopsis longicolla causes wrinkling and a white moldy chalky appearance. Cercospora kikuchii causes a purple seed stain. Fusarium and Alternaria spp. have been associated with red and dark brown lesions, respectively. These pathogens all have been detected with a high degree of accuracy, based on these symptoms in a computer imaging system (48). At Iowa State University, a blotter seed health test is used routinely to detect several pathogens including the four just mentioned, Aspergillus and Penicillium spp., and the bacterium Bacillus subtilis (49). The method requires surface sterilization of seeds in 1.0% sodium hypochlorite for 30 sec followed by a rinse in sterile water, then incubation of damp blotters in plastic boxes at 25 C in the dark for 10 days. Pathogens are identified on colony characteristics on infected seeds. Other seedborne pathogens commonly tested for include soybean mosaic virus by an ELISA procedure (50) and Pseudomonas syringae pv. glycinea by a semiselective media method. Typical management actions that might follow on the results of these tests would include, decisions on the need for seed treatments, issuance of phytosanitary certificates for seed to be exported, cleaning of grain or seed lots to remove physically altered seeds, using heavily diseased seed lots for crushing rather than planted seed, or deciding on the risk of planting seed that could transmit a pathogen to the following crop.

ACKNOWLEDGEMENTS

Journal Paper J-14806 of the Iowa Agriculture and Home Economics Experiment Station. Project 2621.

REFERENCES

1. Gareth Jones, D., _Plant Pathology Principles and Practice._ Prentice Hall, Englewood Cliffs, 1987, pp. 100-4.

2. Backman, P. A., and Jacobsen, B. J., Control of foliage, stem and pod diseases with fungicides. In _World Soybean Research Conference IV Proceedings_, ed. A. J. Pascale, Orientacion Grafica Editora, Buenos Aires, Argentina, 1989, pp 2091-2096.

3. Sinclair, J. B., and Backman, P. A., _Compendium of Soybean Diseases._ 3rd ed., American Phytopathological Society, St. Paul. 1989.

4. Backman, P. A., Mack, T. P., Rodriguez-Kabana, R., and Herbert, D. A., A computerized integrated pest management model (_AUSIMM_) for soybeans grown in the southeastern United States. In _World Soybean Research Conference IV Proceedings._ ed. A. J. Pascale, Orientacion Grafica Editora, Buenos Aires, Argentina, 1989, pp. 1494-1499.

5. Tachibana, H., Prescribed resistant cultivars for controlling brown stem rot of soybean and managing resistance genes. _Plant Dis.._ 1982, **66**, 271-274.

6. Oyekan, P. O., and Naik, D. M., Fungal and bacterial diseases of soybean in the tropics. In _Soybeans for the Tropics: Research. Production and Utilization._ eds. S. R. Singh, K. O. Rachie, and K. E. Dashiell, John Wiley & Sons Ltd, 1987, pp. 47-52.

7. Wyllie, T. D., Charcoal rot of soybeans -- current status. In _Soybean Diseases of the North Central Region._ eds. T. D. Wyllie, and D. H. Scott, American Phytopathological Society, St. Paul, 1988, pp. 106-113.

8. Short, G. E., Wyllie, T. D., and Bristow, P. R., Survival of _Macrophomina phaseolina_ in soil and in residue of soybean. _Phytopathology._ 1980, 70, 13-17.

9. Dhingra, O. D., and Sinclair, J. B., Effect of soil moisture and carbon:nitrogen ratio on survival of _Macrophomina phaseolina_ in soybean stems in soil. _Plant Dis. Rep.._ 1974, **58**, 1034-1037.

10. Gangopadhyay, S., Wyllie, T. D., and Teague, W. R., Effect of bulk density and moisture content of soil on the survival of *Macrophomina phaseolina*. Plant Soil. 1982, **68**, 241-247.

11. Bromfield, K. R., Soybean rust. Monograph 11. American Phytopathological Society, St. Paul, 1984.

12. Melching, J. S., Dowler, W. M., Koogle, D. L., and Royer, M. H., Effects of duration, frequency, and temperature of leaf wetness periods on soybean rust. Plant Dis.. 1989, **73**, 117-122.

13. Keogh, R. C., Notes on the survival of *Phakopsora* *pachyrhizi* on stands of *Kennedia* *rubicunda* in coastal New South Wales. Aust. Plant Pathol.. 1979, **8**, 31-32.

14. Kuchler, F., Duffy, M., Shrum, R. D., and Dowler, W. M., Potential economic consequences of the entry of an exotic fungal pest: the case of soybean rust. Phytopathology. 1984, 74, 916-920.

15. Rowe, R. C., Beute, M. K., and Wells, J. C., Cylindrocladium black rot of peanuts in North Carolina - 1972. Plant Dis. Rep.. 1973, 57, 387-389.

16. Berner, D. K., Berggren, G. T., Snow, J. P., and White, E. P., Distribution and management of red crown rot of soybean in Louisiana Appl. Agric. Res.. 1988, 3, 160-166.

17. Berner, D. K., Berggren, G. T., Snow, J. P., and Pace, M. E., The occurrence of red crown rot in Louisiana. Phytopathology. 1986, 76, 125.

18. Skotland, C. B., Bean pod mottle virus of soybean. Plant Dis. Rep.. 1958, 42, 1155-1156.

19. Walters, H. J., Bean pod mottle virus disease of soybeans. Arkansas Farm Res.. 1970, 19, 2, 8.

20. Horn, N. L., Newsom, L. D., Carver, R. G., and Jensen, R. L., Effects of virus diseases on soybeans in Louisiana. Louisiana Agric.. 1970, 13, 4, 12-13, 15.

21. Moore, B. J., Scott, H. A., and Walters, H. J., *Desmodium* *paniculatum*, a perennial host of bean pod mottle virus in nature. Plant Dis. Rep.. 1969, 53, 154-155.

22. Walters, H. J., and Lee, F. N., Transmission of bean pod mottle virus from *Desmodium* *paniculatum* to soybean by the bean leaf beetle. Plant Dis. Rep.. 1969, 53, 411.

23. Ghabrial, S. A., and Schultz, F. J., Serological detection of bean pod mottle virus in bean leaf beetles. Phytopathology. 1983, 73, 480-483.

24. Hopkins, J. D., and Mueller, A. J., Efficiency of the Ouchterlony gel diffusion test in detecting bean pod mottle virus infection in soybean. Environ. Entomol., 1984, 13, 1135-1137.

25. Ghabrial, S. A., Hershman, D. E., Johnson, D. W., and Yan, D., Distribution of bean pod mottle virus in soybeans in Kentucky. Plant Dis., 1990, 74, 132-134.

26. Kmetz, K. T., Schmitthenner, A. F., and Ellett, C. W., Soybean seed decay: prevalence of infection and symptom expression caused by Phomopsis sp., Diaporthe phaseolorum var. sojae, and D. phaseolorum var. caulivora. Phytopathology, 1978, 68, 836-840.

27. TeKrony, D. M., Egli, D. B., Stuckey, R. E., and Loeffler, T. M., Effect of benomyl applications on soybean seedborne fungi, seed germination and yield. Plant Dis., 1985, 69, 763-765.

28. McGee, D. C., Prediction of Phomopsis seed decay by measuring soybean pod infection. Plant Dis., 1986, 70, 329-333.

29. Stuckey, R. E., Jacques, R. M., Tekrony, D. M., and Egli, D. M., Foliar fungicides can improve seed quality. Kentucky Crop Improvement Association, Lexington, Kentucky, 1981.

30. Garzonio, D. M., and McGee, D. C., Comparison of seeds and crop residues as sources of inoculum for pod and stem blight of soybeans. Plant Dis., 1983, 67, 1374-1376.

31. Balducchi, A. J., and McGee, D. C., Environmental factors influencing infection of soybean seeds by Phomopsis and Diaporthe species during seed maturation. Plant Dis., 1987, 71, 209-212.

32. Tekrony, D. M., Egli, D. B., Stuckey, R. E., and Balles, J. Relationship between weather and soybean infection by Phomopsis sp. Phytopathology, 1983, 73, 914-918.

33. Lamka, G.. Effects of environmental and genotypic factors on Phomopsis infection of soybean pods. MS thesis, 1986, Iowa Sate University.

34. Kmetz, K., Ellett, C. W., and Schmitthenner, A. F., Soybean seed decay: sources of inoculum and nature of infection. Phytopathology, 1979 69, 798-801.

35. Wu, W. S., and Lee, M. C., Occurrence, pathogenicity and control of Phomopsis sojae on soybean. Mem. Coll. Agric. Nat. Taiwan Univ., 1985, 24, 16-26.

36. Spilker, D. A, Schmitthenner, A. F., and Ellett, C. W., Effects of humidity, temperature, fertility and cultivar on the reduction of soybean seed quality by Phomopsis sp. Phytopathology, 1981, 71, 1027-1029.

37. McGee, D. C., and Nyvall, R. F., Pod test for foliar fungicides on soybeans. Coop. Ext. Serv. Iowa State University, Pm-1136, 1984.

38. Tekrony, D. M., Stuckey, R. E., Egli, D. B., and Tomes, L., Effectiveness of a point system for scheduling foliar fungicides in soybean seed fields. Phytopathology. 1985, 69, 962-965.

39. Stuckey, R. E., Foliar fungicides: predictive systems. In World Soybean Research Conference IV Proceedings. ed. A. J. Pascale, Orientacion Grafica Editora, Buenos Aires, Argentina, 1989, pp. 1362-1367.

40. McGee, D. C. Evaluation of current predictive methods for control of Phomopsis seed decay of soybeans. In Soybean Diseases of the North Central Region. eds. T. D. Wyllie, and D. H. Scott, American Phytopathological Society, St. Paul, 1988, pp. 22-25.

41. Wilcox, J. R., Laviolette, F. A., and Martin, R. J., Heritability of purple seed stain resistance in soybeans. Crop Sci.. 1975, 15, 525-526.

42. Suzuki, H., Mode of occurrence and control of purple speck of soybean. JARQ. 1985, 19, 7-12.

43. Nedelchev, N. K., Harmfulness of downy mildew of soybean. Rast. Nauk.. 1978, 15, 116-121.

44. Dorworth, C. E., and Christensen, C. M., Influence of moisture content, temperature, and storage time upon changes in fungus flora, germinability, and fat acidity values of soybeans. Phytopathology. 1968, 58, 1457-1459.

45. Sauer, D. B., Grain quality and grading standards. In Soybean Diseases of the North Central Region. eds. T. D. Wyllie, and D. H. Scott, American Phytopathological Society, St. Paul, 1988, pp. 32-38.

46. Lisker, N., Ben-Efraim, A., and Henis, Y., Involvement of fungi in the increase of free fatty acids in stored soybeans. Can. J. Microbiol.. 1985, 31, 799-803.

47. McGee, D. C., and Christensen, C. W., Storage fungi and fatty acids in seeds held thirty days at moisture contents of fourteen and sixteen per cent. Phytopathology. 1970, 60, 1775-1777.

48. Paulsen, M. R., Using machine vision to inspect oilseeds. INFORM. 1990, 1:50-55.

49. McGee, D. C., and Nyvall, R. F., Soybean seed health. Coop. Ext. Serv. Iowa State University, Pm-990, 1984.

50. Lister, R. M., Application of the enzyme-linked immunosorbent assay for detecting viruses in soybean seed and plants. Phytopathology. 1978, 68, 1393-1400.

MANAGEMENT OF VIRUS DISEASES IN SOYBEAN

O.P. SEHGAL AND G. THOTTAPPILLY

Department of Plant Pathology, University of Missouri, Columbia, MO 65211, USA, and Biotechnology Unit, International Institute of Tropical Agriculture, Ibadan, Nigeria, W. Africa

INTRODUCTION

Soybean, Glycine max, is susceptible to about 50 viruses belonging to 18 groups and is commonly used as an indicator host in plant virus research. Approximately 25 viruses occur naturally on soybean and of these 10 are of economic importance. These viruses, and the groups to which they belong, are: bean pod mottle (BPMV, comovirus), cowpea chlorotic mottle (CCMV, bromovirus), cowpea mild mottle (CMMV, carlavirus), cowpea mosaic (CpMV, comovirus ; syn.= cowpea yellow mosaic), mung bean yellow mosaic (MYMV, geminivirus), soybean chlorotic mottle (SCMV, caulimovirus), soybean dwarf (SDV, luteovirus), soybean mosaic (SMV, potyvirus), tobacco ringspot (TRSV, nepovirus) and tobacco streak (TSV, ilarvirus). Furthermore, it is increasingly being recognized that several undefined viruses or virus-like agents transmitted by whiteflies offer considerable threat to soybeans in tropical parts of Africa and Asia (1).

GEOGRAPHICAL DISTRIBUTION AND DISEASE INCIDENCE

SMV and TRSV are the two most widespread and damaging soybean viruses worldwide and locally serious epiphytotics by these viruses have resulted in considerable yield losses (Table 1).

TABLE 1
Economically important soybean pathogenic viruses[a]

Causal Virus	Geographical distribution	Transmission modes	Economic importance; crop loss
BPMV	USA	Sap, beetle	Locally serious; 15%-40%
CCMV	USA, Costa Rica	Sap, beetle	Moderate
CMMV	Africa, Asia	Sap, whitefly	Moderate
CpMV	Africa, USA	Sap, beetle	Locally serious; 25%-35%
MYMV	Asia	Whitefly	Locally serious;
SCMV	Japan	Sap	Moderate
SDV	Japan, USA Australia	Aphid	Locally serious 15%-25%
SMV	Worldwide	Sap, aphid, seed	Serious; 35%
TRSV	Worldwide	Sap, thrips, seed	Locally destructive; 20%-100%
TSV	Brazil, USA	Sap, thrips, seed	Locally serious in south-eastern Brazil

[a]Alfalfa mosaic virus, bean yellow mosaic virus, and peanut mottle virus are fairly widespread on soybeans but are usually of minor importance.

CCMV, CMMV, CpMV, MYMV and SDV are somewhat restricted in their distributions but infections by these viruses have caused moderate to severe crop losses. TSV, the causal agent of Brazilian soybean bud blight disease, which occurrs primarily in southeastern Brazil, can be especially serious if infection occurs early in the season (2).

BPMV has been reported only from U.S.A. where it has caused several epidemics in the eastern, midwestern and southern states (3). The estimates of yield losses due to BPMV infection range from 5-40%. Occasionally, double infection of soybean with BPMV and SMV occurrs and in such cases, losses are much higher i.e., up to 80%.

SCMV has been recorded only from Aichi Prefacture in Japan where it has caused moderate crop losses (4).

HOST RANGE

BPMV, CCMV, CpMV, MYMV, SCMV, SDV and SMV are rather restricted in their host ranges and infect legumes mostly. CMMV is transmissible to plants in 5 families, besides Leguminosae. TRSV and TSV possess very wide host ranges which includes many dicotyledonous and some monocotyledonous species.

A virus with broad host range is adapted better for survival under divergent ecological and environmental conditions. Additionally, if susceptible hosts include annual or perennial weeds then they can serve as field sources of the viral inocula. Even for viruses with restricted host ranges, susceptible perennial wild plants are important field reservoirs. A critical assessment of the virus host range, therefore, is essential for an understanding of its disease cycle and formulation of appropriate control measures.

SYMPTOMATOLOGY

Symptoms induced by selected soybean viruses are shown in Figure 1. Infection by SDV causes severe puckering of leaves, interveinal yellowing and a marked stunting of the plant (Fig. 1A). SCMV induces chlorotic mottling and mosaic on leaves and a moderate degree of stunting (Fig. 1 B). CpMV and CMMV cause yellow mosaic symptoms which in the former appear as small patches (Fig. 1 D), while in the latter the symptoms are in the form of pronounced vein clearing (Fig. 1 F). BPMV induces mottling on the systemically-invaded leaves (Fig. 1 E) which is usually accompanied by mild puckering; infected pods and seed coats also are mottled.

The earliest sign of SMV infection is the development of vein-clearing and light green mosaic (Fig. 1 C). This is followed by a distortion mosaic on younger leaves and development of dark green bands along the principal veins. In chronic stages of infection, plants are distinctively stunted and produce misshapen pods with few seeds. SMV infection enhances seed discoloration and reduces seed size by 10-20% (1,5). Root nodulation may also be severely affected. In mixed infections by SMV and BPMV, a severe apical necrosis of soybean plants commonly develops (1).

The initial symptom of TRSV infection (bud blight disease) is a characteristic curving of the apical

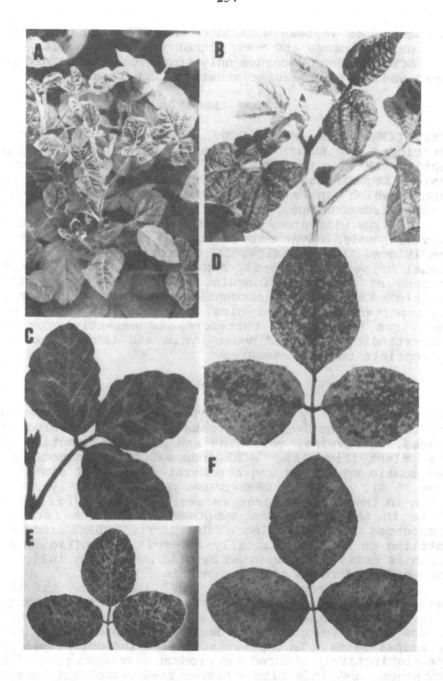

Fig.1. Symptoms induced by selected soybean viruses. A, Soybean dwarf virus; B, Soybean chlorotic mottle virus; C, Soybean mosaic virus, very early symptoms; D, Cowpea mosaic virus; E, Bean pod mottle virus; and F, Cowpea mild mottle virus.

meristem. This is followed by a bronzing of the young leaves and necrosis of the terminal bud. The plant becomes stunted and many lateral buds proliferate giving it a bushy appearance. Pods are generally underdeveloped or completely aborted (2).

MODES OF TRANSMISSION

Except for MYMV and SDV, other soybean viruses are readily transmitted by sap inoculation (Table 1). This signifies that most of these viruses attain a fairly high endogenous concentration and are relatively stable in vitro. However, TSV requires the presence of antioxidants in the extraction buffer to stabilize its infectivity (6). The most common mode of field transmission of soybean viruses is by insects viz., aphids, beetles, thrips and whiteflies (Table 1).

BPMV, CCMV and CpMV are transmitted efficiently by the bean leaf beetle, Ceratoma trifurcata, and their relationship with the vector is of circulative, nonpropagative type; these viruses persist for several days within their vectors (3).

CMMV is acquired and transmitted by the whitefly, Bemisia tabaci, within a few minutes and is lost rapidly by the insects indicating that it is nonpersistent in its vector (2, 6). MYMV exhibits a circulative, nonpropagative type of relationship with B. tabaci and persists for 3-10 days in the vector (6).

SDV remains viable in its aphid vector, Acyrthosiphon solani, for about 40 days but does not multiply in the vector (6). SMV is transmitted in a nonpersistent manner by about 24 different aphid species but its principal natural vectors are Acyrthosiphon pisum, Aphis craccivora, and A. fabae (5).

TRSV can be transmitted experimentally from soybean to soybean by Thrips tabaci and persists in this vector for about 10 days (2). TSV is transmitted by T. tabaci and Frankliniella occidentalis although the precise role of these insects in TSV infection of soybeans is not known (6).

Transmision through seed plays a major role in the field occurrence and dispersal of SMV and TRSV. The world wide incidence of SMV is largely due to the distribution of contaminated seed. The extent of seed transmission of SMV varies from 5-20% while for TRSV it ranges from 30-100% (1,5).

PHYSICOCHEMICAL PROPERTIES

From the physicochemical and structural standpoints, soybean viruses are very divergent (Table 2).

TABLE 2

Some physiochemical properties of soybean viruses

Virus	Genome	Coat protein(s) mass (kDa)	Shape and size (mm)
BPMV	ssRNA bipartite	37,26	Isometric, 30
CCMV	ssRNA tripartite	20	Isometric, 30
CMMV	ssRNA monopartite	36	Flexuous, 650x15
CpMV	ssRNA bipartite	37,22	Isometric, 30
MYMV	ssDNA	28	Geminate, 30x18
SCMV	dsDNA	not determined	Isometric, 50
SDV	ssRNA monopartite	22	Isometric, 25
SMV	ssRNA monopartite	28	Flexuous, 750x15
TRSV	ssRNA bipartite	58	Isometric, 30
TSV	ssRNA tripartite	29	Isometric, 30

MYMV and SCMV contain DNA as their genetic material while the others contain ssRNA which may be in the form of mono- , bi-, or tri-partite genetic elements. Most of these viruses are good immunogens and produce high titred antisera.

BPMV and CCMV are among the few plant viruses which exhibit the property of a progressive decline in their specific infectivities in relation to the age of infection (7). In case of BPMV this is due to a preferential degradation of one of the two genomic RNAs in situ while in CCMV it results from a strengthening of the coat protein:RNA crosslinks. It is unclear if any host constituents play a role in this viral inactivation process.

VIRAL IDENTIFICATION AND MONITORING PROTOCOLS

Accurate and rapid viral identification is essential in formulating appropriate control measures. Furthermore, from epidemiological standpoints it is important to establish identity of the primary and secondary sources of the field inocula. Moreover, the ability to detect viruses within their vectors should be useful in forecasting disease so that necessary precautions can be taken. Serological procedures are quite useful in soybean viral identificatioin and diagnoses. For instance, the use of enzyme-linked immunosorbent assay (ELISA) has permitted rapid detection of TRSV and SMV in leaf tissue and in the seed (8). Similarly, ELISA test of beetle (Ceratoma trifurcata) homogenates have proved reliable in predicting field incidence of BPMV (3). A modification of the ELISA procedure, termed dot immunobinding assay, (DIBA) has proved quite useful in detecting BPMV in field samples (Figure 2), and for quantification of BPMV coat antigen (7).

EPIDEMIOLOGICAL CONSIDERATIONS

Considerable information is avaliable concerning SMV disease cycle (5). Seed-borne SMV is major source of the field inoculum and virus survives from season to season in the seed. Weeds play no significant role in SMV epidemiology. From the contaminated seedlings, SMV is spread by aphids and infected plants can serve as virus source through much of the growing season. SMV spread in the field is largely due to the activity of transient aphids that probe briefly on soybean leaves and then move onto new plants.

Naturally-infected perennial or annual weeds and seed-borne inocula serve as field sources of TRSV and TSV (2, 6). BPMV reservoir hosts include perennial lespedezas, Desmodium spp., crimson clover, and velvet bean (6). The possibility of BPMV overwintering in its beetle vector has also been considered (3). Desmodium scopirus, Calpogonium muconoides, and Centrosema pubescens act as field reservoirs of CMMV (2).

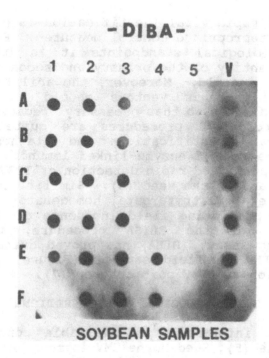

Fig.2. Dot immunobinding assay for detecting bean pod mottle in field-collected soybean leaves. One μl of the appropriately diluted extract of the various leaf samples (columns 1-5, rows A-F) or of the purified virions (column V, rows A-F) were spotted on the nitrocellulose membrane. After blocking the remaining sites, the membrane was treated successively with rabbit polyclonal BPMV antibodies, goat anti-rabbit antibody conjugated to alkaline phosphatase and NBT-BCIP substrate (Ref. No.7). Positive reaction was indicated by the development of purple color.

DISEASE AND VECTOR MANAGEMENT

Certain agricultural practices can be effective in achieving a satisfactory level of control of soybean virus diseases. These include: use of virus-free seed, roguing diseased plants, crop rotation, eliminating weeds that serve as hosts for viruses and their vectors and use of barrier crops to impede vector activity (2). Viruses that are seed-borne in soybean are carried within the embryo and thus cannot be eradicated. Consequently, efforts should be directed in obtaining and utilizing virus-free seed. An attractive alternative is to develop cultivars with an inherently low incidence of seed-transmision (2,6). This approach should be particularly useful for SMV control since seed-borne inoculum plays a crucial role in starting SMV epiphytotics.

Application of insecticides to a crop for preventing virus introduction, especially of the nonpersistently-transmitted viruses (e.g.,SMV) is usually ineffective because chemicals fail to kill the insects fast enough i.e., prior to completion of the inoculation feeding. Use of chemicals for controlling weeds that harbor viruses or their vectors offers a better promise. Viruses that are transmitted in a persistent manner by vectors can be controlled to some extent by insecticidal application of the crop. This is illustrated by the success achieved in controlling MYMV by using a systemic insecticide and mineral oil to kill the whitefly vector (9).

Development and use of genetically resistant varieties is the ideal approach in controlling soybean viruses. Several different types or components of resistance against these viruses have been identified (5,6,10). Furthermore, virus control by utilizing natural genetic resistance against vectors appears to be a desirable strategy.

Recent progress in genetic engineering and plant transformation procedures has permitted formulation of a novel type of disease resistance based upon the expression of viral genes or sequences. Transgenic tobacco plants expressing the coat protein gene of SMV have been obtained and they exhibit a high degree of resistance against infection by two unrelated potyviruses (11). It is now feasible to conduct similar studies with soybean as appropriate methodology in obtaining transgenic soybeans has become available (12).

CONCLUSIONS

Viral infections can drastically affect plant efficiency and productivity. At least 10 different viruses economically impact soybeans worldwide. A basic understanding of the viral disease cycle, including feeding behavior and ecology of vectors, is necessary in formulating appropriate control measures. Pesticides to control vectors at best, should supplement other components of the disease management system. Genetic modification via biotechnology offers attractive alternatives for incorporating viral resistance in soybeans.

ACKNOWLEDGEMENTS

We thank Dr. T. Tamada for providing Figure 1A ; to Dr. M. Kamaya-Iwaki for Figure 1B and to the American Phytopathological Society for permission for its reproduction.

Contribution from Missouri Agricultural Experiment Station. Journal Series No.11598.

REFERENCES

1. Allen, D.J., The Pathology of Tropical Food Legumes, John Wiley and Sons, New York, 1975, 413 p.

2. Thottappilly, G., and Rossel, H.W., Viruses affecting soybean. In Soybeans for the Tropics; Research, Production and Utilization, ed. S.R. Singh, K.O. Rachie and K.E. Dashiell, John Wiley and Sons, New York, 1987, pp 53-68.

3. Ghabrial, S.A., and Schultz, F.J., Serological detection of bean pod mottle virus in bean leaf beetles. Phytopathology, 1983, 73,480-483.

4. Iwaki, M., Isogawa, Y., Tsuzuki, H., and Honda, Y., Soybean chlorotic mottle, a new caulimovirus on soybean. Plant Disease, 1984,68, 1009-1011.

5. Irwin, M.E., and Goodman, R.M., Ecology and control of soybean mosaic virus. In Plant Diseases and Vectors: Ecology and Epidemiology, ed. K. Maramorosch and K.F. Harris, Academic Press, New York, pp. 181-220.

6. Brunt, A., Crabtree, K., and Gibbs, A., Viruses of Tropical Plants, C.A.B. International, Wallingford, Oxon, U.K., 1990, 707 p.

7. Kartaatmadja, S.A., and Sehgal, O.P., Decline of bean pod mottle virus specific infectivity *in vivo* correlates with degradation of encapsidated RNA-1. Phytopathology, 1990, 80, 1182-1189.

8. Lister, R.M., Application of the enzyme-linked immunosorbent assay for detecting viruses in soybean seed and plants. Phytopathology, 1978, 68, 1393-1400.

9. Nene, Y.L., Control of *Bemisia tabaci* Genn, a vector of several plant viruses. Indian J. agric. Sci., 1973, 43, 433-436.

10. Ross, J.P., Registration of four soybean germplasm lines resistant to bean pod mottle virus. Crop Sci., 1986, 6, 20.

11. Stark, D.M., and Beachy, R.N., Protection against potyvirus infection in transgenic plants; evidence for broad spectrum resistance. Biotechnology, 1989, 7, 1257-1262.

12. Hinchee, M.A.W., Connor-Ward, D.V., Newell, C.A., McDonnell, R.E., Sato, S.J., Gasser, C.S., Fischoff, D.A., Re, D.B., Fraley, R.T., and Horsch, R.B., Production of transgenic soybean plants using *Agrobacterium*-mediated DNA transfer. Biotechnology, 1988, 6, 915-922.

WEED MANAGEMENT IN SOYBEANS IN THE UNITED STATES

C. G. MCWHORTER
Plant Physiologist/Application Technology Research Unit
USDA/ARS
Stoneville, MS 38776, USA

ABSTRACT

The total cost of weeds in soybean (<u>Glycine</u> <u>max</u> L.) production in the USA is about $3.0 billion each year. About four dozen weed species contribute to the yield losses and reductions in soybean seed quality. Crop rotations and cultural practices are beneficial to a successful weed management program but herbicides are the primary component of most successful control programs. Practically all soybeans are treated with herbicides each year. Preemergence soil treatments have been used more extensively than postemergence treatments but the usage of postemergence treatments is increasing and there is a slight decline in the use of preemergence treatment. The total amount of herbicide applied annually peaked in the 1980s and has declined slightly thereafter. The trend toward increased use of postemergence treatments will probably continue especially as increased use is made of conservation tillage practices.

INTRODUCTION

The monetary losses due to weeds in soybean production in the USA probably exceed the combined losses due to all other pests (7). In 1984, yield losses due to weeds were approximately $1.9 billion and the estimated cost of weed control was about $1.1 billion bringing the total cost of weeds in soybean production to about $3 billion annually. In the 1970s, losses in yield and quality of soybeans caused by weeds were about 17% of the total production. However, losses were reduced to about 10% of the total production in the 1980s. These reduced losses indicate that the efficiency of weed control technology in soybeans has improved dramatically during the last decade. Even so, losses continue to be high, and this is largely due to the complexity of the ever changing weed control problem confronted by American soybean producers. This paper summarizes the major weed problems confronted by the American soybean producer, the herbicides being used to control weeds, and some of the factors that may affect future weed control in soybeans.

PRESENT WEED PROBLEMS

In 1987, Jordan et al. prepared a list of the 48 worst weeds in soybeans in the USA (7). Seventy-seven percent of these weeds are annuals while 23% are perennials. Sixty-three percent of the major weeds are dicotyledons while 37% are monocotyledons. More than 50% of the major weeds that occur in soybeans have been shown to be allelopathic so it may be possible that allelopathy is more involved in reducing soybean yields than previously thought. Only a few of the dicotyledonous weeds are C_4 (dicarboxylic acid pathway) whereas most of the grasses are C_4. The C_4 photo-synthetic pathway is associated with a large number of anatomical, biochemical, and ultrastructural characteristics in plants as compared to the more conventional C_3 photosynthetic pathway (10). The C_4 method of carbon fixation has been shown to contribute to the "weediness" of plants.

Corn Belt States: Annual weeds are generally much more troublesome in the Corn Belt states than perennials. Predominant annual grass weeds include giant foxtail (Setaria faberi Herrm.), fall panicum (Panicum dichotomiflorum Michx.), barnyardgrass [Echinochloa crus-galli (L.) Beauv.], and crabgrass (Digitaria sp.). Shattercane [Sorghum bicolor (L.) Moench] may also be troublesome in localized situations. Volunteer corn (Zea mays L.) may also be troublesome as a weed in soybeans. It is likely to be more troublesome in reduced-tillage systems when the corn stubble is not buried into the soil.

The most troublesome annual broadleaf weeds include pigweeds (Amaranthus sp.), smartweeds (Polygonum sp.), velvetleaf (Abutilon theophrasti Medicus), and common cocklebur (Xanthium strumarium L.). Less frequently but highly troublesome when they occur, other annual broadleaf weeds include: morningglories (Ipomoea sp.), jimsonweed (Datura stramonium L.), Eastern black nightshade (Solanum ptycanthum Dun.), common ragweed (Ambrosia artemisiifolia L.), common lambsquarters (Chenopodium album L.), giant ragweed (Ambrosia trifida L.), wild mustard [Brassica kaber (D.C.) L.C. Wheeler], and sunflower (Helianthus sp.).

Perennial weeds are not as troublesome on a widespread basis in the Corn Belt as in the southern USA but they may cause significant yield losses in localized situations. Canada thistle [Cirsium arvense (L.) Scop.] and quackgrass [Elytrigia repens (L.) Nevski] are the most troublesome weeds in the northern part of the region while johnsongrass [Sorghum halepense (L.) Pers.] is most troublesome in the southern part of the region. Other troublesome perennials may include honeyvine milkweed [Ampelamus albidus (Nutt.) Britt.], common milkweed (Asclepias syriaca L.), yellow nutsedge (Cyperus esculentus L.), and in the western region, field bindweed (Convolvulus arvensis L.).

Northeastern States: Heavy weed infestations are probably less common in this region than in either the Corn Belt or the southern USA. Competition may be less because of lower air temperatures and a shorter growing season. Annual grasses that are most troublesome are crabgrass, fall panicum, and giant foxtail. Goosegrass [Eleusine indica (L.) Gaertn.] and broadleaf signalgrass [Bracharia platyphylla (Griseb.) Nash] often present problems in the southern portion of this area.

The most troublesome annual broadleaf weeds are pigweeds, smartweeds, common cocklebur, morningglories, Eastern black nightshade, jimsonweed, common ragweed, common lambsquarter, and velvetleaf. Perennial weeds are also less troublesome in this region than in other regions although quackgrass and yellow nutsedge may create heavy infestations on a localized basis.

Southern States: It is generally acknowledged that weeds are more troublesome in this region than in other areas of the USA. There is a wider array of troublesome weeds in this region because of a longer growing season, higher rainfall and, in the Delta States, a higher level of soil fertility. Predominant annual grasses include crabgrass, goosegrass, broadleaf signalgrass, red sprangletop [Leptochloa filiformis (Lam.) Beauv.] and barnyardgrass. Major broadleaf weeds are annual morningglories, common cocklebur, pigweed, sicklepod (Cassia obtusifolia L.), and prickly sida (Sida spinosa L.). Others that may be troublesome in certain areas within the region include crotons (Croton sp.), Florida beggarweed [Desmodium tortuosum (SW.) DC.], bristly starbur (Acanthospermum hispidum DC.), hemp sesbania [Sesbania exaltata (Raf.) Rydb. ex A.W. Hill], ragweeds, smartweeds, wild cucumber [Echinocystis lobata (Michy.)T. & G.], and jimsonweed. Less frequently occurring broadleaf weeds are Florida pusley (Richardia scabra L.) and spurred anoda [Anoda cristata (L.) Schlecht.].

The major perennial weed throughout the region is johnsongrass. Several annual and perennial sedges may be also locally troublesome. Many areas within the region may also be infested with several perennial vines that include the milkweeds, redvine [Brunnichia ovata (Walt.) Shinners] and trumpetcreeper [Campsis radicans (L.) Seem. ex Bureau].

Specific Troublesome Weeds: Sicklepod, a tall growing broadleaf weed, was confined to only a few southeastern states until recent years. It has now developed into a major troublesome weed throughout much of the south and the area that it infests each year continues to increase. It is especially difficult to control with herbicides making it especially troublesome under reduced tillage programs.

Wild-proso millet (Panicum miliaceum L.) is an annual grass confined primarily to four or five Corn Belt states and it can cause severe yield losses in soybeans. It has the potential of spreading throughout much of the soybean producing areas of the USA. Itchgrass [Rottboellia cochinchinensis (Lour.) W. Clayton], another tall growing annual grass, is presently confined to a few localized areas in Gulf Coastal states but, like wild-proso millet, has the potential of spreading over much of the soybean-growing area of the USA.

Crotalaria (Crotalaria sp.), an annual broadleaf weed, was originally grown as a green manure crop. It is presently confined to a few Gulf Coast states and is a major concern because its seeds are highly toxic to animals. Soybean seeds contaminated with crotalaria seed are greatly reduced in value. Balloonvine (Cardiospermum halicacabum L.) is an annual broadleaf weed that produces seed that are the same size as soybean seed making it extremely difficult to separate from soybean seed. While only troublesome in isolated situations in the South, its presence in harvested soybean seed greatly reduces the value of harvested product. Similarly southern cowpea [Vigna unguiculata (L.) Walpers] produces a seed that is similar to soybean seed and is difficult to remove in the cleaning process. Southern cowpea is grown as a cultivated plant and infrequently becomes a weed in localized situations. Its presence in soybean seed greatly reduces the value of harvested soybeans. It is important that these special weeds be controlled with a high level of efficiency to avoid further spread. Recommendations for the best control of these weeds vary on a state-by-state basis.

Weed Management Systems: Preventive control should always be a component of a good weed management system. This includes planting weed-free seed and the cleaning of all field equipment thoroughly before moving from an infested area to an uninfested area. Reproductive portions of weeds including seeds, rhizomes, bulbs, and tubers are frequently transported to uninfested area, especially by combines. Although this report is primarily concerned with the use of herbicides for weed control, there are other components of good weed management systems that will be mentioned briefly.

Crop Rotation: Crop rotations are often beneficial in controlling individual weed species. Rotations of both crops and herbicides are especially helpful in controlling perennial weeds and certain broadleaf annual weeds in soybeans. Corn is the predominant crop rotated with soybeans in the Corn Belt. This has been an excellent rotation because this permits the rotation of herbicides which aides in more efficient weed control programs in both crops. Winter wheat has also been used in the cropping sequence in the southern half of the Corn Belt. This too has been beneficial because the wheat competes with the summer annual weeds that are common in both corn and soybeans.

In the southern USA, soybeans may be rotated with rice, cotton, corn or peanuts. The presence of red rice (Oryza sativa L.) in rice has been a major problem in a rotation that includes soybeans even though there are now several herbicides that effectively control red rice in soybeans (9). Farmers in the South have also rotated soybeans with a grain crop grown during the winter months. This has been effective in controlling many troublesome weeds, especially perennials.

Tillage: Until the 1980s both preplanting and postemergence tillage was used in most soybean production. More recently, increased usage of herbicides such as paraquat and glyphosate have reduced or replaced preplanting tillage. Possibly as much as 30% of the total soybean production is now with conservation tillage that may include various forms of no-tillage, mulch-tillage, reduced-tillage, or minimum-tillage. Poor weed control has often been cited as the primary reason for farmers being reluctant to adopt various forms of conservation-tillage. Even so, the practice is increasing not only because it is widely advocated by many government and non-government organizations, but also because it reduces soil erosion and conserves soil moisture.

The use of postemergence tillage in soybean production has been reduced significantly during the last decade although some type of postemergence tillage is probably still used on more than 65% of the soybeans produced. Planting soybeans in narrow rows that cannot be cultivated reduces cultivation costs but often the costs of the herbicide treatments that are used are greater than the use of cultivation. Perennial weeds, especially in the southern USA, tend to increase in reduced tillage systems, and the cost of controlling these with herbicides increases the cost of soybean production.

ethalfluralin, alachlor, pendimethalin and metolachlor (trade names and chemical names are presented in Table 1). Use of trifluralin has decreased slightly during the last decade as sales of pendimethalin increased. Up to 50 or 60% of the soybeans in the Corn Belt states receive treatment with one of this these herbicides while they are probably used even more frequently in southern states. Use of alachlor declined during the past decade as use of metolachlor increased. The use of metolachlor, alachlor, and ethalfluralin is more frequent in northern states than in southern states.

The use of selective herbicides for postemergence grass control in soybeans with materials such as sethoxydim and fluazifop-P has not reached expectations. The increased use of these since introduction in the early 1980s has been slow. Use of these probably does not exceed 10 to 15% of the production in the Corn Belt states or 15 to 20% of soybeans in southern states. These herbicides are very selective and highly effective when applied properly, but their reduced effectiveness in late season and their comparatively high cost per hectare treated has probably slowed their expanded use. Future use of these and related herbicides will probably increase slowly as further reductions in tillage occur.

The most commonly used preemergence herbicides for broadleaf weed control during the last decade have been metribuzin, linuron, chloramben, clomazone, chlorimuron, imazaquin, imazethapyr, and thifensulfuron. Chloramben was used on a large percentage of the soybeans in the 1970s but its use declined and sales were terminated in 1990. The use of linuron has also declined. Sales of metribuzin were extensive from the late 1970s to the late 1980s, but its use decreased after the introduction of clomazone, chlorimuron, imazaquin, imazethapyr and thifensulfuron in the mid 1980s.

Imazaquin was marketed aggressively following its introduction in the mid 1980s but a severe drought in 1988 contributed to widespread residual injury in cotton and corn in 1989. This resulted in changes in its registration, and its use has been generally replaced by imazethapyr in the northern portion of the corn belt.

Predominant herbicides used for postemergence control of broadleaf weeds during the last decade have been bentazon, acifluorfen, chlorimuron, imazaquin and imazethapyr. The extent of the area treated varies from state to state, but in general, 30 to 40% of the soybean hectarage may be treated in northern states while 50 to 60% of the hectarage may be treated postemergence in southern states each year. The use of postemergence treatments will probably continue to increase slightly while the use of preemergence treatments will probably decrease.

A large number of herbicides are available (Table 2) and soybean producers will continue to be dependent on the use of herbicides for some years to come. Even if expanded use of conservation tillage does not appreciably affect the total amount of herbicides used annually, it is almost certain to change use patterns of individual herbicides. Reduced tillage will likely result in increased sales of materials more effective for perennial weed control and reduce sales of herbicides applied for preemergence weed control. Other important variables that could affect use patterns would be future ecological shift of weeds and worldwide soybean production. The increased production of soybeans in other

There will be continued interest in the USA in reducing cultivation whenever possible in soybean production, especially where soil erosion is troublesome. Postemergence tillage will, however, continue to be an important part of soybean production until more effective and economical weed control practices are developed for the control of both annual and perennial weeds.

Mycoherbicides: Two mycoherbicides, Collego and DeVine, have been used commercially in the USA for the last few years (3). Collego has been used to control northern jointvetch [Aeschynomene virginica (L.) B.S.P.] in rice and soybean fields, primarily in Arkansas. Grower acceptance of this pathogen has been good, but its use potential is limited because the weed is not widespread. DeVine has been used commercially in Florida but is not used in soybeans.

In the early 1980s, it was discovered that Alternaria cassiae was highly effective as a mycoherbicide for control of sicklepod, coffee senna (Cassia occidentalis L.) and crotalaria (2). This discovery helped provide impetus for additional research on mycoherbicides and several others were discovered that appeared promising at the time (3). Alternaria cassiae was licensed to the Mycogen Corporation for commercial development under the trade name CASST (3) but, after several years of research, further development ceased. It does not appear that this mycoherbicide will be made available commercially. Likewise further development of other mycoherbicides for weed control in soybeans has slowed or stopped, and it does not appear that mycoherbicides will be made available for weed control in soybeans before the year 2000. Possibly the first new pathogen to be marketed as a mycoherbicide in soybeans will be Colletotrichum truncatum, an anthracnose disease that controls hemp sesbania and several other weeds. It is not likely that mycoherbicides will have an appreciable impact on the sale of herbicides in the United States for the next two decades unless more efficacious organisms are discovered or unless more efficient production techniques are developed.

Cultural Practices: Narrow-row spacings of soybeans are more competitive to weeds than wide-row spacings (1, 11). Better weed control will be obtained even with herbicides when soybeans are grown in narrow rows. Also, certain soybean cultivars are more competitive than others, so proper selection of the soybean cultivar may be important in obtaining maximum control (6).

Herbicides: Herbicides are widely used by soybean farmers. Herbicide use increased from less than five percent of the soybeans produced in the early 1950s to more than 95% of the present production. The total volume of herbicides used annually probably reached a maximum in the mid 1980s and has declined slightly since that time. Even though there has been some decline in the total amount of herbicides used, the percent of the total soybean production treated with herbicides has not changed appreciably.

Herbicides are available that will control most individual weed species although failure to apply herbicides on a timely basis or under favorable environmental conditions may greatly reduce the effectiveness of individual treatments. Soil applied herbicides used most frequently for control of grasses during the past decade include trifluralin, countries has resulted in comparatively low soybean prices in the USA. A 20 to 30% increase in the price of harvested soybeans in the United States would probably result in increased herbicide sales because one of the factors relating to reduced sales in recent years has been the severe economic plight of many farmers.

The increased availability of computer programs will enable producers to be more efficient in developing weed management systems. An increasing number of states such as North Carolina now have computer programs available for personal computers in all extension offices to permit extension personnel and growers to select optimized programs for their own situations (8). This trend will continue as more herbicide options become available to producers. Many states now recommend many dozens of combinations of herbicides and application methods. Computer programs will be necessary to sort through the many herbicide combinations available for the many different weed complexes.

The number of herbicide-resistant weed biotypes has increased rapidly during the last two decades (5). There has been speculation that the development of herbicide-resistant weeds could significantly affect herbicide usage in soybeans in the United States. To date, there have been no cases of herbicide-resistance reported that would significantly affect overall herbicide use in soybeans in the United States. The wide variety of chemical alternatives and the widely varying environmental and geographic conditions throughout the United States will probably reduce the impact of any individual herbicide-resistant weed biotype that may be reported in the future. Also a wide variety of options are available in crop rotations, herbicide rotations, and the use of non-chemical cultural techniques to reduce the impact of future resistant biotypes in soybean production.

Several companies are conducting research to develop herbicide-resistant soybeans cultivars and it has been predicted that the availability of these could significantly alter herbicide use patterns. A recent report on herbicide-resistant crops suggests that, even though herbicide-resistance in soybean cultivars would be advantageous in some situations, its potential impact has been over emphasized (4). The report concluded that although this technology is feasible this does not mean that it will play a major role in weed management in soybeans during the next decade.

TABLE 1
Common, trade, and chemical names of herbicides used for weed control in soybeans in the USA.

Common Name	Trade Name	Chemical Name
acifluorfen	Blazer	5-[2-chloro-4-(trifluoromethyl)phenoxy]-2-nitrobenzoic acid
alachlor	Lasso	2-chloro-$\underline{N}$-(2,6-diethylphenyl)-$\underline{N}$-(methoxymethyl)acetamide
bentazon	Basagran	3-(1-methylethyl)-($1\underline{H}$)-2,1,3-benzothiadiazin-4($3\underline{H}$)-one 2,2-dioxide

TABLE 1 continued
Common, trade, and chemical names of herbicides used for weed control
in soybeans in the USA.

Common Name	Trade Name	Chemical Name
chloramben	Amiben	3-amino-2,5-dichlorobenzoic acid
chlorimuron	Classic	2-[[[[4-chloro-6-methoxy-2-pyrimidinyl)amino]carbonyl]amino]sulfonyl] benzoic acid
clomazone	Command	2-[(2-chlorophenyl)methyl]-4,4-dimethyl-3-isoxazolidinone
ethalfluralin	Sonalan	N-ethyl-N-(2-methyl-2-propenyl)-2,6-dinitro-4-(trifluoromethyl)benzenamine
fluazifop - P	Fusilade 2000	(R)-2-[4-[[5-(trifluoromethyl)-2-pyridinyl]oxy]phenoxy]propanoic acid
glyphosate	Roundup	N-(phosphonomethyl)glycine
imazaquin	Scepter	2-[4,5-dihydro-4-methyl-4-(1-methylethyl)-5-oxo-1H-imidazol-2-yl]-3-quinolinecarboxylic acid
imazethapyr	Pursuit	(±)-2-[4,5-dihydro-r-methyl-4-(1-methylethyl)1H-imidazol-2-yl]-5-ethyl-3-pyridinecarboxylic acid
linuron	Lorox	N'-(3,4-dichlorophenyl)-N-methoxy-N-methylurea
metalachlor	Dual	2-chloro-N-(2-ethyl-6-methylphenyl)-N-(2-methoxy-1-methylethyl)acetamide
metribuzin	Sencor Lexone	4-amino-6-(1,1-dimethylethyl)-3-(methylthio)-1,2,4-triazin-5(4H)-one
paraquat	Gramoxone	1,1'-dimethyl-4,4'-bipyridinium ion
pendimethalin	Prowl	N-(1-ethylpropyl)-3,4-dimethyl-2,6-dinitrobenzenamine
sethoxydim	Poast	2-[1-(ethoxyimino)butyl]-5-[2-(ethylthio)propyl]-3-hydroxy-2-cyclohexen-1-one
thifensulfuron	Pinnacle	3-[[[[(4-methoxy-6-methyl-1,3,5-triazin-2-yl)amino]carbonyl]amino]sulfonyl]-2-thiophenecarboxylic acid
trifluralin	Treflan	2,6-dinitro-N,N-dipropyl-4-(trifluoromethyl) benzenamine

TABLE 2

Estimated level of control of individual weeds following the proper use of herbicides in soybeans (adopted from 1991 Weed Control Guidelines for Mississippi, Miss. State Univ., Miss. State, MS 39762, 238 p.). Control scale is 0 = no control, 10 = perfect control. PPI is preplanting-incorporated, OT is over-the-top spray and DIR is directed sprays).

Weeds

Herbicides	Barnyardgrass	Broadleaf signalgrass	Crabgrass	Goosegrass	Seedling johnsongrass	Rhizome johnsongrass	Fall panicum	Cocklebur	Entireleaf morningglory	Pitted morningglory	Palmleaf morningglory	Smallflower morningglory	Purple moonflower	Purslane	P. smartweed	Hemp sesbania	Prickly sida	Spurred anoda	Pigweed	Balloonvine	Texas gourd	Sicklepod	Cutleaf groundcherry	Common ragweed	Yellow nutsedge	Annual Sedge	Velvetleaf	Jimsonweed	Red rice	Spotted spurge	Hophornbeam copperleaf	Showy crotolaria	Wild poinsettia	Crop tolerance (G=Good, F=Fair)
Preplant - PPI																																		
Canopy	7	6	7	7	7	3	7	9	8	8	8	9	6	10	9	9	9	8	9	8	8	7	9	9	5	9	7	9	8	9	8	·	8	F
Command	9	9	9	9	9	3	8	7	4	7	6	6	3	9	8	4	9	8	7	·	·	0	·	·	·	·	10	8	6	8	8	·	9	G
Scepter	7	7	7	5	6	2	5	9	6	8	8	8	5	9	9	3	8	7	10	4	9	5	8	9	4	9	6	8	5	9	7	·	7	G
Sonalan	9	9	9	9	2	8	0	2	2	2	2	1	9	2	0	0	0	8	·	·	0	8	2	0	·	3	·	·	·	·	·	·	·	G
Prowl or Treflan	9	9	9	9	9	3	9	0	2	2	2	2	1	9	2	0	0	0	8	0	0	2	0	3	0	3	2	3	4	2	0	0	0	G
Treflan-2X or Prowl-2X	10	10	10	10	9	7	9	0	4	4	4	4	3	10	2	1	2	0	9	0	5	3	1	4	0	4	3	4	8	3	0	0	0	G
Prowl or Treflan + Scepter	9	9	9	9	9	4	8	9	6	8	8	8	5	9	9	0	8	7	10	4	9	5	8	9	4	9	6	8	5	9	7	·	7	G
Prowl or Treflan + Sencor/Lexone	9	9	9	9	9	3	9	5	2	6	7	8	5	9	9	9	9	9	10	9	6	6	9	9	2	9	8	7	8	9	9	6	6	F
Dual	8	8	9	9	6	0	9	0	2	2	2	2	9	5	2	6	3	9	1	5	5	9	7	9	9	4	5	9	4	5	·	·	3	G
Lasso 3.5 to 4 qt	8	8	9	9	6	0	9	0	2	2	2	2	9	5	2	6	3	9	1	5	6	9	7	7	9	4	5	9	4	5	·	·	3	G
Tri-Scept																																		
Squadron	9	9	9	9	9	4	8	9	6	8	8	8	5	9	9	0	8	7	10	4	9	5	8	9	4	9	6	8	5	9	7	·	7	G
Freedom	8	8	8	8	6	0	8	0	2	2	2	2	9	5	2	3	2	9	0	4	6	9	8	7	9	4	5	8	4	6	·	·	·	G
Commence	9	9	9	9	9	0	8	5	3	7	6	6	2	9	7	3	8	7	8	5	0	0	0	3	0	3	9	7	6	7	7	·	8	G
Turbo	8	8	8	8	6	0	8	5	3	7	7	8	5	9	9	9	9	9	9	9	6	7	9	9	7	9	8	8	9	9	9	6	6	F
Salute	8	8	8	8	9	0	8	5	2	7	7	8	5	9	9	8	9	9	8	9	6	9	0	9	·	7	7	7	8	9	6	·	6	F
Preemergence																																		
Canopy	7	6	7	7	7	3	7	9	8	8	8	9	6	10	9	9	9	8	9	8	8	7	9	9	5	9	7	9	8	9	8	·	8	F
Command	9	9	9	9	9	3	8	6	4	7	6	6	3	9	8	4	9	9	7	5	0	0	0	5	·	·	10	8	7	8	8	·	9	G
Dual	8	8	9	9	7	0	9	0	0	0	0	0	9	4	0	4	0	9	·	·	3	9	5	7	9	3	4	8	3	·	·	·	3	G
Lasso	8	7	9	9	5	0	9	0	0	0	0	0	9	4	0	4	0	9	1	3	4	9	5	5	9	3	4	8	3	5	·	·	3	G
Scepter	7	7	7	5	7	2	5	9	6	8	8	8	5	9	9	3	9	7	10	5	9	5	·	9	4	9	6	8	5	9	7	·	·	G
Sencor/Lexone	8	6	8	7	5	0	7	6	2	7	2	8	6	9	9	9	9	9	9	9	9	7	8	9	9	2	9	8	8	4	9	9	7	F
Zorial	8	8	9	8	7	2	7	4	4	5	5	4	4	9	5	4	8	7	8	·	5	·	4	9	7	·	8	·	·	·	·	·	·	F
Zorial (Split)	8	8	9	8	3	8	4	5	6	6	5	5	9	6	5	9	8	9	9	·	5	·	·	5	9	7	·	7	·	·	8	·	·	F
Squadron	7	7	7	6	8	0	7	9	6	8	8	8	5	9	9	0	8	6	10	4	9	5	8	9	4	9	6	8	5	9	7	·	7	G
Turbo	8	8	8	8	6	0	8	5	3	7	7	8	5	9	9	9	9	9	9	9	6	7	9	9	7	9	8	8	9	9	9	6	6	F
Postemergence-OT																																		
Classic	0	0	0	0	0	0	9	8	9	9	8	9	5	9	8	9	2	4	10	5	6	7	·	8	6	8	8	9	0	0	4	·	8	G
Cobra	4	4	4	4	3	2	3	8	8	9	8	8	9	9	6	9	8	6	9	9	8	5	9	8	3	6	8	9	0	8	8	9	8	F
Scepter	2	2	3	3	6	5	5	10	5	6	6	7	5	9	7	2	3	2	10	0	6	3	·	8	5	7	3	0	2	4	3	0	8	G
Basagran	0	0	0	0	0	0	9	2	6	7	9	3	7	9	4	8	8	5	8	0	0	6	9	6	8	9	8	·	0	0	0	5	G
Basagran +2,4-DB	0	0	0	0	0	0	9	5	8	9	9	5	7	9	5	8	9	5	8	0	0	6	9	6	·	9	8	·	0	0	0	6	F
Blazer	3	4	3	3	2	2	5	8	9	9	8	9	8	7	9	1	2	8	8	7	3	9	8	3	5	·	8	2	7	8	9	7	G
Blazer +2,4-DB	3	4	3	3	2	2	7	8	9	9	8	9	8	7	9	1	·	8	8	7	3	9	8	·	·	8	2	7	8	9	8	F
Reflex	3	3	3	3	3	2	8	8	9	8	8	9	8	7	9	2	2	9	8	·	3	9	8	6	7	·	9	0	5	8	9	8	G
Fusilade	8	8	8	9	9	9	8	0	0	0	0	0	0	0	0	0	0	0	0	0	0	0	0	0	0	0	0	8	0	0	0	·	G
Poast	8	9	9	9	9	9	0	0	0	0	0	0	0	0	0	0	0	0	0	0	0	0	0	0	0	0	0	8	0	0	0	·	G
Whip	8	9	7	8	9	8	8	0	0	0	0	0	0	0	0	0	0	0	0	0	0	0	0	0	0	0	0	7	0	0	0	·	G
Assure	8	9	8	8	9	9	9	0	0	0	0	0	0	0	0	0	0	0	0	0	0	0	0	0	0	0	0	6	0	0	0	·	G
Pursuit	7	7	7	5	8	6	7	9	7	9	8	9	6	·	7	0	6	6	10	4	4	0	·	6	7	8	7	9	4	8	2	0	9	G
Scepter-OT	3	4	3	3	3	2	4	10	6	8	8	8	7	9	7	9	3	2	10	8	7	3	9	8	4	7	3	8	0	6	7	8	8	G
Storm	3	4	3	3	3	0	2	9	8	9	9	9	7	8	8	9	7	7	8	8	7	2	9	9	6	8	8	8	0	6	7	9	6	G
Postemergence-Dir																																		
2,4-DB	0	0	0	0	0	0	9	9	9	9	9	9	3	0	3	3	2	2	1	2	0	0	1	0	·	3	4	0	0	2	·	3	G
Gramoxone	9	9	9	8	8	0	8	4	5	4	6	7	4	8	5	1	4	3	8	2	2	8	7	8	3	·	6	7	9	5	7	·	8	G
Lorox	7	7	8	7	7	0	7	7	8	8	8	8	7	8	7	8	8	8	8	·	7	8	8	·	·	6	7	6	7	7	·	8	G
Lorox + 2,4-DB	7	7	8	7	7	0	7	9	10	9	9	10	9	9	7	8	8	8	9	9	5	9	10	9	2	·	7	8	6	7	9	·	8	G
Sencor/Lexone	7	7	8	7	7	0	·	8	7	7	7	7	·	7	7	8	8	8	8	8	8	7	7	0	·	8	·	8	4	·	·	5	G
Sencor + 2,4-DB	7	7	8	7	7	0	·	9	9	9	9	9	8	3	7	7	8	8	8	9	8	9	8	8	0	·	8	7	8	4	8	·	7	G

271

REFERENCES

1. Bendixen, L.E., Soybean (Glycine max.) competition helps herbicides control johnsongrass (Sorghum halepense). Weed Technol., 1988, 2, 46-48.

2. Boyette, C.D., Host range and virulence of Colletotrichum truncatum, a potential mycoherbicide for hemp sesbania (Sesbania exaltata). Plant Disease, 1991, 75, 62-64.

3. Charudattan, R., The mycoherbicide approach with plant pathogens. In Microbial Control of Weeds, ed. David O. Tebeest, Chapman and Hall, New York, NY, 1991, pp. 24-57.

4. Duke, S.O., Christy, A.L., Hess, F.D. and Holt, J.S., Herbicide-resistant crops. Council for Agricultural Science and Technology, Ames, IA, May, 1991, pp. 24.

5. Holt, J.S. and LeBaron, H.M., Significance and distribution of herbicide resistance. Weed Technol., 1990, 4, 141-49.

6. James, K.L., Banks, P.A. and Karnok, K.J., Interference of soybean, Glycine max, cultivars with sicklepod, Cassia obtusifolia. Weed Technol., 1988, 2, 404-9.

7. Jordan, T.N., Coble, H.D. and Wax, L.M., Weed control. In Soybeans: Improvement, Production, and Uses. Second Edition, ed. J.R. Wilcox, American Society of Agronomy, Inc., Madison, WI, 1987, pp. 429-60.

8. Linker, H.M., York, A.C. and Wilhite, D.R., WEEDS - A system for developing a computer-based herbicide recommendation program. Weed Technol., 1990, 4, 380-85.

9. Salzman, F.P., Smith, R.J. and Talbert, R.E., Control and seedhead suppression of red rice (Oryza sativa) in soybeans (Glycine max). Weed Technol., 1989, 3, 238-43.

10. Stowe, L.G. and Teerio, J.A., The geographic distribution of C_4 species of the dicotyledonae in relation to climate. Am. Nat., 1978, 112, 609-23.

11. Yelverton, F.H. and Coble, H.D., Narrow row spacing and canopy formation reduces weed resurgence in soybeans (Glycine max). Weed Technol., 1991, 5, 169-74.

SOYBEAN WEED PROBLEMS IN ARGENTINA AND THEIR CONTROL

AGUSTIN MITIDIERI
Est. Exp. Agrop. de San Pedro
Instituto Nacional de Tecnología Agropecuaria
C.C. 43 (2930) San Pedro, Argentina

ABSTRACT

The 10 most important soybean weeds in Argentina are: **Sorghum halepense,
Amaranthus hybridus, Datura ferox, Ipomoea purpurea, Chenopodium album,
Anoda cristata, Cynodon dactylon, Cyperus rotundus, Echinochloa colonum**
and **Digitaria sanguinalis**. Losses are estimated to be above 400 million dollars yearly; about half of which is the cost of control. Intended control is achieved by mechanical and chemical methods. Good contribution is attained through cultural practices and somewhat by natural
agents. Research projects are aimed at reducing both the cost of control
and losses in yield and harvest operation. About 2/3 of the chemical
treated area corresponds to post-emergence products. The most used
herbicides for PPI, PRE, broadleaf POST and grass POST treatments are:
trifluralin, metribuzin, bentazon and haloxyfop-methyl and fluazifop-
P-butyl, respectively. Recent important introductions have been imazaquin, imazethapyr, chlorimuron-ethyl, clethodim, propaquizafop and
quizalofop-P-tefuryl.

INTRODUCTION

The soybean crop has experienced a great expansion in the last 20 years
in Argentina, from 80,000 hectares in 1971 to about 5 millions in 1990.
At the beginning weed problems were little and only mechanical control
was purposely used. Along with the increase of the planted area, weeds
became more and more troublesome, because of monoculture, lack of very
effective broad spectrum herbicides, high cost of the new specialized
chemicals, and changes in the cropping systems. The only products available for weed control in soybeans were prometryne, 2,4-DB and trifluralin but their use was limited.

In 1974, bentazon was registered for controlling broadlead weeds.
Up to that time, no summer crops could be cultivated in a very infested
area with johnsongrass (**Sorghum halepense**). In 1975, the use of 2X
rate of trifluralin was adopted for rhizome johnsongrass control. Two

years later, pirifenop (the first aryloxy-phenoxy grasskiller) was introduced for controlling seedling and rhizome johnsongrass. Since then, new chemicals were continuously introduced and about 30 herbicides are now directly or indirectly used for the management of weeds in soybeans in Argentina.

WEED PROBLEMS

About 60 species of weeds have been registered in soybeans in Argentina that cause from very little to very severe problems. The 10 main weeds are:

Sorghum halepense (L.) Pers.	JOHNSONGRASS
Amaranthus hybridus L.	SMOOTH PIGWEED
Datura ferox L.	LARGE THORNAPPLE
Ipomoea purpurea Lam.	TALL MORNINGGLORY
Chenopodium album L.	COMMON LAMBSQUARTERS
Anoda cristata (L.) Schlecht	SPURRED ANODA
Cynodon dactylon (L.) Pers.	BERMUDAGRASS
Cyperus rotundus L.	PURPLE NUTSEDGE
Echinochloa colonum (L.) Link	JUNGLE RICE
Digitaria sanguinalis (L.) Scop.	LARGE CRABGRASS

Less diffussed weeds or those of local importance are:

Portulaca oleracea L.	COMMON PURSLANE
Tagetes minuta L.	WILD MARIGOLD
Xanthium cavanillesii Schouw	LARGE COCKLEBUR
Xanthium strumarium L.	COMMON COCKLEBUR
Bidens subalternans DC	BEGGARTICKS
Euphorbia heterophylla L.	WILD POINSETTIA
Eleusine indica (L.) Gaertn.	GOOSEGRASS
Convolvulus arvensis L.	FIELD BINDWEED
Brassica campestris L.	WILD MUSTARD
Nicandra physaloides (L.) Gaertn.	APPLE-OF-PERU

Weeds, as is a general rule, cause two types of economic damage: a) indirectly, because they add to the cost of production, and b) directly, because they reduce yield when they are not adequately controlled, make harvest difficult and increase the cost of commercialization. It has been estimated that soybean growers spend in mechanical and chemical control practices about 150 and 100 million dollars per year, respectively. On the other hand, reduction in yield and troubles in the harvest operation and commercialization cause losses between 150 and 200 million dollars yearly. Both types of damage sum up above 400 million dollars.

WEED CONTROL PRACTICES

As weed problems increased, soybean growers became more dependent on chemical control. While in 1977/78 only 19 % of the planted area with soybeans was treated with herbicides (228,000 hectares), in 1989/90 with a planted area of about 4.8 million hectares, 72 % received one to three chemical treatments (Figure 1). Without the use of chemical control, very severe losses would have occurred in spite of any mayor involvement of mechanical control. In such situation in a few years, yield reduction, hindering of harvest and increase in commercialization expenses due to weeds would probably cause losses of more than 40 per cent.

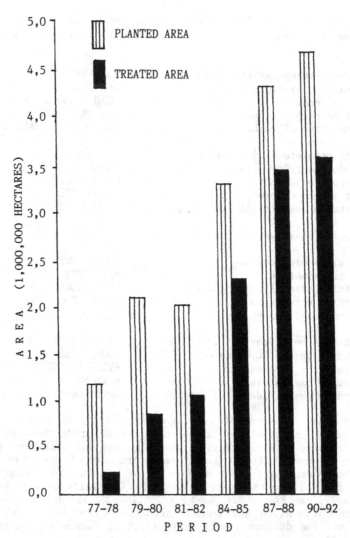

Figure 1. Evolution of planted area with soybeans and area that was treated once to three times with herbicides, in Argentina.

Mechanical control

Mechanical control is performed in soybeans with different tools and implements. Moldboard, chisel and disk plows are used for primary tillage. In many circumstances at least one cultivation with one of these implements is done for weed control purposes. Disk and/or spring-tooth harrows and field cultivators are used for secondary tillage either to attain additional seedbed preparation or to control weeds or both. One to two of these operations are aimed at controlling weeds.

Selective tillage is performed pre- and/or post-emergence of soybeans with broadcast implements, such as light springtooth harrows or rotary hoes; these operations, either for breaking the soil crust or for controlling weeds are done no later than a week after soybean emergence. One of the most effective and commonly used mechanical control of weeds is achieved with interrow tillage. Several types of row-crop cultivators are used, mainly alone but in some cases they are equipped with a sprayer to perform band, post-emergence treatments in the same operation.

Special implements have been manufactured with the purpose of controlling perennial weeds, such as johnsongrass and bermudagrass. They look and/or work like rotary hoes, but with longer straight or curved teeth. The tillage is performed during the fallow season and by repeated operations 50 % or even more of the rhizomes are taken to soil surface were they die by desiccation or frost damage.

Weed control is frequently complemented by hand operations using hoe or machete, mainly when large plant species, such as smooth pigweed, large thornapple, large cocklebur and common lambsquarters have escaped from other control practices. It costs between 4 to 5 dollars per hectare, and it is mainly implemented for preventing seed production and for cosmetic purposes other than to prevent yield reduction.

Chemical control

Improved technology in chemical control has made possible in the last 10 years to increase or keep a very good standard in the average yield of soybeans. Chemical control is performed in six different situations:

1. Pre-plant treatments with post-emergence herbicides mainly for perennial weed control. The main chemical used is glyphosate (Table 1).

2. Pre-plant incorporated herbicides. Trifluralin is extensively used for controlling annual grasses and some broadleaf weeds. The 2X rate for rhizome johnsongrass control is not used anymore. (Table 2).

3. Pre-emergence treatments. The main chemical used is metribuzin either broadcast or band applied (Table 3).

4. Post-emergence herbicides for broadleaf weeds. The most used product is bentazon among several contact herbicides. Recently translocated chemicals such as chlorimuron-ethyl and imazetapyr have been introduced (Table 4).

5. Post-emergence grasskillers. There are two types of treatments:

a) selective herbicides for over-the-top applications to control annual and perennial grasses.
b) glyphosate used with roller or rope-wick applicators for controlling rhizome johnsongrass (Table 5).

6. Pre-harvest treatments in order to facilitate the harvest operation. There is very little use of 2,4 D and/or paraquat for this purpose.

The main use recommendations are presented in the tables that follow.

TABLE 1
Pre-plant treatments with post-emergence herbicides

Herbicide treatments	Rate (kg/ha)	Controlled weeds
Glyphosate[a]	0,96 - 1,20	Johnsongrass
	2,40 - 2,88	Bermudagrass, purple nut-sedge, field bindweed
Dalapon	5,0 - 7,0	Bermudagrass, johnsongrass
MSMA	2,5 - 3,5	Johnsongrass
Aminotriazole	2,5 - 3,0	"Yuyo sapo" (**Wedelia glauca** (Ort.) Hoff.

[a]The rate if gkyphosate can be reducwed 30 to 40 % by adding adjuvants based on fatty amine ethoxylate at 0,4 per cent.

TABLE 2
Pre-plant incorporated herbicides

Herbicide treatments	Rate (kg/ha)	Controlled weeds
Trifluralin	0,72 - 0,96	Annual grasses and some broadleaf weeds
Pendimethalin	1,00 - 1,20	Idem
Vernolate	2,10 - 2,80	Nutsedges, annual grasses and some broadleaf weeds
Metolachlor	2,10 - 2,80	Idem
Imazaquin	0,16 - 2,00	Broad spectrum weed control including nutsedges

Incorporation is generally implemented with disk harrows. Imazaquin may be applied from one month before up to planting time. Vernolate has to be incorporated soon after application, whereas trifluralin incorporation can be delayed four hours.

To improve the control of some annual grasses, mainly large crab-grass and goosegrass, imazaquin may be tank mixed with trifluralin or pendimethalin.

TABLE 3
Pre-emergence treatments

Herbicide treatments[a]	Rate (kg/ha)[b]	Controlled weeds
Metribuzin	0,25 - 0,50	Several annual broadleaf and some grass weeds
Alachlor	1,50 - 2,00	
Metolachlor	1,50 - 2,00	Annual grasses and some broadleaf weeds
Acetolachlor	2,00 - 2,50	
Imazaquin	0,16 - 0,20	Broad spectrum weed control Poor in some annual grasses
Imazethapyr	0,16 - 0,20	Broad spectrum weed control Fairly good in annual gras-ses

[a]Several formulated and tank mixtures are also used in order to broaden the spectrum of weed control; some of them are metribuzin + alachlor, metribuzin + metolachlor, prometryn + linuron and alachlor + linuron.

[b]Low rates are recommended for light soils. Do not use pre-emergence treatments in very light soils with low organic matter content.

For pre-emergence herbicides to be effective soil moisture is required within a week after application; if no rain occurs shallow incorporation is recommended with springtooth harrows or rotary hoes.

As rainfall is not frequent after planting double crop soybeans, it is not convenient to rely on pre-emergence treatments in such situation.

Band application is possible with pre-emergence treatments. Treatment cost and parallel chemical contamination can be reduced 50 to 60 per cent. If the chemical has a long period of residual effect, the reduction of the herbicide quantity will decrease the danger to the following crop.

Band application to be effective must be performed along with the planting operation.

TABLE 4
Post-emergence treatments for broadleaf weed control

Herbicide treatments	Rate (kg/ha)	Controlled weeds
Bentazon[a]	0,25 - 0,30	Large thornapple
	0,48 - 0,75	Several broadleaf weeds
Bentazon	0,48 - 0,75	Several broadleaf weeds,
+ imazaquin	0,03 - 0,04	including smooth pigweed
Acifluorfen[a]	0,11 - 0,15	Large thornapple
	0,18 - 0,25	Several broadleaf weeds
Fluoroglycofen[a,b]	0,024 - 0,036	Large thornapple
	0,048 - 0,060	Several broadleaf weeds
Fomesafen[a]	0,125 - 0,175	Large thornapple
	0,200 - 0,325	Several broadleaf weeds
Benazolin	0,30 - 0,50	Some broadleaf weeds
Imazethapyr[c]	0,08 - 0,10	Broad spectrum weed control including annual grasses and nutsedges
Chlorimuron-ethyl	0,007 - 0,008	Smooth pigweed
	0,010 - 0,015	Several broadleaf weeds
Imazaquin[d]	0,06 - 0,08	Smooth pigweed

[a]Low rates can be used in good to very good soil moisture conditions
[b]Weeds must be very small (less than 5 cm tall)
[c]It is recommended to apply imazethapyr in very early post-emergence
[d]These rates are effective when smooth pigweeds are small (less   than 10 cm tall. When the weed is bigger the rate must be increased up to 120 - 200 g/ha

Several formulations of surfactants are used for broadleaf treatments at 0,1 to 0,25 % of the commercial product, following label instructions. For controlling spurred anoda, common lambsquarters and tall morningglory, 2,4 DB at 25 to 32 g/ha is added to bentazon and fomesafen.

The mixture of selective post-emergence grasskillers with acifluorfen, fluoroglycofen, fomesafen and chlorimuron-ethyl is not recommended because grass control may be reduced.

TABLE 5
Post-emergence treatments for grass weed control

Herbicide treatments	Rate (kg/ha)	Controlled weeds
A) Selective herbicides[a]		
Fluazifop-P-butyl	0,06 - 0,09 0,10 - 0,15	
Haloxyfop-methyl	0,05 - 0,06 0,08 - 0,10	
Haloxyfop-P-methyl	0,03 - 0,04 0,04 - 0,05	
Quizalofp-ethyl	0,04 - 0,05 0,08 - 0,10	
Quizalofop-P-ethyl	0,02 - 0,03 0,04 - 0,05	Johnsongrass and annual grasses
Fenozaprof-P-ethyl	0,10 - 0,15	
Propaquizafop	0,03 - 0,04 0,05 - 0,07	
Quizalofop-P-tefuryl	0,03 - 0,04 0,05 - 0,07	
Clethodim[b]	0,06 - 0,09 0,15 - 0,24	
Sethoxydun	0,50 - 0,75 0,18 - 0,24	Johnsongrass Annual grasses
A) Glyphosate with rope-wick applicators		
Glyphosate: 12 % (acid equivalent) (Treat when johnsongrass is at boot stage or later)		Rhizome johnsongrass, only when infestation is light to moderate

[a]Low rates can be used in good to very good soil conditions and in conventional tillage systems.

[b]Rates of clethodim can be reduced 50 % by adding dilauril ester polyethylenglycol ester at 2 L/ha.

Bermudagrass can be controlled with the selective grasskillers by increasing 40 to 60 % the normal rates.

The performance of post-emergence herbicides for controlling johnsongrass, depend on several factors: i.e., weed density, rainfall pattern before and after planting soybeans, stage of johnsongrass, soil moisture, relative humidity, size of rhizomes, mixture with other chemicals, adyuvants (type and rate), biotype, and quality of application. In some cases insolation during midday applications will adversely affect the results.

By knowing and managing many of these factors, a better decision can be made on the election of the adequate chemical and the proper rate. Consequently, better results and/or a reduction in the cost and/or lesser risk of polluting the environment, may be achieved.

On this line, the main objectives of research projects are to study aspects of weed bioecology (weed density and yield reduction, weed-free period, population dynamics, seed bank), to look for new products that have less impact in the environment, to study how different factors affect chemical control, and search for new alternative methods of managing weeds.

ECONOMIC IMPORTANCE

The weed problem in soybeans in Argentina has evolved in the last 10 years, leading to a significant increase of the herbicide market. The most used herbicides belong to post-emergence groups (Table 6). In 1990 24.4 % of the planted area with soybeans was treated with broadleaf post-emergence products whereas 41.8 % was applied with post-emergence grasskillers. Of the 75.7 % treated area, 44.6 % received 2 to 3 treatments, either tank mixed or sequential.

The main combinations are: a pre-plant soil incorporated or a pre-emergence herbicide followed by post-emergence treatments, and mixtures of an herbicide for broadleaf weeds and a selective post-emergence grasskiller.

Only about 300,000 hectares of soybeans are in double-cropping no tillage systems. The main treatments used are post-emergence herbicides applied before planting (paraquat and glyphosate), followed by a residual product or selective post-emergence herbicides.

TABLE 6
Treated area with herbicides in soybeans in different periods
(1,000 hectares)

Type of treatment	1982–83	1984–85	1988–89	1990–91
Pre-plant incorporated	568	555	899	1,080 (20 %)
Pre-emergence	193	231	873	752 (14 %)
Post-emergence (broadleaf)	215	780	1,473	1,323 (24 %)
Post-emergence (grasses)	378	1,204	1,375	2,265 (42 %)
Total	1,354	2,820	4,620	5,420 (100 %)
Treated 2 to 3 times	266	293	1,428	1,573 (34 %)
Treated at least once	1,088	2,527	3,192	3,447 (76 %)
Planted area	2,226	3,300	4,500	4,950 (100 %)

The economic importance of weeds in soybeans in Argentina can be appreciated in Table 7. Again, the most used chemical treatments are those of post-emergence herbicides. In 1990, 32 % of what was spent by the soybean growers for chemical weed control was for broadleaf post-emergence treatments, whereas 43 % was for post-emergence grasskiller applications. An important role was also played by the mechanical control. In 1990 about 60 % was spent of the total cost of weed control was spent in mechanical practices.

TABLE 7

Cost of weed control in soybeans in Argentina in different periods
(1,000 dollars)

Type of treatment	1982-83	1984-85	1988-89	1990-91
Pre-plant incorporated	7,851	7,116	12,390	14,374 (13 %)
Pre-emergence	4,191	5,171	14,910	12,402 (12 %)
Post-emergence (broadleaf)	4,937	23,868	38,830	33,677 (32 %)
Post-emergence (grasses)	9,489	13,455	29,210	46,491 (43 %)
Chemical control[a]	26,468	49,610	95,340	106,944 (40 %)
Mechanical control	75,684	112,200	147,900	157,148 (60 %)
T O T A L	102,152	161,810	243,240	264,092 (100 %)

[a] Chemical control includes the herbicide price paid by the growers and the application cost.

Herbicides share the main part of the pesticide market in Argentina. In 1986 the total market was of 175.2 million dollars: 54.6 % were herbicides, whereas 27.8 % and 11.5 % were insecticides and fungicides, respectively. In 1990, the total pesticide market reached 244 million dollars: 67.1 % were herbicides (half of which were used in soybeans), whereas 20 % and 7.5 % were insecticides and fungicides, respectively.

On the other hand, several natural agents (insects and pathogens) affect the growth of many weeds in soybeans in Argentina. There is no evaluation of the importance of this contribution. Some of these agents are soybean pests; then, there are complex interactions among them. It is an important aspect to be taken into account in future research projects.

ACKNOWLEDGMENTS

Thanks are due to BASF ARGENTINA S.A., CYANAMID DE ARGENTINA S.A., DOW-ELANCO ARGENTINA S.A., UNIROYAL QUIMICA DE ARGENTINA S.A.C.I. and Mr. Kief SEMINARIO for the information they provided on the herbicide market.

WEED-SOYBEAN INTERFERENCE STUDIES IN BRAZIL

ROBINSON ANTONIO PITELLI
Faculty of Agriculture and Veterinary Sciences/UNESP
14870, Jaboticabal, São Paulo, Brazil

ABSTRACTS

In Brazil, there is a great number of weed species which are important in soybean crop. Nowadays, an expressive weed shifting has occurred in some soybean growing regions, as a result of the tillage system change and herbicide use. The weed communities are very diversified and promote a strong interference on soybean growth and productivity, especially when the shading of superior leaves of the canopy occurs. Some changes on certain agriculture practices, such as cultivars, row spacing and sowing density, were studied and were considered able to be used for the establishment of the integrated weed managment system under Brazilian conditions.

INTRODUCTION

The soybean area in Brazil (1990/91) was about 9.5 million of hectares, of which 60 % were located in the South region with a mild climate and high fertility soils. The remainder 40 % were located in the Center-West region with a sub-tropical climate and very acid soils (Cerrado areas)

The main weeds in the South Region of Brazil are Euphorbia heterophylla and Brachiaria plantaginea. Such weeds were selected after a long period of residual herbicide use, especially the combination of metribuzin and trifluralin. Other very important species are Bidens pilosa, Commelina benghalensis, Sida spp (mainly S. rhombifolia), Digitaria spp (mainly D. horizontalis) and Raphanus raphanistrum.

There was an increase in the diversity of soybean herbicides in the last years. As a result of this diversity, the weed selection was changed and other species are becoming more important nowadays such as Cardiospermum halicacabum, Ipomoea spp (mainly I. purpurea and I. aristolochiaefolia) and Acanthospermum hispidum.

Another factor that contributes to the weed shifting in the South region of Brazil is the adoption of No Tillage system in some areas. Some species, such as Digitaria insularis, Borreria alata and Erigeron bonariensis, that were less important in the conventional system are now very important in the No Tillage cultivation areas. It is accepted that the lack of soil disturbance and the limitation in the use of certain herbicides are the main factors of such behavior. The recent introduction of Sorghum halepense is another preoccupying fact in the South region of Brazil.

A massive plantation of soybean in the Central-West region of Brazil is still recent. That is why there is a strong influence of previous crops (especially upland rice, reforestation and pastures) in the weed species composition. The low technological level used in some areas made possible the formation of different zones of the weed seeds introductions. Besides, there was no sufficient time so that the use of herbicides promote an expressive weed shifting and, as a result of this, a reduction of the number of weed species and a increase of its areas of geographic distribution has occurred.

For this reason, there is not a group of weeds which are dominant in all Central-West region of Brazil. There are some species which are very important in the main plantation areas such as Digitaria horizontalis, Bidens pilosa, Acanthospermum australe, Acanthospermum hispidum, Cenchrus echinatus, Senna tora, Ipomoea spp (mainly I. acuminata and I. aristolochiaefolia) and Ageratum conyzoides.

Other weeds are becoming very important species in the "Cerrado areas", mainly due to the quick population growth, herbicide resistance and high dispersal capacity, such as Hyptis suaveolens, Tridax procumbens and Nicandra physaloides.

In all soybean growing areas in Brazil, the selection of species of the Fabales order (Leguminosae), are becoming more and more strong, specially in the warm regions. The main species are Senna tora, Desmodium tortuosum and Chamaecrista nictitans.

WEED INTERFERENCE ON SOYBEAN GROWTH AND PRODUCTIVITY

Usually the term interference is used to refer to the group of environmental stresses suffered by the soybean crops due to the presence of the weeds in the agroecosystem. The main direct forms of weed interference are competition, allelopathy and parasitism. The occurrence of parasitic weeds in soybean crop is not considerable in Brazil.

The acquisition of any resource for the plant development is susceptible of weed interference. Nevertheless, really the interference is important only for light, nutrients and water.

The light interception caused by fast weed growth may be the most important form of interference on the soybean crop. Many authors, such as Pitelli & Neves [01], Chemalle & Fleck [02], Pitelli et al [03], Gazziero et al [04], Velini [05] and Spadotto [06], observed a high reduction on the soybean leaf area when the crop was submitted to weed interference. Soybean growth analysis studies carried out by Pitelli & Neves [01], Velini [05] and Spadotto [06] , demonstrated that the leaf area reduction is especially due to a premature senescence of leaves when the light compensation point was reached, as a consequence of the intense shading caused by the weeds.

An interesting aspect observed by Velini [05] and Rossi [07] is that, at early stages of soybean cycle (40 - 60 days),an expressive leaf area reduction occurs without, above all, causing damages to the production of grains. A possible explanation for such behavior is that the senescence occurred only in the basal leaves of the plant. At this time,normally, these basal leaves were submitted to self-shading.

At an advanced stage of the soybean cycle, the weeds begin to shade the superior leaves which are photosynthetically more active. After this, the reflexes over the crop productivity will be expressive, as related by Rossi [07].

Another important aspect related to the light interception and the senescence of parts of the plants is the reduction of lateral branching of soybean plant, as observed by Pitelli & Neves [01], Machado Neto [08] and Velini [05]. This behavior appears to be fundamental once it can be related to the main production characteristic affected by the weed interference, which is the number of pods per plant [02, 05, 07, 08, 09]. The reduction in the number of pods can be consequence of a smaller number of floral buds, due to the reduction of lateral branches [05].

In addition, it is important to mention that the number of seeds per pod is not affected by the weed interference [02, 05, 07, 08, 09, 10]. Usually, the seed size is not affected by weed interference [02, 08, 10]. However Velini [05] observed an expressive reduction on the soybean seed size as a result of the weed interference. Pitelli [11] commented that a low uniformity in the distribution of weed species in the crop area can provide a great lack of uniformity in the flowering and ripening of soybean seeds. Thus, it is possible that during the harvest some seeds could be picked in the formation process, reflecting in the results of grain average weight.

The water competition did not deserve notability in the Brazilian research, probably because the soybean crop is conduced during the rainy season.

The competition for nutrients between soybean and Cyperus rotundus was studied by Pitelli et al [03] who observed reductions in the soybean dry matter accumulation and in the nitrogen, phosphorus, potassium, calcium and magnesium contents. They also observed that a reduction occurred in the number and size of Rhizobium sp nodules on soybean roots.

Rassini [12] observed that the weed interference reduced the nitrogen and phosphorus contents in soybean leaves, but did not change the contents of potassium, calcium and magnesium. Durigan et al [13] observed that the effects of weed interference on the protein contents of soybean seeds were not expressive.

Usually the Brazilian soils have low phosphorus contents, but the fertilization located in the seed row and the immobility of this nutrient in the soil, practically eliminate the possibilities of a strong competition for the element.

In Brazil, the researches involving the allelopathic interference of weeds on the soybean crop are scarce. There is a number of papers studying the effects of aqueous extracts of weed plants parts on the

285

soybean germination and growth. Such studies were carried out involving
Brachiaria plantaginea [14], Mabea sp [15] and Nicandra physaloides [16].
Gastal & Casela [17] and Almeida et al [14] studied the effects of
Brachiaria plantaginea and Cyperus rotundus residues soil incorporations.
All these publications showed reduction effects on the soybean germination
and early growth. Almeida et al [14] observed the need of cumulative
effects of the allelopathic-like substance in soil.

FACTORS AFFECTING THE WEED INTERFERENCE DEGREE

Among the factors which affect the interference degree between the soybean
crop and the weed communities (Figure 01) the studies related to the crop
itself and to the periods that are critical in the weed-crop relationships
were those that deserved special attention by the Brazilian research
programs.

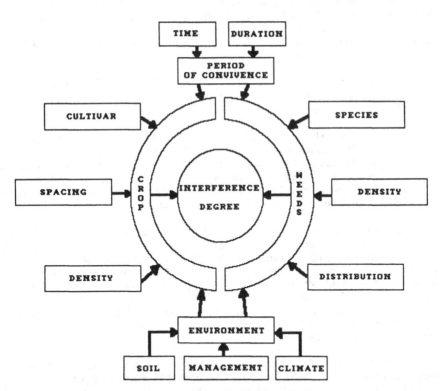

Figure 01. Factors affecting the weed-crop interference degree. From
Pitelli [18] (adapted from Bleasdale [19] and Blanco [20]).

There are great differences concerning competitive ability of the
soybean cultivars sowing in Brazil. Chemalle & Fleck [02] comparing six
soybean cultivars observed that cv. 'Prata' was the most susceptible to
the Euphorbia heterophylla interference. This high susceptibility was
attributed to the characteristics of a low height and low leaf area index.
The least susceptible soybean cultivar was 'BR-1', which exhibited high
height, rapid initial growth and a high leaf area index.

Rassini [12] observed a larger weed control effectiveness proportionated by sub-doses of lactofen and fomesafen on plots planted with 'IAC-8' when compared with 'Foscarin'. This difference in the herbicides efficacy was attributed to the faster initial growth and larger shading proportioned by the IAC-8 cultivar.

The row spacing and seeding rate are very important factors influencing the quickness and the degree of shading promoted by the soybean crop. Maia et al [09] showed that the weed interference effects on soybean crop were reduced as the soybean plant population increased. Rassini [12] observed that sub-doses of lactofen and fomesafen promoted a more effective weed control at soybean row spacing of 0.30 m, when compared with the spacing of 0.60 m.

In the Brazilian soybean growing areas the weed communities are very diversified. For this reason, there are few papers studying the interference involving soybean plants and a particular weed species. Chemalle & Fleck [02] studied the Euphorbia heterophylla interference on soybean crop. The effects were influenced by the weed presence period and by the weed density. The reduction of the grain production was 43% for a weed population of 54 plants/m .

Pitelli et al [03], at greenhouse conditions, observed that the Cyperus rotundus interference reduced the soybean leaf area, dry matter accumulation and nutrient uptake.

Gazziero et al [04] studied the effects of a Sorghum halepense population of 48 shoots/m (average) on the soybean productivity. The reductions in the productions observed were 23%, 45% and 64% for periods of weed presence from the soybean emergence until 28, 42 and 56 days, respectively. These results explain the great care about the recent appearance of this species in the South of Brazil.

Field studies trying to obtain correlations between weed density or weed dry matter and soybean productivity have not reached satisfactory results. The possible explanation for such a thing is the great lack in uniformity that occurs in the species composition, geographic distribution and in the growth of weed species even in small areas. Chemalle & Fleck [02] observed that the soybean productivity reduction occurred in proportion to the Euphorbia heterophylla density. They observed productivity reductions of the 19% and 43% for weed densities of 12 and 54 plants/m , respectively.

Machado Neto [08] noted that certain growth regulators such as giberelic acid and clomerquat chloride promoted morphological alterations on soybean plants, although such alterations were not able to change the soybean competitive ability.

Studies involving periods considered critical for the weed presence or for the weed control on soybean productivity were those which received a better attention from the Brazilian research programs.

The period that is more studied is the period from the soybean emergence which the crop must be kept free of the presence of weeds so that the crop productivity will not be affected by the interference process. This period was denominated by Pitelli & Durigan [21] as "Total Period of Weed Interference Prevention".

As a whole, the experimental models used were the traditional ones involving crescent periods of weed control and with results interpretation based on Multiple Comparison Tests (Tukey or Duncan). More recently it has been adopted the tendency models studies especially the exponential [05] and the Broken Stick models [06].

The studies made in Brazil showed low values for the total period of weed interference prevention [05, 06, 07, 09, 10, 22], probably due to the germinative behavior of the weeds at our climate conditions. Durigan et al [10], Rossi [07], Velini [05] and Spadotto [06] observed that during the period of 0-30 days after the soybean sowing occurred the emergence of almost all the weed community. After this period, the weed emergence was always small. Probably, for this reason, the majority of the experimental models carried out in Brazil has studied periods of control reaching even 60 days after sowing, which harms the data analysis through the exponential models, as observed on Figure 02.

Some kinds of critics that usually takes place concerning this type of study is that only the effects on the productivity are analyzed and not those on the work done during harvest. Hirata [23] observed that in areas with a poor weed control serious damages occur while handling a reaper machine or damages with the grain harvested.

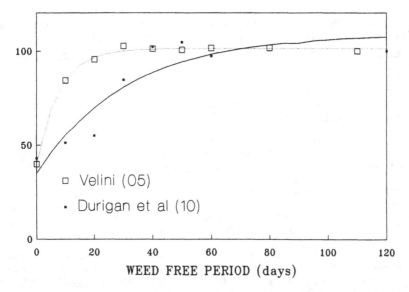

WEED FREE PERIOD (days)

Figure 02. Effects of weed free period on the soybean grain production. Adapted from Velini [05] and Durigan et al [10].

The period in which the crop can live together with the weed community before a significant damage occurs on its productivity, was also studied by various authors. Usually these periods are very long and are not very useful to give orientation on control measures to be applied at post-emergence situations and are denominated "Pre-interference Periods" [21].

The comparison between these two types of periods are very useful in weed control orientation. When the Pre-interference Periods are longer than those of the Total Period of Weed Interference Prevention, the farmer has an adequate situation to use only a single post-emergence weed control measure. Otherwise, when the pre-interference period is shorter than the total period of weed interference prevention, there is a critical period of interference prevention, being necessary the use of subsequent weed control measures or a measure with residual effects [18].

FINAL CONSIDERATIONS

Recently a great number of researches about weed interference on soybean crop have been carried out in Brazil, especially effects of tillage practices, such as fertilization, row spacing and crop cultivars influences, economic thresholds levels and other studies which can be useful in the establishment of integrated weed control programs. Studies about the effects of long term herbicide use on the weed species composition are being carried out in some regions of Brazil.

REFERENCES

01. Pitelli, R.A., Neves, A.S., Efeitos da competição das plantas daninhas sobre algumas características morfológicas e agronômicas de plantas de soja. In: Seminário Brasileiro de Herbicidas e Ervas Daninhas, 12th, Fortaleza, 1978, Abstracts, p.104.

02. Chemalle, V.M., Fleck, N.G., Avaliação de cultivares de soja em competição com Euphorbia heterophylla sob tres densidades e dois períodos de ocorrência. Planta Daninha, 1982, 05, 36-45.

03. Pitelli, R.A., Durigan, J.C., Benedetti, N.J., Estudos de competição inter e intraespecífica envolvendo Glycine max (L.) Merril e Cyperus rotundus L., em condições de casa de vegetação. Planta Daninha, 1983, 06, 129-137.

04. Gazziero, D.L.P., Ulbrich, A.V., Voll, E., Pitelli, R.A., Estudos dos efeitos do período de convivência do capim massambara (Sorghum halepense (L.) Pers.) com o crescimento e a produção da cultura da soja. In: Brasil, EMBRAPA, 1989, Annual Report 88/89, p.306-308.

05. Velini, .D. Avaliação dos efeitos do comunidades infestantes naturais, controladas por diferentes períodos, sobre o crescimento e produtividade da cultura da soja (Glycine max (L.) Merril). Jaboticabal, FCAVJ/UNESP, 1989. MSc Dissertation, 153p.

06. Spadotto, C.A. Determinação do período crítico para prevenção da interferência e avaliação de parâmetros para o controle monitorizado de plantas daninhas de folhas largas na cultura da soja (Glycine max (L.) Merril). Botucatu, FCA/UNESP, 1991, MSc Dissertation, 83p..

07. Rossi, C.A., Efeitos de períodos de controle e convivência de plantas daninhas na cultura da soja (Glycine max (L.) Merril). Jaboticabal, FCAVJ/UNESP, 1985, Graduation Thesis, 40p.

08. Machado Neto, J.G., Estudos preliminares dos efeitos de fitorreguladores nas plantas daninhas, na soja cv. Santa Rosa e nas relações competitivas entre a cultura e a comunidade infestante. Jaboticabal, FCAVJ/UNESP, 1981, Graduation Thesis, 71p.

09. Maia, A.C., Machado, A.M., Laca-Buendia, J.P., Efeito de espaçamento e população de plantas no controle de plantas daninhas na cultura da soja (Glycine max (L.) Merril) em solo de cerrado. In: Seminário Nacional de Pesquisa de Soja, 2nd, Brasilia, 1981, Annals, p.331-338.

10. Durigan, J.C., Victoria Filho, R., Matuo T., Pitelli, R.A., Períodos de matocompetição na cultura da soja [Glycine max (L.) Merril], cultivares Santa Rosa e IAC-2. I- Efeitos sobre os parâmetros de produção. Planta Daninha, 1983, 06, 86-100.

11. Pitelli, R.A. Efeitos do período de competição das plantas daninhas sobre a produtividade do amendoim (Arachis hypogaea) e o teor de macronutrientes em suas sementes. In: Piracicaba, ESALQ/USP, 1980, MSc Dissertation, 89 p..

12. Rassini, J.B., Integração de práticas culturais e baixas dosagens de herbicidas em pos-emergência para o controle de plantas daninhas na cultura da soja (Glycine max). Jaboticabal, FCAVJ/UNESP, 1988, PhD Thesis, 115p.

13. Durigan, J.C., Victoria Filho, R., Matuo, T., Pitelli, R.A., Períodos de matocompetição na cultura da soja [Glycine max (L.) Merril], cultivares Santa Rosa e IAC-2. II- Efeitos sobre características morfológicas das plantas e constituição química dos grãos. Planta Daninha, 1983, 06, 101-114.

14. Almeida, F.S., Rodrigues, B.N., Voss, M., Efeitos alelopáticos e competição da Brachiaria plantaginea na soja. In: Congresso Brasileiro de Herbicidas e Plantas Daninhas, 16th, Campo Grande, 1986. Abstracts, p. 5-6.

15. Souza, I.F., Potencial alelopático do jambreiro (Mabea sp) sobre a cultura da soja. In: Congresso Brasileiro de Herbicidas e Plantas Daninhas, 16th, Campo Grande, 1986, Abstracts, p. 12.

16. Braga, P.E.T., Pereira, R.C., Propriedades alelopáticas do joá de capote (Nicandra physaloides) em tomate, soja e milho. In: Congresso Brasileiro de Herbicidas e Plantas Daninhas, 18th, Brasilia, 1991. Abstracts, p.18.

17. Gastal, M.F., Casela, C.R., Influência alelopática de Cyperus rotundus L. em Glycine max (L.) Merril). In: Seminario Nacional de Pesquisa de Soja, 2nd, Brasilia, 1981. Annals, p.366-370.

18. Pitelli, R.A., Interferências das plantas daninhas em culturas agrícolas. Informe Agropecuário, 1985, 11, 16-27.

19. Bleasdale, J.K.A., Studies on plant competition. In: The Biology of Weeds, ed. J.L. Harper, Blackwell Scientific Publications, Oxford, 1960, p.1323-42.

20. Blanco,H.G., A importância dos estudos ecológicos nos programas de controle de plantas daninhas. O Biológico, 1972, 38, 343-50.

21. Pitelli, R.A., Durigan, J.C., Terminologia para períodos de convivência e controle das plantas daninhas em culturas anuais e bianuais. In: Seminário Brasileiro de Herbicidas e Plantas Daninhas, 15th, Belo Horizonte, 1984, Abstracts, p.73-74.

22. Blanco, H.G., Oliveira, D.A., Araujo, J.B.M., Grassi, N., Observações sobre o período em que as plantas daninhas competem com a soja (Glycine max (L.) Merril). O Biologico, 1972, 38, 33-35.

23. Hirata, M.A.K., Efeito do manejo de plantas daninhas sobre a quantidade e qualidade de produção da soja. Jaboticabal, FCAVJ/UNESP, 1986, Graduation Thesis, 61p.

INTEGRATED METHODS FOR WEED MANAGEMENT IN SOYBEANS IN BRAZIL

DIONISIO LUIZ PISA GAZZIERO

EMBRAPA-Centro Nacional de Pesquisa de Soja (CNPSo)

Cx. Postal 1061, 86.001, Londrina, PR, Brasil.

ABSTRACT

Although many other options to the use of herbicides are already available, the chemical control of weeds in soybeans is the method most used by Brazilian farmers, because it is fast and apparently easy to perform. The combination of physical and cultural control with the chemical control is common, especially on the small and medium size farms. In the no-tillage system, the remaining crop residues in the soil surface has lead to modifications in the spectrum of weed infestation and reduced the germination of some species, such as Brachiaria plantaginea, a serious weed problem in Southern Brazil. There are no standard models to be followed in the establishment of integrated weed control systems. An agronomist is always required for the analysis of the weed problems at each farm as well as for the stablishment of an adequate alternative. Integrated weed control methods have permitted to increase the efficiency and to reduce the cost and other problems created by the use of chemical products.

INTRODUCTION

Soybeans are grown in Brazil in two regions. The first one located in the south, the traditional region, has a normal rainfall distribution all year round, and allow to grow summer and winter crop species. During the 1990/91 growing season, soybeans were grown over an area of 5.8 mill.ha in this region. The other area, located in the central-west Savanna area of Brazil, the expanding region, has an irregular rainfall distribution which is concentrated in the summer and permits only one crop a year. In this region soybeans were grown over an area of 3,7 mill.ha during the 1990/91 growing season. A third region, located in the north and north-east part of the country, could be characterized as a potential for soybean production area, although in this region, the area occupied by soybeans is still very small (less than 50,000 ha).

Several systems are used in Brasil for soil preparation. With the conventional system, one disc plowing followed by two disc harrows are used.

In the reduced system, which has been most frequently used, a heavy disc harrow followed by two disc harrows are used. The heavy disc harrow associated with inadequate soil and farm management have lead to serious problems of soil compaction and erosion. The use of the chisel plow has increased in Brasil, replacing the disc plow and heavy disc harrow.

In the southern region many farmers are using the no-tillage system. This system is already in used in more than 1 mill.ha with several field crops, mainly soybeans, wheat and maize. Despite its advantages in preventing erosion this system turns weed control into a very difficult task.

In several ways, the weed species are similar in both production regions. Although some species have specific requirements in terms of soil and climatic conditions and are not adapted to all production areas. Losses due to weeds lead farmers to use different methods of control, mainly chemical control. Many problems may arise with the use of chemicals products for it is very difficult to conciliate efficiency with low cost.

In this paper, the major weeds as well as the alternatives to the integrated management are discussed.

WEEDS

To determine the effect of weed competition with soybeans many experiments were conducted with different weed species (1). Results indicated that the yield can be reduced from 40% to 90%, depending on the weed species and that the critical competion period was the first 45 days after soybean emergence. The broad-leaf weeds were generally less aggressive than grasses. Weeds not only decrease yield but also affect the efficiency of harvest and the quality of the final product (2).

The range of weed species infesting the soybean crop is wide and a list of the major weeds that occur in Brazil is presented in Table 1. The importance of each weed vary from place to place. Species such as Cassia tora and Desmodium purpureum, although present in some areas of the traditional region, are more frequent in the expanding region, where they are considered very important weeds for the soybean crop. Some other species like Rottboelia exaltata were already identified in Brasil, but still did not reach economic importance for soybeans. In the some way, Cyperus rotundus, which is very common throughout the country, is not important for the crop. Other examples are Brachiaria plantaginea, which is the biggest problem in the south, and Cenchrus echinatus, which is the worst problem in the central-west region.

WEED MANAGEMENT

Methods of Control

Although the chemical control is most common, several other methods are available to the farmers. It is very important to know each one of them in order to be able to establish a suitable system which would be adapted to the weed problems of each farm.

Prevention

Prevention is one alternative to avoid the introduction of a weed species where it is still not prevent. When soybeans were introduced in the expanding region, several weed species were introduced from the south region. Different methods were suggested to present weed dissemination (3, 4). The use of certified seeds, a thorough cleaning of equipment and machinery, crop rotation, good planning for the use of the farm land as well as training of technicians and farmers are good examples of how to prevent introduction of weed in to new areas. Unfortunately in Brazil, prevention does not receive adequate attention and the problems with weeds are becoming more important, much more difficult and expensive to control. Euphorbia heterophylla was rapidly disseminated throughout the Southern region of the country, when soybeans started to be grown extensively. Nowadays Cassia tora and Desmodium purpureum, among others, have been disseminated in an alarming way.

Considering these examples, it is clear that prevention is a very important method of control since erradication is in practice, technically difficult and a economically impossible for extensive areas.

Cultural Control

The Brazilian farmers generally use the technology recomended by research and extension services, thus permitting an adequate development for the crop. These cultural practices also help in promoting good conditions for soybeans to compete with weeds.

Crop rotation, mainly with maize, has been used by many farmers. Research results (5) indicated

TABLE 1
Most important weeds in soybean, in Brazil.

Scientific name	Common name	Family
Acanthospermum australe	Carrapicho-rasteiro	Compositae
Acanthospermum hispidum	Carrapicho-de-carneiro	Compositae
Ageratum conyzoides	Mentrasto	Compositae
Alternanthera tenella	Apaga-fogo	Amaranthaceae
Amaranthus hybridus	Caruru	Amranthaceae
Amaranthus viridis	Caruru-de-mancha	Amaranthaceae
Bidens pilosa	Picão-preto	Compositae
Brachiaria plantaginea	Capim marmelada	Gramineae
Brachiaria decumbens	Capim braquiaria	Gramineae
Cassia tora	Fedegoso	Leguminosae
Cenchrus echinatus	Capim carrapicho	Gramineae
Commelina benghalensis	Trapoeraba	Commelinaceae
Desmodium purpureum	Carrapicho beiço-de-boi	Leguminosae
Digitaria horizontalis	Capim colchão	Gramineae
Echinochioa crus-galli	Capim-arroz	Gramineae
Eleusine indica	Capim pé-de-galinha	Gramineae
Emilia sonchifolia	Serralha	Compositae
Eupatorium pauciflorum	Botão azul	Compositae
Euphorbia heterophylla	Amendoim bravo (leiteiro)	Euphorbiaceae
Galinsoga parviflora	Fazendeiro	Compositae
Hyptis lophanta	Fazendeiro/cheirosa	Labiatae
Hyptis suaveolens	Cheirosa	Labiatae
Ipomoea aristolochiafolia	Corda-de-viola	Convolvulaceae
Mitracarpus hirtus	Poaia-da-praia	Rubiaceae
Nicandra physaloides	Joá-de-capote	Solanaceae
Pennisetum setosum	Capim custódio	Gramineae
Portulaca oleraceae	Beldroega	Portulacaceae
Raphanus raphanistrum	Nabiça	Cruciferae
Richardia brasiliensis	Poaia-branca	Rubiaceae
Sida rhombifolia	Guanxuma	Malvaceae
Solanum americanum	Maria-pretinha	Solanaceae
Solanum sisymbriifolium	Joá-bravo	Solanaceae
Sorghum halepense	Capim massambar	Gramineae
Tridax procumbens	Erva de touro	Compositae
Vigna sinensis	Feijão miúdo	Leguminosae

that after a five year rotation period, the number of broad-leaf weeds in soybeans decreased in both conventional and no-tillage systems. Similar results were also observed with the wheat crop (5).

In areas highly infested with Cassia tora, crop rotation with maize was the best control method since the herbicides used for maize highly decreased the population of that weed once the weed population is low, herbicides which do not achieve satisfactory level of control under normal conditions, could be effectively used. The reduction of the spacing between rows to 35-40 cm is another cultural practice adopted by the farmers. The early application of postemergence herbicides associated with narrow spacing had permitted dosage reduction of the herbicides fluazifop-butyl and setoxidim in controling Echinochloa spp. and Oryza sativa (6). Field trials using row spacings varyng from 0,20 m to 0,80 m (7) demonstrated that weed control increased as row spacing decreased.

Physical Control

The physical control has been adopted mainly in the small size farms of the traditional growing region. Hand weeding is the method most commonly used. Besides its efficiency this method has also a very important social function since it provides work for many low income field workers.

The use of pre-plant incorporation of trifluralin followed by mechanical cultivation is also commonly used. Several different models of cultivators are available on the Brazilian market.

The association of pre-emergence herbicide applied in the row and mechanical weeding between rows was studied in field trials (8). Results indicated that the level of control was the same obtained by the use of herbicides applied to the total area, with significant reduction in costs. Although efficient, this practice presents some problems which prevent its wide adoption. One of these problems is the fairly sharp slopes of the fields in the southern region. Another one is the poor control achieved by the pre-emergence herbicides in areas highly infested with Brachiaria plantaginea. This method has been used in maize.

In the expanding region, the constant rainfall during the growing season, the size of the farms and the lack of labour has prevented the extensive use of physical control.

Chemical Control

In Brazil, the herbicides represent 50% of the total chemicals used in agriculture, and from this figure 23% represent the herbicides used in soybeans (9). The Brazilian market was always promising and stimulated the introduction of different chemical products. In 1970, in the technical recomendations for the Soybean Crop (10) only two herbicides were recommended for the control of three weeds which were considered of economical importance. Nowadays more than 33 products and mixtures are recommended for the control of approximately 40 different weeds in both conventional and no-tillage systems. A list of the major herbicides used in the crop is presented in the Table 2.

These figures demonstrate the preference of the farmers for chemical control, which is explained by its low cost in labour and quickness in application. Despite these advantages, many problems have also to be taken into consideration; 1) Contamination of the environment; 2) adjusting of equipment; 3) wrong choice of dosage and product; and 4) the appropriate time for application.

Trifluralin has been the most widely used herbicide in controlling grass weeds, not only for its efficiency but also for its low cost as well. For broad-leaf weeds, imazaquin is the most used herbicide. Besides its broad spectrum, this herbicide provides levels up to 80% control of Euphorbia heterophylla.

The post-emergence herbicides have been extensively used, although presenting inconvenience of strict dependance on weather conditions and limitations in controlling broad-leaf weeds as far as development stages are concerned.

Tank mixtures are frequently used by farmers. The most common mixtures are the ones including residual products for grass weeds with products specific for the control of broad-leaf weeds. Mixtures of two post-emergence products for the control of broad-leaf weeds aiming to increase efficiency and spectrum, are also used.

When the efficiency of control is not well achieved by the farmer, during crop development or when foliar retention occurs due to physiological problems, paraquat is used as dessicant at harvest time. Results of field trials concerning chemical residues in soybean seeds, meal and oil (11) indicated residue levels below those recomended by FAO and the Braziliam Ministry of Agriculture, since the seven day interval between application and harvest is respected.

In the no-tillage system the weeds that germinate before planting are controlled by the use of paraquat + diuron in one or two applications or with glyphosate in a single application. In both cases, a mixture with 2,4-D is generally used to control broad-leaf weeds and to reduce costs. This product has to be used with a 10 days interval between application and planting to avoid damages to the crop. Fig. 1.

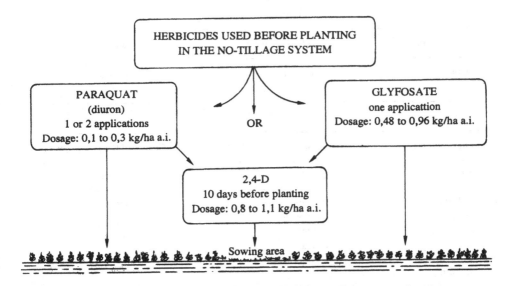

Figure 1. Desiccant herbicides used alone or in combination in the control of weeds that germinate before planting summer or winter crops.

After desiccation, the chemicals used in weed control are pratically the same used in the conventional system except that the pre-plant incorporated and some pre- emergent herbicides such as cyanazine can cause phytotoxicity. In this system the germination of some weeds may increase, mainly the perennials like Sida rombifolia, Senecio brasilensis, and Erigeron bonariensis. The annuals such as Brachiaria plantaginea tend to be reduced. Field observations were confirmed by trials (12) which demonstrated variations in the weed germination from 13% to 60% in the conventional system and from 3% to 5% in the notillage system. Other experiments (13) indicated reduction in the number of Brachiaria plantaginea from 61 to 21 plants/m2 when soil covered with wheat straw was compared with soil covered with oat straw at planting time. This can be explaned by the physical, chemical and biological characteristics of the soil which is greatly changed in the no-tillage system (14) interfering with seed dormancy. Besides, this depletion of the seed bank may occur. These factors in addition to allelopathic activity are responsable for the changes in infestation. Allellopathic effects have been largely studied in Brazil (15), especially with winter crops, such as Avena strigosa, Secale cereale, Lolium multiflorium and others which keep the soil completely clean for long periods. For this reason, in south Brazil, the use of muching is widely used by farmers which adopted the no-tillage system.

In the northern and eastern areas of the state of Parana, southern Brazil, it is possible to anticipate the planting date of soybean in the no-tillage system in 15 to 20 days as compared with the normal planting date. This practice has also reduced application of dessicants. In the no-tillage it is very important to adequately manage the weeds keeping all crops constantly celan. The success of the system greatly depends on an adequate weed control.

TABLE 2
Herbicides used in soybean production in Brazil.

Pre-Plant Incorporated		
Grass control	Broad-leaf control	Grass and broad-leaf control
– trifluralin	– imazaquin – metribuzin	– trifluralin + metribuzin

Pre-Emergence		
Grass control	Broad-leaf control	Grass and broad-leaf control

– alachlor
– alachlor
  + trifluralin
– clomazone
– metolachlor
– oryzalin
– pendimethalin
– trifluralin[1]
	– chlorimuron - ethyl[2]
	  + diuron
	– imazaquin
	– metribuzin
	– linuron
	– cyanazine[2]
		– cyanazine[2]
		  + metolachlor
		– metribuzin
		  + metolachlor

Post-Emergence		
Grass control	Broad-leaf control	Grass and broad-leaf control

– clethodim
– fenoxaprop-p-ethyl
– fluazifop-p-butyl
– haloxyfop-methyl
– sethoxydim
	– acifluorfen-sodium
	– bentazon
	– bentazon
	  + acifluorfen
	– chlorimuron-ethyl
	– fomesafen
	– lactofen
		-fluazifop-p-butyl
		  + fomesafen
		– imazethapyr

[1] Formulated for pre emergence

[2] Used only for no-tillage

Other Alternatives

Weed Science is relatively recent in Brazil. Even though the technical information already available is reasonable. Many researchers have been conducting studies in basic areas such as biology, competition and populational dynamics of weeds.

Biological control has been studied (17) with good perspectives for the control of Euphorbia heterophylla by the fungus Helminthosporium sp. (= Bipolaris euphorbiae). This pathogen, which is specific for this weed, was isolated from diseased field plants, studied in the laboratory and formulated as a biological herbicide.

Studies have been conducted in the savanna area of central Brazil in order to achieve better soil and weed management (18). Under these conditions some species such as Panicum maximum, Stilozobium sp. and others, present promising features to be used in mulching.

Integrated Methods

In a country like Brazil, with large production areas with different and specific problems and characteristics it is impossible to stablish standard rules for weed management. Besides, plant communities composed by different species, each one with its own characteristics, are generally found.

Each problem and each farm has to be individually analysed before a weed control recommendation is issued. An agronomist has to analyse each situation to be able to schedule the recomendations about the most adequate alternative for the integrated weed management.

Although further studies are needed, many changes have already been observed in Brazil, as far as the philosophy of the weed control is concerned. At the research level, the search for the improvement of weed management methods has increased. At farmer and extensionist levels the concern about technical, economic and environmental issues of weed control has also increased.

REFERENCES

1. Blanco, H. G. Ecologia das plantas daninhas: competição de plantas daninhas em culturas brasileira. In: Controle Integrado de Plantas Daninhas, Conselho Regional de Engenharia, Arquitetura e Agronomia, São Paulo, 1982, p. 43.

2. Hoffmann-Campo, C.B., Gazziero, D.L.P. and Barreto, J.N., Estudos de competição de amendoim bravo (Euphorbia heterophylla L.) e a soja. In: Resultados de Pesquisa de Soja 1982/83. EMBRAPA-CNPSo, Londrina, 1983, pp. 160-61.

3. Gazziero, D.L.P. and Guimarães, S.C., Disseminação das plantas daninhas na cultura da soja cultivada em áreas de cerrado. EMBRAPA-CNPSo, Londrina, 1984, pp.1-4. (Comunicado Técnico, 26).

4. Gazziero, D.L.P., Guimarães, S.C. and Pereira, F.A.R., Plantas Daninhas: cuidado com a disseminação. EMBRAPA-CNPSo, Londrina, 1989, (Folder).

5. Ruedell, J., Efeito do manejo do solo e da rotação de culturas. In: Primeros Jornadas BiNacionales de Cero Labranza. Sociedad de Conservacion de Suelos de Chile, 1990, pp. 169-82.

6. Pinto, J.O., Estudo e controle de capim-arroz e arroz-vermelho na soja cultivada em terras-de-arroz. EMBRAPA-CPATB, Pelotas, 1991. (EMBRAPA, PNPSoja. Projeto 00589009-0. Projeto em andamento.

7. Ruedell, J., Sediyama, T. and Barni, N.A., Resposta da soja (Glycine max (L.) Merrill) ao efeito conjugado de arranjo de plantas e herbicidas. I. Controle de Plantas daninhas e rendimento de grãos, Agron. Sulnogr., 1981, 17, 95-106.

8. Mesquita, C.M., Gazziero, D.L.P. and Roessing, A.C., Adaptação de equipamentos de pulverização em semeadora-adubadora para aplicação de herbicidas pré-emergentes em faixas. In: Resultados de Pesquisa de Soja 1981/82. EMBRAPA-CNPSo, Londrina, 1982. pp.84-89.

9. ANDEF. Vendas de defensivos agrícolas por destinação - 1989/90. ANDEF/SINDAF, 1991, pp.1-2.

10. Brasil. Ministério da Agricultura, Soja no Paraná. IPEAME. Curitiba, 1971, 24 pp. (Circular, 9).

11. Gazziero, D.L.P., Skora Neto, F, Rodrigues, B.N., Almeida, F.L.S., Vidal, R.A, Avaliação do resíduo de paraquat em grãos, farelo e óleo de soja. In: Décimo Oitavo Congresso Brasileiro de Herbicidas e Plantas Daninhas, Piracicaba, 1988, pp. 381-82.

12. Voll, E., Karan, D. and Gazziero, D.L.P, Dinâmica de populações de plantas daninhas em soja. In: Resultados de Pesquisa de Soja 1989/90. EMBRAPA-CNPSo, Londrina, (no prelo).

13. Gazziero, D.L.P., Karan, D., and Voll, E., Impacto de produtos herbicidas sobre Comunidade infestante da cultura da soja. EMBRAPA-CNPSo, Londrina, 1991. 1 v. (EMBRAPA. PNPSoja. Projeto 590021-2). Projeto em andamento.

14. Fancelli, A.L., Torrado, P. V. and Machado, J., Atualização em plantio direto. Fundação Cargil, Campinas, 1985, pp.343.

15. Almeida, F.S., A alelopatia e as plantas, IAPAR, Londrina, 1988, 60 pp. (circular, 53).

16. Almeida, F.S., Controle de plantas daninhas em plantio direto. IAPAR, Londrina, 1991, 34 pp. (IAPAR. Circular, 67).

17. Gazziero, D.L.P., Calçavara, P.R. Sarlo, S.M.Z., Yorinori, J.T., Biological Control of Wild Poinsetia: Current Status of Research. In: XII International Plant Protection Congress, Rio de Janeiro 1991.

18. Pereira, F. de A.R., Gazziero, D.L.P. and Bonamigo, L.A., Avaliação de espécies com potencial para produção de cobertura morta em áreas do cerrado Plantio Direto, 1988, 6, 6.7.

UPLAND WEEDS AND THEIR MANAGEMENT IN SOYBEAN IN JAPAN

K. NOGUCHI
Head, Upland Weed Laboratory, National Agriculture Research Center
Tsukuba, Ibaraki 305 Japan

ABSTRACT

Weed growth was extremely inhibited by 80 - 90% shading conditions. The relative light intensity under the canopy of soybean was reduced to below 10% at about 73 days after seeding, and at that time the interrow space up to the height of about 50cm was placed under the same light intensity below 10%. On the other hand, a period of 40 days or more was necessary for _Digitaria ciliaris_ to grow to about 50cm in plant height. Therefore, a period of 33 days (73 - 40 days) of weed-free condition is theoretically required to be adopted after the seeding of soybean in the community of soybean and _Digitaria ciliaris_.

INTRODUCTION .

Japan is an island country located near to Asia Continent. It consists of four principal island, i.e., Hokkaido, Honshu, Shikoku and Kyushu. It belongs to the temperate zone except Okinawa Prefecture which belongs to the subtropical zone. As Japan is situated in the monsoon belt, the rainfall is abundant, averaging between one or two thousand mm per year. Due to abundant rainfall and high temperature in the summer season, upland weeds flourish markedly in Japan, causing a high cost for weed management in upland fields including soybean cultivation. Nowadays herbicides are widely used in paddy and upland fields in Japan. Although the herbicide application is useful in stabilizing crop production and saving labor, too much use of herbicides causes several problems such as adverse effects on environment(1). Therefore, there is an urgent need to establish reasonable weed control methods for soybean cultivation.

PRODUCTION OF SOYBEAN IN JAPAN

Changes of planted area, production and yield of soybean in Japan was shown in Figure 1. The peak of planted area was during 1900 - 1920. Planted area was gradually reduced from 1920 to 1945 due to war. It was increased after the World War II, but has remarkably reduced recently. A

little increase after 1980's is due to adjustment of rice production.
The production of soybean has been followed the tendency of planted area.
The yield per ha has been increased gradually and about 1.8 t/ha has
achieved recently(2).

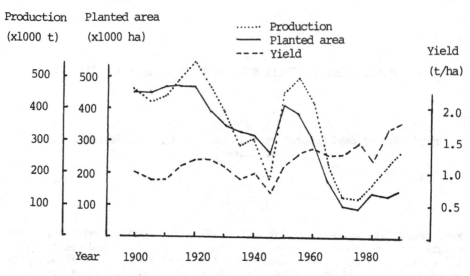

Figure 1.   Changes of planted area, production and yield of soybean in
Japan.

MAIN UPLAND WEED IN SOYBEAN

In Japan, Kasahara(3) listed 417 weed species in 43 family. The number
of weed species growing in paddy field, upland field and both site are
191, 302 and 76, respectively. About a third of paddy weeds and a half

TABLE 1
List of main weeds of soybean fields in Japan

No.	Weed species	Name in Japanese	a/p*
1	Digitaria ciliaris	mehishiba	a
2	Echinochloa crus-galli	himeinubie	a
3	Eleusine indica	ohishiba	a
4	Cyperus microiria	kayatsurigusa	a
5	Persicaria longiseta	inutade	a
6	Chenopodium album	shiroza	a
7	Portulaca oleracea	suberihiyu	a
8	Amaranthus lividus	inubiyu	a
9	Amaranthus viridis	honagainubiyu	a
10	Commelina communis	tsuyukusa	a
11	Equisetum arvense	sugina	p

* : a : annual, p : perennial

of upland weeds are perennials. Main weeds of the soybean fields are shown in Table 1. Digitaria ciliaris is the most dominant and serious weed in upland fields in Japan.

DAMAGE DUE TO WEED AND ESTABLISHMENT OF PERIOD FOR WEED FREE MAINTENANCE

Crop yield losses due to weeds

Weed damage, insect injury and disease injury are factors that lead to crop yield losses or deterioration of quality. Weeds injure crops slowly compared with insect or disease, but losses due to weeds are remarkable. Crop plants grow in competition with weeds. When crop plants grow vigorously, weed growth is retarded. On the other hand, when crops lack vigor, weeds flourish. Major environmental factors in plant competition are water, light and mineral nutrients. Weeds compete directly with crops for their utilization(4).

Wide range of yield loss of upland crops due to weeds were observed. About 90, 60, 30 and 3% of crop yield compared with weeded field are obtained in non-weeded field in corn, soybean, upland rice and peanuts, respectively(5). Only a few percent of crop yield can be obtained under non-weeded conditions in these kinds of crops. Therefore, there is a need to conduct reasonable weed control for stability of crop production.

Effect of shading on growth of weeds

Plant length or main stem length of weeds increased at slight shading condition, but it decreased in heavy shading condition. Top dry weight and the number of tillers or branches of each weed decreased at all shading conditions. It was understood that weed growth was extremely inhibited by 80% or more shading from Portulaca oleracea and Cyperus microiria, by 80 - 90 shading from Persicaria lapathifolia, and by 90% or more shading from Digitaris ciliaris and Chenopodium album, respectively(6).

Changes of light environment in crop canopies and growth of competing weeds

Changes of light environment in crop canopies were different among crops, but it was shown that the relative light intensity was reduced below 10% someday after crop seeding. High negative correlation was found between logarithms of relative light intensity on interrow ground surface and the number of days after seeding of soybean, and then as shown in Figure 2, the regression line was obtained. From this regression line, the number of days when relative light intensity is reduced to lower than 20 or 10% was estimated, i.e., about 63 or 73 days after seeding of soybean, respectively(7).

The linear regression was obtained between LAI(leaf area index) of soybean and logarithm of relative light intensity on interrow ground surface(Figure 3). From this regression line, LAI at the time when relative light intensity is reduced to below 20 or 10% was estimated at 2.3 or 3.3, respectively(7).

At the time when relative light intensity at the interrow ground surface was reduced to about 20 or 10% in soybean field, the interrow space up to the height of about 60 or 50cm from ground surface showed the same relative light intensity as that of interrow ground surface, i.e., 20 or 10%, respectively(7).

In early summer season when relative light intensity was reduced by crop canopies, plant length or main stem length of main competing weeds

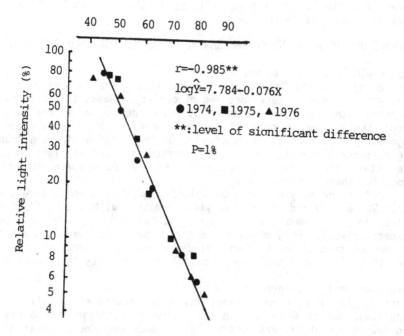

Figure 2. Relationship between the relative light intensity on the ground surface and the number of days after soybean seeding.

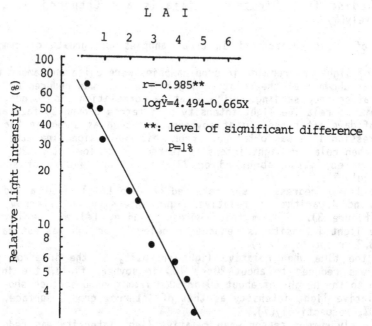

Figure 3. Relationship between the relative light intensity on the ground surface and LAI of soybean.

in this season were 1-2, 7-13, 16-28, 39-52cm in <u>Digitaria ciliaris</u>, 1-2, 6-12, 12-26, 20-43cm in <u>Cyperus microiria</u>, and 1>, 4-14, 9-25, 17-33cm in <u>Portulaca oleracea</u> at 10, 20, 30 and 40 days after seeding, respectively(7).

A concept of weed-free field condition
On the basis of the above results, a concept of introducing a certain period of weed-free condition into crop management system was developed, aiming at avoiding efficiently crop yield reduction due to competing weeds. The procedure is shown in Figure 4. Growth of <u>Digitaria ciliaris</u> was greatly diminished by relative light intensity lower than 10%. The relative light intensity under the canopy of soybean was reduced to below 10% at about 73 days after seeding, and at that time the interrow space up to the height of 50cm was placed under the same light intensity below 10%. On the other hand, a period of 40 days or more was necessary for Digitaria ciliaris to grow to about 50cm in plant length. Therefore, a period of 33 days (73 - 40 days) of weed-free condition is theoretically required to be adopted after the seeding of soybean in the community of soybean and <u>Digitaria ciliaris</u>. It was expected that growth of <u>Digitaria ciliaris</u> which emerged after the 33 days from crop seeding would be greatly inhibited by shading of crop canopies.

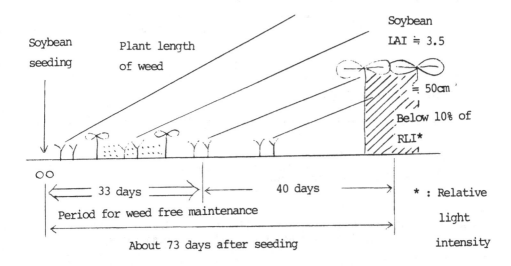

Figure 4. Procedure for estimating weed-free field condition of soybean.

As shown in Figure 5, the experimental field of soybean was treated with different number of days of weed-free condition. It was observed that the weed grew beyond the canopy of the soybean when no treatment was given or only 30 days of the treatment was given, although the relative light intensity under the canopy became lower than 10%. On the contrary, the weed growth was markedly suppressed by the treatment of 47 or 58 days of weed-free condition. The weed was not able to outgrow the canopy.

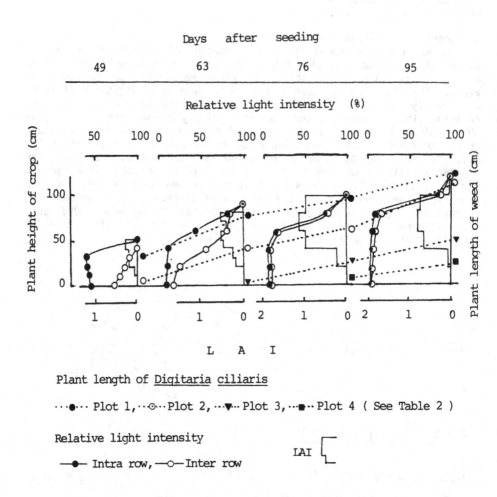

Figure 5. Plant length of <u>Digitaria</u> <u>ciliaris</u> as influenced by the relative light intensity and LAI in soybean canopy.

Growth and yield of soybean as influenced by the treatment are shown in Table 2. The yield of soybean was reduced in the plots without the treatment(0 days) or with 30 days of the treatment, because the weed growth was not diminished greatly by crop canopies. On the contrary, yield reduction did not occur in the plots with 47 and 58 days of treatment of weed-free condition, because the weed growth was greatly suppressed by the crop canopy. From the above results, the proposed hypothesis on the effectiveness of introducing a certain period of weed-free condition after seeding into the cultural management of soybean was proved to be applicable to practical use to avoid crop yield reduction caused by competing weeds(5).

TABLE 2
Effect of different post-seeding periods of weed-free condition
on soybean yield

Plot	Days of weed-free condition	Stem	Bean	Total
		%		
1	0	66	39	52
2	30	80	73	76
3	47	95	99	96
4	58	98	98	98
5	Full season	100	100	100
l.s.d. 5%		11.3	13.9	12.1
Actual weight in Plot 5(kg/ha)		1,770	2,090	4,940

WEED CONTROL SYSTEM

Weed control system is able to be established on the basis of various factors such as ecological and physiological growth characteristics both crops and weeds, mutual interaction(competition) between crops and weeds, mode of action or method for application of herbicide, cost for application, etc..

Wide-scale use of herbicides began in 1947, after the discovery of 2,4-D. Recently, Chemical weeding is the most popular and effective method in Japan. Chemicals have released farmers from painful weeding work especially in summer season. Figure 6 showed changes of working hours for soybean cultivation. In 1955, it was necessary for soybean cultivation to spend a time more than 220 hours per ha, but in 1985, it was reduced to only 60 hours(2). This mainly due to development of weed science.

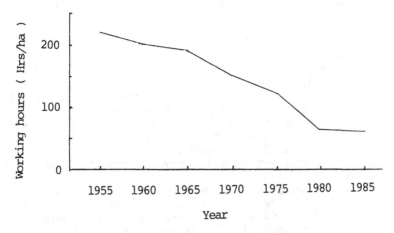

Figure 6. Changes of working hours in soybean in Japan.

A weed control system is established by applying the concept of post-seeding period of weed-free condition. This would be changed by condition of cropping, i.e., variety, cropping site, cropping pattern, etc.. Weed control system for soybean are described as follows;

```
                        (a)          (b) (c) (d)
(1)Rotary tillage-----Seeding-----Emerging---------------Harvesting
   For cold or cool region(Hokkaido, Tohoku, Weed-free period : 50
   days)

                        (a)              (c) (d)
(2)Rotary tillage-----Seeding-----Emerging---------------Harvesting
   For mild region(Honshu, Weed-free period : 30-35 days)

                        (a)                   (d)
(3)Rotary tillage-----Seeding-----Emerging---------------Harvesting
   For warm region(Kyushu, Weed-free period : 20 days)

           (e)          (a)          (b) (c) (d)
(4)Non tillage--------Seeding-----Emerging---------------Harvesting
   For non tillage cultivation in mild region
```

(a):Herbicides for soil treatment such as trifluralin, alachlor, metolachlor, benthiocarb-prometryn, linuron etc.
(b):Herbicide for foliage treatment such as alloxydim, sethoxydim, fluazifop-butyl, quizalofop-ethyl etc.
(c)Intertillage using rotary cultivator
(d)Picking weed off by hand
(e)Herbicide for foliage treatment such as glufosinate, bialaphos

CONCLUSION

Weed control can not be considered in isolation from the other plant production system. There are several possible strategies and tactics to adopt in weed control methods. It is very difficult to accomplish efficient weed control by using any one control method such as herbicide application. Integrated management is necessary for efficient weed control.

REFERENCES

1. Yamada, T., Fate and safe of herbicides. Weed Res. Japan, 1975, 20, 1-7(in Japanese).
2. Ministry of Agriculture, Forestry and Fisheries, Crop Statistics, 32, Norin Tokei Kyokai, Tokyo, 1991, pp.1-535(in Japanese).
3. Kasahara, Y., The species grouping of wild herbs, ruderals, naturalized plants, weeds and crops on the basis of their habitats. Weed Res. Japan, 1971, 12, 23-27(in Japanese).
4. Crafts, A.S., Modern weed control. Univ. of Carifornia Press, Carifornia, 1975, pp.67-100.
5. Noguchi, K. and Nakayama, K., Studies on competition between upland crops and weeds. V. The period for weed-free maintenance. Japan Jour. Crop Sci., 1978, 47, 637-643(in Japanese with English summary).
6. Noguchi, K. and Nakayama, K., Studies on competition between upland crops and weeds. III. Effect of shade on growth of weeds. Japan Jour.

Crop Sci., 1978, 47, 56-62(in Japanese with English summary).
7. Noguchi, K. and Nakayama, K., Studies on competition between upland crops and weeds. IV. Changes of light environment in crop canopies and hypothesis about the period for weed-free maintenance. Japan Jour. Crop Sci., 1978, 47, 381-387(in Japanese with English summary).

IMPLEMENTATION OF ECONOMIC THRESHOLDS FOR WEEDS IN SOYBEAN

HAROLD D. COBLE
Professor of Crop Science
North Carolina State University
Raleigh, North Carolina, 27695, USA

ABSTRACT

Herbicide selection for soybean is a complicated task, and consideration of the weed species present, weed density, crop value, potential yield of the field, herbicide and application cost, and the grower's management style are all necessary in making the best decision. Many growers are now making use of remedial tactics for weed control, including the use of postemergence herbicides. The best use of these remedial treatments is after careful consideration of the potential losses from a weed population compared to the potential economic gain from treatment, or economic thresholds. Field competition studies have been used to develop economic thresholds for weeds in soybean, and these thresholds have been used to create a computerized decision aid for herbicide selection in that crop based on net economic return from the selected treatment. This computer software program (HERB™) is currently being sold in the U.S., and a similar program for other crops is under development.

INTRODUCTION

Four distinct pest management strategies generally recognized by specialists are: avoidance, prevention, suppression, and eradication. Avoiding a weed population may be possible under certain circumstances, such as using crop rotation to avoid a particular weed problem, or planting the crop after most weeds of a particular species have germinated. However, this strategy is very limited with weeds since almost all fields are infested with a potentially economically damaging level of weed seeds. Prevention, simply stated, means not allowing a weed population to become established in a field. This is a lofty goal because of the propensity of weed seeds to spread from place to place. However, prevention should be an underlying strategy with every grower and practiced when feasible as a general sanitation measure. Suppression is the strategy employed by essentially all growers in an effort to produce their crop while keeping weed populations below an economically damaging level. It is this strategy that should receive most attention with respect to choosing and placing into practice the tactics for carrying out the plan. Eradication has very limited usefulness to most growers, since it is practically impossible except on small, isolated weed infestations.

The tactics for carrying out the suppression strategy include competition from the crop, cultivation, and herbicides. Herbicide selection is a fairly complicated task, with the first decision, the philosophical approach, a critical one. There are basically three ways to approach herbicide use: prophylactic (preventive), remedial (wait-and-see), and a combination of the previous two. The prophylactic approach has as its basis the assumption that weeds will be a problem and they can best be controlled with a preplant or preemergence herbicide. In most instances this assumption is correct, and success with the approach depends on selection of the correct herbicide for the weeds present and adequate weather conditions for herbicide activity to be expressed. The down side of this approach is that there are times when weed populations do not develop, even though the potential is present, and herbicide use in these instances results in zero return on the investment. Because there are times when weed populations are not expressed, the remedial approach allows the greatest potential for reductions in weed control cost and herbicide use. With the right weather conditions and the crop grown in rows, cultivation may be all that is necessary for adequate weed control. With this approach, even if the weed population does develop, the availability of selective , highly efficacious herbicides for postemergence use allows remedial control efforts to be as effective as preventive measures.

Another advantage of the remedial approach lies in the ability to make economic judgments on whether or not the weed population is high enough to require treatment, implying the existence of an economic threshold. Threshold is defined as the point at which a stimulus is just strong enough to produce a response [1]. For most cropland situations, weed density is the most easily quantified variable for stimulus measurement, requiring only the ability to count and identify weeds in a finite sampling area. Commonly measured responses with crops include total biomass, yield of the raw agricultural commodity, crop quality, and net economic returns [6, 7, 15, 18, 20]. Other responses, although difficult to quantify, may result from aesthetic perceptions and sociological pressures [23].

An economic threshold for weeds may be defined as the weed population at which the cost of control is equal to value of crop yield attributable to that control [10]. Mathematically, the economic threshold for an individual crop may be defined as:

$$t_E = (Ch + Ca)/(YPLH) \qquad (1)$$

where t_E is the economic threshold, Ch is the herbicide cost, Ca is application cost, Y is the weed-free crop yield, P is the value per unit of crop, L is the proportional loss per unit weed density, and H is the proportional reduction in weed density by the herbicide treatment [12].

It is apparent from this equation that any increase in herbicide or application cost will increase economic thresholds, other factors being constant. On the other hand, any increase in crop yield, value, degree of weed control, or crop loss per unit weed density will lower economic thresholds, other factors being constant. Three of the factors involved in economic threshold calculations, herbicide cost, application cost, and crop value per unit can be estimated fairly accurately by individual growers. However, potential crop yield, proportional loss per unit weed density, and herbicide efficacy are more difficult to estimate because of the variability associated with weather, weed species makeup, weed size, and cropping systems effects on these variables [2, 4, 17, 18, 21].

Economic thresholds generally refer to in-season decisions during a single crop year, and do not include a cost factor associated with possible increases in the soil seedbank. Cousens [13] has used the term economic optimum threshold (EOT) to include the impact of seedbank dynamics on long-term profitability of weed management decisions. Predicting the seed rain from various weed populations would not be a difficult task. Seed production by weeds grown in competition with the crop has been quantified for several weeds in soybean [Glycine max (L.) Merr.] [5] and winter wheat (Triticum aestivum L.) [13, 14]. In these same studies, simulation models were developed to estimate the EOT. For each case, the EOT weed population was significantly smaller than the economic threshold population. Although seed production and weed interference have been accurately quantified, estimates of seedbank mortality, seed dispersal, and seedling recruitment remain largely untested [16, 19]. Attempts at relating seedbank population size to resulting seedling population size have been for the most part unsuccessful [3]. To date, these EOT studies [5, 13, 14] have helped to identify important gaps in our understanding of the influence of management practices on weed population dynamics. Much more work is needed in this area to enhance our understanding of threshold theory in weed science.

More growers are likely to use modifications of economic thresholds called action thresholds. Action thresholds may be defined as the weed population at which a grower decides to institute a control tactic, and may be above, below, or at the economic threshold level. For most growers, risk aversion is a more compelling philosophy than profit maximization, and they are likely to take control actions at weed populations below the true economic threshold [22]. These growers see risk aversion as a means of sustainability of profit over years. However, even those growers who choose action thresholds below true economic thresholds would benefit from knowing what the economic thresholds are in order to have a beginning point for making action threshold choices. The best action thresholds should still be based on some sort of economic justification that may be mitigated by other factors [14, 23].

FUTURE DIRECTIONS

Because of increased emphasis on profitability at the farm level and more public pressure to reduce pesticide use, greater reliance on remedial weed control measures seems certain. Profitability is most often increased with the wait-and-see approach to weed control, since herbicide use may not be required. If herbicide use is required, it may only be necessary on part of the field in question, so that over fields and over years, a reduction in herbicide use is a near certainty. In addition, most of the new postemergence herbicides are used at rates 10 to 100 times lower than the more traditional preventive treatments. Although the rate of application of active ingredient is not the only important consideration in reduced pesticide use, it certainly is a major factor.

Changes in tillage practices with emphasis on leaving a high percentage of crop residue on the soil surface are likely to occur in the near future so that growers can conform to soil conservation practice guidelines. These changes will most likely be toward reduced tillage, in effect eliminating widespread use of preplant herbicides which must be incorporated into the soil through tillage operations. In most instances where tillage is reduced, water infiltration through the soil is increased. This increased movement of water through the profile may increase the possibility of some of the standard soil applied herbicides ending up as groundwater contaminants. For these reasons as well as the aforementioned economic ones, postemergence herbicides used on an "as needed" basis will no doubt play an increasingly important role in weed management in the future.

IMPLEMENTATION PROGRAMS

Utilization of the economic threshold concept in determining the need for weed control has been on the increase in soybean since the development of several highly effective, selective postemergence herbicides for the crop. A similar trend is almost certain for other crops. One of the keys to successful implementation of this new system of weed control lies in the ability of growers to determine when weed populations exceed economic threshold levels, thus requiring treatment. Research at North Carolina State University has been successful in developing multispecies economic thresholds for weeds in soybean [9]. The approach used in this research was to establish an index for comparing the competitive ability of the different weed species infesting the crop. At low populations, where weed areas of influence do not overlap, effect of weed density on crop yield is best described by the linear regression equation, $Y = a + bx$ [8]. In this equation, **a** equates to crop yield in the absence of weed interference, **b** is the slope of the regression line, or loss per weed per unit area of land, and **x** is the number of weeds per unit of land area. Since the effect of yield loss is negative, the equation becomes:

$$Y = a - bx \qquad (2)$$

One way to examine the competitive ability of different weed species is to examine the amount of crop yield loss caused by an individual weed as a proportion of total yield. This can be done by dividing the slope (**b**) of the regression equation by the y intercept (**a**). Thus, competitive ability can be expressed as **b/a**.

The competitive abilities of different weed species can be compared effectively by using an indexing system. Usually, the most competitive species is assigned some arbitrary value, such as 10, and all other weeds are then ranked relative to that most competitive species. Then, the competitive index equation may be written:

$$CI_1 = \frac{b_1/a_1}{b_i/a_i} \; X \; K \qquad (3)$$

For an index based on a 0 to 10 scale, the value for K would be 10.

This indexing procedure allows the summation of the effects of multiple weeds on crop yield to determine the competitive load (CL). Competitive load is defined as the product of the number of weeds of a single species per unit area multiplied by that species' competitive index (CI) value:

$$CL_1 = CI_1 \; X \; \text{number of species 1 per unit area} \qquad (4)$$

This procedure is repeated for each species in the weed population. Then the total competitive load (TCL) is calculated by summing the CL values for the individual species:

$$TCL = CL_1 + CL_2 + CL_3 + \ldots + CL_n \qquad (5)$$

As weed populations increase, and the areas of influence of individual weeds begin to overlap, the linear function no longer is a meaningful description. For a wide range of weed populations, the rectangular hyperbola serves as an appropriate model [11]:

$$Y_L = \frac{Id}{1 + \dfrac{Id}{A}} \qquad (6)$$

where: Y_L = Yield loss

I = yield loss per weed as d -> 0

d = weed density

A = yield loss as d -> max

For soybean, the value for I is approximately 0.5, and A is approximately 80. Since our TCL values are derived from density values multiplied by an index, the TCL values may be substituted for "d" in the equation, thus giving:

$$Y_L = \frac{0.5(TCL)}{1 + \dfrac{0.5(TCL)}{80}} \tag{7}$$

Implementation of information generated in these studies was through a microcomputer based economic decision model named HERB™ [24]. This model uses the multispecies weed competitive index to determine if a weed population is above the economic threshold based on efficacy and cost of the control tactic to be employed (equation 1). The soybean portion of HERB™ has undergone field verification for three years and is being distributed commercially with over 300 copies out to growers and decision makers in the US. The corn portion of the model is presently in verification trials. A new version of the HERB™ is under development that will allow experts in different geographic areas to access the databases to create regionalized versions to more closely fit local conditions.

HERB™ verification trials have shown the model to be very accurate when used with crops grown in traditional 75- to 100-cm row spacings, and the crop planted at the normal planting date. There is enough flexibility in the programming of HERB™ to accommodate inputs for row spacing and planting date, providing the database is available. Such a database is not available at present to allow the expansion of the program to include crops grown in narrow rows or later than normal planting dates. We intend to develop the database necessary to expand HERB™ to narrow rows and late plantings soon.

REFERENCES

1. Anonymous. 1980. Webster's new world dictionary of the American language. Second college edition. Simon and Schuster Publ. New York, NY.

2. Auld, B. A., and C. A. Tisdell. 1987. Economic thresholds and response to uncertainty in weed control. Agric. Syst. 25:219-227.

3. Ball, D. A., and S. D. Miller. 1989. A comparison of techniques for estimation of arable soil seedbanks and their relationship to weed flora. Weed Res. 29:365-373.

4. Bauer, T. A., D. A. Mortensen, and G. A. Wicks. 1990. Environmental variability and economic thresholds for soybeans. Weed Sci. Soc. Am. Abstr. 30:53.

5. Bauer, T. A., and D. A. Mortensen. 1991. A comparison of economic and economic optimum thresholds for two annual weeds in soybeans. Weed Tech. (in press).

6. Burnside, O. C. , C. R. Fenster, L. L. Evetts, and R. F. Mumm. 1981. Germination of exhumed weed seed in Nebraska. Weed Sci. 29:577-586.

7. Chisaka, H. 1977. Weed damage to crops: Yield loss due to weed competition. Pages 1-16 in Integrated Control of Weeds, J. D. Fryer and S. Matsunaka, eds. Univ. of Tokyo Press, Tokyo.

8. Coble, H. D. 1985. Development and implementation of economic thresholds for soybean. p. 295-307 in R. E. Frisbee and P. L. Adkisson, eds. Integrated Pest Management of Major Agricultural Systems., Texas A&M University.

9. Coble, H. D., F. M. Williams, and R. L. Ritter. 1981. Common ragweed interference in soybeans. Weed Sci. 29:339-342.

10. Cousens, R. 1987. Theory and reality of weed control thresholds. Plant Prot. Q. 2:13-20.

11. Cousens, R. 1985. A simple model relating yield loss to weed density. Ann. Appl. Biol. 107:239-252.

12. Cousens, R., B. J. Wilson, and G. W. Cussans. 1985. To spray or not to spray: the theory behind the practice. Proc. 1985 Brit. Crop Prot. Conf. - Weeds. p. 671-678.

13. Cousens, R., C. J. Doyle, B. J. Wilson, and G. W. Cussans. 1986. Modelling the economics of controlling Avena fatua in winter wheat. Pest. Sci. 17:1-12.

14. Doyle, C. J., R. Cousens, S. R. Moss. 1986. A model of the economics of controlling Alopecuris myosuroides in winter wheat. Crop Prot. 5:143-150.

15. Eaton, B. J., O. G. Russ, and K. C. Feltner. 1973. Venice mallow competition in soybeans. Weed Sci. 21:89-94.

16. Egley, G. H. and J. M. Chandler. 1983. Longevity of weed seeds after 5.5 years in the Stoneville 50-year buried seed study. Weed Sci. 31:264-270.

17. Forcella, F. 1990. Breeding soybeans tolerant to weed competition. Weed Sci. Soc. Am. Abstr. 30:51.

18. Legere, A., and M. M. Schreiber. 1989. Competition and canopy architecture   as affected by soybean (Glycine max) row width and density of redroot pigweed (Amaranthus retroflexus). Weed Sci. 37:84-92.

19. Lueschen, W. E., and R. L. Andersen. 1980. Longevity of velvetleaf (Abutilon theophrasti) seeds in soil under agricultural practices. Weed Sci. 28:341-346.

20. McWhorter, C. G., and J. M. Anderson. 1979. Hemp sesbania (Sesbania exaltata) competition in soybeans (Glycine max). Weed Sci. 27:58-63.

21. Mortensen, D. A., and H. D. Coble. 1989. The influence of soil water content on common cocklebur (Xanthium strumarium) interference in soybeans (Glycine max). Weed Sci. 37:76-83.

22. Reichelderfer, K. H. 1980. Economics of integrated pest management: Discussion. Am. J. Agric. Econ. 62:1012-1013.

23. Stoller, E. W., S. K. Harrison, L. M. Wax, E. E. Regnier, and E. D. Nafzigher. 1987. Weed interference in soybeans (Glycine max). Rev. Weed Sci. 3:155-181.

24. Wilkerson, G. G., S. A. Modena, and H. D. Coble. 1988. HERB v2.0: Herbicide decision model for postemergence weed control in soybeans. Users manual. Bull. No. 113, Crop Sci. Dep., N. C. State Univ., Raleigh, NC.

S-23031, A NEW POST-EMERGENCE HERBICIDE FOR SOYBEANS

KATSUZO KAMOSHITA, EIKI NAGANO, KAZUO SAITO, MASAHARU SAKAKI,
RYO YOSHIDA, RYO SATO, HIROMICHI OSHIO
Sumitomo Chemical Company Ltd.,
Takarazaka Research Center,
Agricultural Science Research Laboratory,
4-2-1, Takatsukasa, Takarazuka, Hyogo 665, Japan

ABSTRACT

S-23031, pentyl 2-chloro-4-fluoro-5-[(3,4,5,6-tetrahydro)phthalimido]phenoxyacetate, is a new post-emergence herbicide for broad-leaved weed control in soybeans. S-23031 is a fast acting herbicide and effective against various kinds of broad-leaved weeds with good crop safety at 8 - 125 g a.i. / ha in post-emergence treatment. Weed control spectrum of S-23031 includes troublesome weeds such as velvetleaf, prickly sida, jimsonweed and common lambsquarters. S-23031 shows excellent tank-mix compatibility with other soybean herbicides such as imazethapyr, bentazone and clethodim.

INTRODUCTION

N-phenyl tetrahydrophthalimides have been known as herbicidal compounds, i.e., S-23121, *N*-[4-chloro-2-fluoro-5-[(1-methyl-2-propynyl)oxy]phenyl]-3,4,5,6-tetrahydrophthalimide, as a pre and post-emergence herbicide for cereals (1), and S-53482, 7-fluoro-6-[(3,4,5,6-tetrahydro)phthalimido]-4-(2-propynyl)-1,4-benzoxazin-3(2*H*)-one, as a pre-emergence herbicide for soybeans (2).

S-23031, pentyl 2-chloro-4-fluoro-5-[(3,4,5,6-tetrahydro)phthalimido]phenoxyacetate, is a new post-emergence herbicide for the control of broad-leaved weeds in soybeans. The main objective of this paper is to describe the biological activity of S-23031 in laboratory and field studies.

S-23031 is a fast acting herbicide. When applied to foliage, it is readily absorbed into susceptible plant tissue and causes characteristic herbicidal symptoms such as desiccation, bleaching, browning, wilting or necrosis. Most symptoms are often observed within a day under bright sunlight.

S-23142, one of N-phenyl tetrahydrophthalimide herbicides, expresses its herbicidal activity in the presence of light and oxygen. S-23142 induces massive accumulation of porphyrins. The photosensitizing action of accumulated porphyrins is likely to be one of the causes of membrane lipid peroxidation. The peroxidation of membrane lipids leads to irreversible damage of the membrane functions and the structure of susceptible plants (3,4,5). It is suggested that S-23031 has the same mode of action as S-23142 because the chemical structure and the herbicidal symptoms of S-23031 are similar to that of S-23142.

MATERIALS AND METHODS

Synthesis of S-23031

S-23031 was synthesized by the reaction of 2-fluoro-4-chloro-5-pentyloxycarbonylmethoxy anilin and 2,3,4,5-tetrahydrophthalic anhydride as shown below.

S-23031

Physical and chemical properties

Chemical name   : pentyl 2-chloro-4-fluoro-5-[(3,4,5,6-tetrahydro)phthalimido]=
                       phenoxyacetate
Common name   : Not given yet
Code number   : S-23031, V-23031
Molecular formula   : $C_{21}H_{23}ClFNO_5$
Molecular weight   : 423.87
Melting point   : 88.87 - 90.13 °C
Vapor pressure   : < 1.0×10^{-7} mmHg at 22.4 °C

Density	: 1.3316 g/ml at 20 ℃	
Solubility at 25 ℃	: Water	; 0.189 mg/L
	Methanol	; 47.8 g/L
	Hexane	; 3.28 g/L
	Acetone	; 590 g/L
	Acetonitrile	; 589 g/L

Toxicology of Technical Material

Acute oral LD50	rat male	: >5000 mg/kg
	rat female	: >5000 mg/kg
Acute dermal LD50	rat male	: >2000 mg/kg
	rat female	: >2000 mg/kg
Eye irritation	rabbit	: Minimal irritant
Dermal irritation	rabbit	: Not irritant
Mutagenicity	Ames test	: Negative

Formulation

10 % (w/v) emulsifiable concentrate (EC) of S-23031 was used for all tests.

Greenhouse Tests

Weed spectrum test : In pre-emergence treatment, aqueous solutions of S-23031 without adjuvants were sprayed on the soil surface at the rate of 1000 L/ha. In post-emergence treatment, S-23031 was applied on the foliage of the plants which were at 1-4 leaf stages, 20-30 days after planting. Visual assessment of herbicidal activity and crop phytotoxicity was made 20-30 days after treatment (DAT) with a scale of 0 (no effect) - 10(complete kill). On this basis, scores 0-3 and 7-10 represent acceptable crop phytotoxicity and acceptable weed control, respectively. Weed spectrum tests were repeated 10-15 times.

Crop spectrum test : S-23031 was applied on the foliage of the test crops which were at 1-3.5 leaf stages (14 or 28 days after planted). Nonionic surfactant (X-77) 0.1 % (v/v) was added to spray solutions. S-23031 was applied by a sprayer with an even flat fan tip at the pressure of 2 kg/cm^2. Spray volume was 243 L/ha. Visual assessment of crop phytotoxicity was made at 21 or 28 DAT with the scale of 0 (no effect) - 10 (complete kill).

Field Trials

S-23031 was applied by a sprayer with flat fan tips at the pressure of 2 kg/cm^2. The sprayer was carbon dioxide pressured small hand-held type. Spray volume was 200-240 L/ha. Nonionic surfactant (X-77) 0.25% or crop oil concentrate (Agridex) 1 % was added to spray solution . Crop phytotoxicity and weed control efficacy were visually assessed by a scale of 0 (no effect) - 100 (complete kill).

RESULTS

Greenhouse Tests

Weed spectrum test (Table 1): In post-emergence application, S-23031 at 31 g a.i./ha showed acceptable soybean phytotoxicity, with acceptable control of all tested broad-leaved weeds such as *Abutilon theophrasti*, *Solanum nigrum*, *Xanthium strumarium*, *Ipomoea spp.* and *Galium aparine*. Soybean phytotoxicity was acceptable at higher dosage, that is, 125 g a.i./ha. Efficacy against tested grass weeds, however, was insufficient. In pre-emergence application at 31 to 125 g a.i./ha, S-23031 did not show sufficient efficacies against any tested weeds.

Crop spectrum test (Table 2) : In broadleaf crops, soybeans were the most tolerant to S-23031 in post-emergence application, and their phytotoxicity was acceptable at the dosage of 50 g a.i./ha. Other leguminous crops than soybeans were also tolerant to S-23031. S-23031 at 50 g a.i./ha showed marginally acceptable phytotoxicity against pea and adzuki beans. Kidney beans phytotoxicity was acceptable at 12.5 g a.i./ha. Gramineous crops were tolerant to S-23031. At 50 g a.i./ha of S-23031, phytotoxicities of field corn, rice, wheat and grain sorghum were acceptable.

Field Trials

Post-emergence efficacy of S-23031 alone and tank-mixed with broad-leaved weed killer (Table 3) : S-23031 at 63 g a.i./ha showed excellent control against most of tested weeds, such as *Xanthium strumarium*, *Abutilon theophrasti*, *Sida spinosa*, *Sesbania exaltata* and *Datura stramonium*, and showed good control against *Chenopodium album*. At this dosage, soybean phytotoxicity was acceptable. S-23031 at the lower dosage (31 g a.i./ha) showed also good to excellent control against the above weeds except for *Sida spinosa* and *Chenopodium album*. S-23031 at 31 g a.i./ha combined with imazethapyr at 31 g a.i./ha or bentazone at 600 g a.i./ha showed better weed control than S-23031 alone with very slight phytotoxicity. S-23031 with bentazone excellently controlled all the tested weeds. No antagonistic effect was observed in each combination.

Post-emergence efficacy of S-23031 tank-mixed with grass killer (Table 4) : Post-emergence application of S-23031 at 31 to 63 g a.i./ha showed excellent *Abutilon thephrasti* control, however, efficacies against *Echinochloa crus-galli* and *Sorghum halepense* were poor. S-23031 combined with clethodim, a grass killer, of 100 g a.i./ha showed excellent control against these grass weeds. No antagonistic effect was observed.

Post-emergence efficacy of S-23031 on velvetleaf at various stages (Table 5) : Post-emergence application of S-23031 at 12 to 24 g a.i./ha showed excellent velvetleaf (*Abutilon theophrasti*) control which was superior to bentazone at 800 g a.i./ha at every stage of velvetleaf. S-23031 at 24 g a.i./ha excellently controlled 7 leaf-stage velvetleaf.

TABLE 1
Herbicidal activity of S-23031, evaluated 20 - 30 days after treatment (mean values of 10 - 15 greenhouse trials)

Species	*Phytotoxicity scores		
	8 (g a.i. / ha)	31	125
POST-EMERGENCE			
Soybeans	0.5	1.6	2.5
Abutilon theophrasti	8.9	9.6	9.8
Solanum nigrum	7.6	8.9	8.6
Xanthium strumarium	5.7	7.4	8.4
Ipomoea spp.	8.9	9.6	9.8
Galium aparine	4.8	8.0	9.0
Sorghum halepense	1.6	3.7	4.8
Setaria faberi	0.6	2.5	2.8
Alopecurus myosuroides	0.2	0.3	1.8
PRE-EMERGENCE			
Soybeans		0	0.6
Abutilon theophrasti		0	3.4
Solanum nigrum		0	1.0
Xanthium strumarium		0	0.3
Ipomoea spp.		0	1.3
Galium aparine		0	0
Sorghum halepense		0	0
Setaria faberi		0	0
Alopecurus myosuroides		0	0

*Score : 0 = no effect ; 10 = complete kill
        0-3 = acceptable crop injury ; 7-10 = acceptable herbicidal activity
No adjuvant

TABLE 2
Post-emergence crop injury of S-23031 on various crops,
examined in greenhouse trials

Species	*Phytotoxicity scores		
	12.5 (g a.i. / ha)	25	50
Soybeans	1.0	1.0	1.5
Kidney beans	2.5	3.5	4.0
Pea	1.0	2.5	3.0
Adzuki beans	1.5	2.5	3.0
Cotton	9.0	9.0	10
Sugar beet	6.0	6.0	7.0
Sunflower	5.5	6.0	7.0
Grain sorghum	1.0	2.0	3.0
Rice	0.0	0.0	0.0
Field corn	0.0	1.0	2.0
Wheat	0.0	0.0	1.5

*Score : 0 = no effect ; 10 = complete kill
  0-3 = acceptable crop injury
X-77 0.1%

TABLE 3
Post-emergence performance of S-23031 alone and tank-mixed with imazethapyr or
bentazone, evaluated 20 days after treatment in a field trial in Kasai, Japan (mean values of 3
replications)

Compounds(g a.i./ha)		GLYMA	XANST	ABUTH	SIDSP	SEBEX	DATST	CHEAL
S-23031	(31)	*7	87	100	70	93	100	70
S-23031	(63)	16	100	95	92	93	100	83
S-23031 +imazethapyr	(31) (31)	8	93	100	83	93	100	85
S-23031 +bentazone	(31) (600)	6	100	100	98	95	100	100

*Score : 0 = no effect ; 100 = complete kill
X-77 0.25 %
GLYMA, soybeans: XANST, *Xanthium strumarium*: ABUTH, *Abutilon theophrasti*:
SIDSP, *Sida spinosa*: SEBEX, *Sesbania exaltata*: DATST, *Datura stramonium*: CHEAL,
Chenopodium album

TABLE 4

Post-emergence performance of S-23031 alone and tank-mixed with clethodim, evaluated
28 days after treatment in a field trial in Kasai, Japan (mean values of 3 replications)

Compounds (g a.i./ha)		GLYMA	ABUTH	ECHCG	SORHA
S-23031	(31)	*3	100	0	0
S-23031	(63)	5	100	0	20
S-23031 + clethodim	(31) (100)	7	100	97	97

*Score : 0 = no effect ; 100 = complete kill
Agridex 1 %
GLYMA, soybeans: ABUTH, *Abutilon theophrasti*: ECHCG, *Echinochloa crus-galli*:
SORHA, *Sorghum halepense*

TABLE 5

Effect of post-emergence application of S-23031 on velvetleaf (*Abutilon theophrasti*) at
various stages, evaluated 21 days after treatment in a field trial in Kasai, Japan

Compounds	(g a.i./ha)	Leaf stage of velvetleaf			
		4 L	5 L	6 L	7 L
S-23031	(12)	*85	95	93	N.A.
	(24)	N.A.	93	97	97
bentazone	(800)	78	92	85	67

*Score : 0 = no effect ; 100= complete kill   N.A.= data not available
X-77 0.25 %

CONCLUSIONS

S-23031 is a new N-phenyl tetrahydrophtalimide herbicide for soybeans. When applied in post-emergence, S-23031 controlled many troublesome broad-leaved weeds in soybean field, such as *Abutilon theophrasti, Sida spinosa, Datura stramonium, Sesbania exaltata, Chenopodium album* and *Xanthium strumarium*, at dosages of 31 and 63 g a.i./ha. Particularly, S-23031 was very effective against velvetleaf (*Abutilon theophrasti*). S-23031 at 24 g a.i./ha excellently controlled velvetleaf at the 7 leaf stage. Grass weed control of S-23031 was insufficient. S-23031 showed excellent tank-mix compatibility with soybean herbicides, such as imazethapyr, bentazone and clethodim.

ACKNOWLEDGEMENTS

We would like to thank our colleagues of Plant Physiology Laboratory and Plant Protection and Cultivation Laboratory in Takarazuka Research Center, Sumitomo Chemical Company Ltd., for their valuable discussions and helpful collaboration for greenhouse and field trials.

REFERENCES

1. Hamada, T., Yoshida, R., Nagano, E., Oshio, H. and Kamoshita, K., S-23121 - a new cereal herbicide for broadleaved weed control. Brighton Crop Protection Conference Weeds-1989, 1989, 41-46.

2. Yoshida, R., Sakaki, M., Sato, R., Haga, T., Nagano, E., Oshio, H. and Kamoshita, K., S-53482 - a new N-phenyl phthalimide herbicide. Brighton Crop Protection Conference Weeds-1991, 1991, 69-75.

3. Sato, R., Nagano, E., Oshio, H. and Kamoshita, K. Diphenylether-like physiological and biochemical action of S-23142, a novel N-phenylimide herbicide. Pesticide Biochemistry and Physiology, 1987, 28, 194-200.

4. Sato, R., Nagano, E., Oshio, H., Kamoshita, K. and Furuya, M. Wavelength effect on the action of a N-phenylimide S-23142 and a diphenylether acifluorfen-ethyl in cotyledons of cucumber(Cucumis sativus L.) seedlings. Plant Physiology, 1987, 85, 1146-1150.

5. Sato, R. Mechanism of action of diphenylether-type herbicides. Chemical Regulation in Plants, 1990, 25, 68-78.

NEW WEED CONTROL STRATEGIES FOR SOYBEANS

STEPHEN R. PADGETTE AND JAMES C. GRAHAM
NEW PRODUCTS DIVISION
MONSANTO AGRICULTURAL COMPANY
ST. LOUIS, MO. - U.S.A.

ABSTRACT

Selective herbicide discovery and development has historically been done by combining a chemical synthesis program to a biological whole plant screen. This procedure has generated many good soybean herbicides. Using genetic engineering techniques, it is now possible to develop physiological crop selectivity independent of a chemical synthesis program. The procedures used to develop glyphosate-tolerant soybeans are described as a model for such an approach.

INTRODUCTION

The practice of chemical weed control in soybeans is very sophisticated and growers have a wide

range of options. Current herbicides can be used preplant, preemergence or postemergence to

the soybean corp to the weeds. These products offer grass, broadleaf and sedge control. They

control annual and perennial weeds. In fact, growers have a choice of over twenty active

ingredients. Currently available herbicides will control most weeds that are competitive with

soybeans.

Past Discovery Approaches

The approach used to discover and develop the current herbicides was essentially the same despite being done by over ten independent agro-chemical companies. This approach combines a chemical synthesis program and screening of representative plant species, both weeds and soybeans. In this system, chemicals are applied to weeds and soybeans to select candidates showing control of the economically important weeds. While any chemical will control a number of weeds, none will control all the weeds associated with soybeans. Thus, economic weed control in soybeans usually requires mixtures or sequential applications. Since weed control in soybeans has evolved to its present statues of many products providing nearly complete control, when do companies continue to look for new products? The major factor driving new weed control/discovery programs are:

- evolving weed spectra
- weed resistance to current products
- environmental issues associated with current chemistry
- products having more agronomic flexibility

New Strategies

Since there is a continuing interest in new products for weed control in soybeans, there may be different approaches to the discovery and development of such products. In discovering a new herbicide, considerable synthesis effort and biological evaluations are expended in gaining the necessary selectivity. One is always looking for new techniques that might reduce these.

Herbicide selectivity in crops is based on physical avoidance, physiological selectivity and safeners. The most consistent crop safety is usually based on physiological selectivity. This avoids the need for placement selectivity, safeners or time of application to gain the needed crop safety.

The mechanisms by which plants are physiologically resistant to herbicides include the following (1):

- inhibition of uptake
- degradation
- conjunction
- sequestration
- over-production of a competing substrate or metabolite

- over-production of the target protein
- alteration of the target protein

Degradation is the most common mechanism. Consider the case where one could focus synthesis programs on weed control environmental properties and cost of goods and not be concerned about selectivity. This increases the chance to optimize the other properties. In this system, genetic modification of the crop could be used to infer crop selectivity.

Genetic modification of plants for pest control is not new. Identification of sources of resistance to diseases, nematodes and insects has been a major activity in soybean breeding programs (2). Tolerance has been bred into soybeans varieties for resistance to 2,4-DB and metribuzin (3,4).

Prior to incorporating genetic resistance to a herbicide into a plant one should know the mode of action of the herbicide. The mode of action of many of the newer herbicides is being determined well in advance of their commercialization. Knowing the mode of action or the desired mode of action is required for a rational synthesis program. These programs are usually based on screening isolated enzymes with confirmation evaluation in vivo.

Another approach to diversifying soybean weed control strategies is to use genetic engineering to make non-selective herbicides selective. The rationale for this approach has been described (5).

- There are economic advantages to farmers and agriculture in using presently available herbicides.
- The number of new herbicides with unique modes of action being developed commercially has been decreasing.
- Crops resistant to a herbicide having long soil residual could be rotated easier than today.

Monsanto has employed both strategies in our herbicide discovery programs. We have used the approach of modifying plants' resistance to new classes of chemistry and altering plants' resistance to an existing non-selective herbicides. This paper will focus on the modification of soybeans to confer resistance to the herbicide glyphosate. The principles used here are applicable to the development of a new herbicide family.

Environmental Features of Glyphosate

A number of studies have shown very tight absorption of glyphosate by a wide variety of soils (6). Thus, there should be no contamination of surface or ground water as a result of glyphosate run-off or leaching. Also, once glyphosate reaches the soil and binds to the soil particles, it is unavailable for uptake into plants and has no phytotoxic activity. In soil, glyphosate is rapidly degraded by soil microbes. The initial sole major metabolite of glyphosate has been identified as aminomethylphosphonic acid (AMPA), following incubation of 14_c-glyphosate with soil (7). AMPA is further degraded to CO_2 in the soil.

Plant-glyphosate Interactions

Glyphosate is systemic herbicide and to produce phytotoxic effects, it must first penetrate the plant foliage and be translocated to the target site. The absorption and transport of glyphosate has been reviewed (8). Translocation following cuticle penetration involves cell-to-cell movement followed by long distance transport via the vascular tissues. Movement is severely reduced or abolished in plants which are in the dormant state. There is no convincing evidence that glyphosate is actually metabolized by plants (6).

Glyphosate Mode of Action

Early studies on the mechanism of action of glyphosate's herbicidal activity indicated that growth inhibition of plants and bacteria by glyphosate could be partially alleviated by either L-phenylalanine and/or L-tyrosine (9). These results suggested that glyphosate was interfering with aromatic amino acid biosynthesis in both plants and bacteria. Subsequent studies showed the glyphosate blocked the incorporation of 14_c-shikimic acid into aromatic amino acids in plant cells (10) provided the first evidence that glyphosate specifically inhibits the enzyme 5-enolpyruvylshikimate-3-phosphate synthase (EPSP synthase, EPSPS, 3-phosphoshikimate 1-carboxyvinyl-transferase; EC 2.51.19), using cell-free extracts of *K. pneumoniae*. EPSPS catalyzes the reversible condensation of shikimate-3-phosphate (S3P) and phosphoenolpyruvate (PEP) to give EPSP and inorganic phosphate (Pi). Glyphosate is competitive inhibitor with respect to PEP or EPSPS, and interacts with either E•S3P complex (12). Of the several known PEP-dependent enzymatic reactions, EPSPS is the only enzyme inhibited by glyphosate.

Roundup®-tolerant Soybeans

We have used the target-site modification approach to develop Roundup-tolerant soybeans. This work has been previously reviewed (13). The principle of the work has been to identify EPSPS enzymes which efficiently catalyze the formation of EPSP, even in the presence of glyphosate. The gene for the glyphosate-tolerant EPSPS is then inserted into the soybean DNA, and upon regeneration the genetically modified soybean now expresses two different kinds of EPSPS enzyme: the natural endogenous EPSPS and the glyphosate-tolerant EPSPS. Thus, when the genetically modified soybean is sprayed with Roundup, the endogenous enzyme is inhibited, but the tolerant EPSPS still functions and the plant does not die due to starvation from the products of the EPSPS reaction.

We have isolated an EPSPS which is highly tolerant to glyphosate by selection of *E. coli* for growth in the presence of glyphosate (14). The aroA gene encoding this mutant (SM-1) has been cloned and shown to contain a single amino acid substitution of alanine for glycine at amino acid residue 96. Kinetic studies indicate that the K_i for glyphosate for this enzyme is 4 mM (compared to 0.5 µM for the wild-type *E. coli* enzyme). The mutant enzyme thus has an 8,000-fold decreased glyphosate sensitivity, and the K_m for PEP increased significantly (from 17 µM to 220 µM). This increase in K_m for PEP is not unexpected since glyphosate is a competitive inhibitor with respect to PEP. Alignment of the amino acid sequences of EPSPS from petunia, tomato, *Arabidopsis*, *B. napus*, soybean, maize, *E. coli*, *S. typhimurium*, *A. nidulans*, *S. cerevisiae*, and *B. pertussis* revealed that the Gly[96] residue is located in a highly conserved region of these enzymes. We have shown that this glycine to alanine substitution (G96A) also imparts glyphosate tolerance to five additional EPSPS enzymes (15). This region is therefore a critical part of an EPSPS active site highly conserved between plant and bacterial enzymes and is critical for the interaction of EPSPS with glyphosate and PEP.

Because the G101A petunia EPSPS (which corresponds to G96A *E. coli* EPSPS) also has reduced PEP binding relative to the wild-type petunia enzyme (K_m(PEP)=210µM, K_i(glyphosate)= 2 mM, we undertook random mutagenesis experiments to attempt to identify glyphosate-tolerant EPSPS enzymes which had better PEP binding. Better PEP binding would likely translate into a more efficient glyphosate-tolerant EPSPS in the presence of glyphosate. To this end, mutations were identified which when combined with the G96A mutation, lowered the K_m (PEP) of 54 µM, and a K_i(glyphosate) of 670 µM (16).

Genetically modified soybean plants have been made which express the petunia G101A, A192T double variant EPSPS under control of constitutive promoter element. These plants have been field tested in replicated yield trials in the United States during the summer of 1991. Results show these soybeans to be tolerant to twice the normal expected use rates for glyphosate.

Work is on-going to introduce this genetic material into commercial soybeans varieties. Concurrently, residue work is being done to establish tolerances for glyphosate use on these tolerant soybeans. Commercialization is expected in the mid to late 90's.

REFERENCES

1. Thompson, G. A., Hiatt, W. R., Facciootti, D., Stalker, D. M., and Comai, L. (1987) "Expression in plants of a bacterial gene coding for glyphosate resistance." Weed Science. 35 (Suppl 1):19-23.

2. Hartwig, E. E. (1972). "Utilization of soybean germplasm strains in soybean improvement programs." Crop Science, 12:856-859.

3. Hartwig, E. E. (1974). "Registration of tracy soybeans." Crop Science, 14:777.

4. Hartwig, E. E., Barrentine, and Edwards, C. J. (1980). "Registration of tracy m. soybeans." Crop Science, 20:825.

5. LeBaron, H. (1987). "Introduction of genetic engineering symposium." Weed Science, 35 (Suppl 1),:2,3.

6. Malik, J., Barry, G., and Kishore, G. (1989). "The herbicide glyphosate." Bio-Factors 2, pp. 17-25.

7. Rueppel, M. L., Brightwell, C. C., Schaefer, J., and Marvel, J. T. (1977). J. Agric. Food Chem, 25:517-528.

8. Casle, J. C., and Coupland, D. (1985). In Grossbard, E., and Atkinson, D. (eds.), "The herbicide glyphosate." Butterworths, London, pp. 82-123.

9. Jaworski, E. G. (1972). "Mode of action of n-phosphonomethyl glycine: inhibition of aromatic amino acid biosynthesis." J. Agric. Food Chem., 20:1195-1198.

10. Höllander, H., and Amrhein, N. (1980). "The site of the inhibition of the shikimate pathway by glyphosate." Plant Physiology, 66:823-829.

11. Steinrücken, H. C., and Amrhein, N. (1980). "The herbicide glyphosate is a potent inhibitor of 5-enolpyruvyl-shikimic acid 3-phosphate synthase." Biochem. Biophys. Res. Commun., 94:1207-1212.

12. Boocock, M. R., and Coggins, J. R. (1983). "Kinetics of 5-enolpyruvyl-shikimate-3-phosphate synthase inhibition by glyphosate." F.E.B.S. Lett., 154:127-133.

13. Padgette, S. R., della-Cioppa, G., Shah, D. M., Fraley, R. T., and Kishore, G. M. (1989). Selective Herbicide Tolerance Through Protein Engineering in Cell Culture and Somatic Cell Genetics of Plants. (Schell, J. and Vasil, I., eds.). 6:441-476, Academic Press, New York.

14. Kishore, G. M., Brundage, L., Kolk, K., Padgette, S. R., Rochester, D., Huynh, Q. K., and della-Cioppa, G. (1986) "Isolation, purification and characterization of a glyphosate-tolerant mutant e. coli EPSP synthase." Fed. Prac., 45:1506.

15. Padgette, S. R., Re, D. B., Gasser, C. S., Eichholtz, D. A., Grazier, R. B., Hironaka, C. M., Levine, E. B., Shah, D. M., Fraley, R. T., and Kishore, G. M. (1991). "Site-directed mutagenesis of a conserved region of the 5-enolpyruvlshikimate-3-phosphate synthase active site." _J. Biol. Chem._, 266:22364-22369.

16. Eichholtz, D. A., Gasser, C. S., Padgette, S. R., Re, D. B., Kishore, G. M., and Fraley, R. T. (1990). "A novel selection scheme in _E. coli_ for identifying glyphosate tolerant variants of petunia EPSP synthase." _Genetics Society of America_, July 18-21, San Francisco, CA.

THE CONTROL OF MILK WEED (*Euphorbia heterophylla*) IN SOYBEAN
WITH A MYCOHERBICIDE

JOSE T. YORINORI and DIONISIO L.P. GAZZIERO
Centro Nacional de Pesquisa de Soja
EMBRAPA-CNPSo
Caixa postal 1061, 86001, Londrina, PR, Brazil

ABSTRACT

Milk weed (wild poinsettia)(*Euphorbia heterophylla*) may be a classical example of a weed species that, in few years, evolved to a major problem, being selectively favored by the herbicides that controlled the other weeds but left it unaffected. An alternative biological method of control has shown great potential for practical use. The fungus *Helminthosporium* sp. has demonstrated high virulence and host specificity, ability to sporulate on simple media, relatively long shelf life, compatibility with herbicides and insecticides, tolerance to application under dry conditions, is effective on post-application regrowth and delayed germination of weed, and greater flexibility in time of application in relation to the age of the weed. Some of the drawbacks that need further investigation are related to the high density of the spores that result in quick settling, the occurrence of localized resistant biotypes of the weed, development of an industrial process for commercial production and refinement of the formulation.

INTRODUCTION

In 1976, Dr. Glen Davis, then a USDA-AID weed specialist and consultant to the National Soybean Research Project in Brazil, predicted that milk weed (wild poinsettia) (*Euphorbia heterophylla* L.) was going to be a troublesome weed in the future. His field observations had shown that the rapid expansion of soybeans with the intensive use of herbicides was selectively favoring the milk weed which was not affected by the chemicals used at the time. He did not stay in Brazil long enough to see his prediction fulfilled. By 1980, milk weed had indeed become a serious problem in certain areas of the states of Parana and Santa Catarina, in southern Brazil. It is currently one of the most difficult weed to control and is disseminated to most of the soybean producing regions of the country (1).

The only herbicide available for many years, acifluorfen, applied postemergence, gave only reasonable control, especially due to the long germination period of the weed and the short time the seedlings are vulnerable to the herbicide. With enough moisture in the soil, the weed vegetates and produces seed all year round, infesting the field in few crop seasons. Plants beyond three to four leaf stages, if not killed, usually regrow and produce seeds normally. More recently, two postemergent (lactofen and imazethapyr) and one preemergent-incorporated (imazaquin) herbicides were made available (2).

In 1987/88, an estimated 200,000 hectares of crop land, mostly soybean fields were infested, requiring ca. US$ 3.7 million worth in chemicals (3). Of about 8 million hectares of soybeans grown in the 1991/92 crop season, at least 2 million hectares required herbicides for the control of milk weed. Weed control is estimated to represent about 20% of the production cost of soybeans and most often control is by use of herbicides. The chemicals used to control milk weed are generally more expensive than those used on other weeds.

In an attempt to find an alternative, nonchemical method that could be integrated with chemical control, a research project on biological control with pathogens was initiated in 1981 at the National Soybean Research Center (CNPSo) at Londrina (4).

A disease survey made on *E. heterophylla* populations in various regions showed that the weed was naturally affected by the following diseases: 1. Euphorbia mosaic (E. mosaic virus-EMV); 2. rust (*Uromyces euphorbiae* Cke. et PK.); 3. scab [anam. *Sphaceloma krugii* Bitanc. & Jenkins (?), teleom. *Elsinoe brasiliensis* Bitanc. & Jenkins); 4. leaf spot (*Helminthosporium sp.*); 5. stem canker (*Alternaria* sp.); 6. damping- off and root rot (*Rhizoctonia solani* Kuhn); 7. white mold [*Sclerotinia sclerotiorum (L.)* de Bary]; 8. powdery mildew [*Microsphaeria euphorbiae* Pk. (?)] (4,5).

Euphorbia mosaic, rust, scab, leaf spot and stem canker were most commonly associated with *E. heterophylla*, and often occurred simultaneously. Euphorbia mosaic usually appears by mid-January, frequently affecting 100% of the population but the plants have adapted to the virus infection and seed production does not seem to be much affected. The virus is not seed-transmitted. EMV seems to cause little damage to the weed plants but is transmitted from the milk weed to soybeans by the white fly (*Bemisia tabaci*), causing severe stunting , especially when soybeans are planted as a second crop in late February to March. The rust fungus occurs mostly from the middle to the end of the soybean growing season and usually allows normal seed production of milk weed. Being an obligate parasite, the rust fungus is difficult to manipulate. The scab disease has been quite aggressive on milk weed but its occurrence is unpredictable and much dependent upon weather condition. It is possible to artificially grow the scab fungus but its growth is very restricted. *R. solani* and *S. sclerotiorum* occur in certain regions but are also very destructive to soybeans and other important crops. Powdery mildew was found in the cooler region of southern Parana and is capable of causing considerable defoliation. Its occurrence is unpredictable and usually appears too late, when the weed is at late flowering to fruiting stages.

Under experimental conditions, only *Alternaria* sp. and *Helminthosporium* sp. were found most efficient and promising. Both pathogens are widely disseminated but are generally found causing minor diseases. Nevertheless, they have the advantage of being easily cultured under artificial conditions (4).

Alternaria sp. did not cause defoliation but was effective when infecting the stem of young plants. One stem infection was enough to kill the plant. It sporulated very poorly on any artifical media used and, therefore, was not considered as of practical use at the moment. The isolates of *Helminthosporium* sp. *behaved similarly to the Alternaria* isolates. They grew well on artificial media but sporulated very poorly. Among several isolates obtained, one from Rio Grande do Sul, originally very poor in sporulation, happened to produce sectors of black, massive quantities of conidia, without aerial mycilia. Spores from these sectors, when single-spored and cultured on tomato-agar remained stable, henceforth giving uniform growth of sporulating colonies. The isolate was subcultured many times for spore production, with periodic reisolation from greenhouse inoculated plants.

After the pathogenicity of *Helminthosporium* sp. was established, further studies were pursued under laboratory, greenhouse and field conditions. The following is an account of the work done on biological control of milk weed with *Helminthosporium* sp. since 1981, at CNPSo.

MATERIALS AND METHODS

Laboratory studies involved: a. spore production on artificial media using PDA, V-8 juice, tomato-agar, tomato broth, autoclaved grain sorghum, squash-agar and squash-broth; incubation was done under normal laboratory conditions; b. selection of appropriate container (petri dishes, Erlenmeyer flasks, and aluminum lunch-box measuring 6cm x 11cm x 17cm) for spore production; c. formulation in powder using kaolin and lactose at the rate of about 70% of inert material and 30% of spore mass obtained from the broth culture; d. longevity of the spores by maintaining pure spores and lactose formulated conidia under room temperature and in the refrigerator, at 10C ± 1C, for 6 to 14 months; e. and compatibility with herbicides and adjuvants (bentazon, bentazon + assist, sethoxydin + assist, assist, fomesafen + Energic, fomesafen and Energic alone) and insecticides (*Bacillus subtillis*, *Baculovirus anticarsiae* and carbaryl), based on germination of spores suspended in recommended dosages of each herbicide or insecticide, and plated on water-agar. The greenhouse studies involved: a. incubation time: spray-inoculated plants were exposed to different time under moisture saturation by covering with plastic bags; 32-day-old plants were inoculated with a spore concentration of 1.0x 10^4 conidia/ml; b. test of filtered liquid extract of culture media for phytotoxicity; c. host range of *Helminthosporium* sp.; d. variability of reaction to *Helminthosporium* sp. among populations of *E. heterophylla*. The field studies involved: a. time of inoculation (8:00, 12:00 and 16:00) under bright, sunny day and temperatures varying from 23.8 C to 32.2 C and relative humidity from 88% to 64%; assessment of the effects were based on % defoliation and disease severity at 6, 11 and 27 days after inoculation, and fresh weight; b. inoculum concentration of 50, 100, 200, and 300 thousand conidia/ml with disease severity assessed at 8, 26 and 36 days after inoculation.

RESULTS

Laboratory studies

Spore production: The sporulating isolate of *Helminthosporium* sp. produced abundant conidia on any culture media used but was best on V-8 juice, tomato or squash, either on solid or in broth media. V-8 juice was imported and because of high cost, its use was discontinued. Tomato-agar and squash-agar were used when pure dry spores were produced.

Culture container: The aluminum lunch-box was the most practical for spore production and has been routinely used in broth culture.

Spore longevity: Pure and dry conidia stored under room temperature and in the refrigerator for 14 months had 83% and 78% germination, respectively, after 24 hours incubation on water-agar, at room temperature. With three hours of incubation, only 1.4% (spores kept at room temperature) and 0.2% (spores kept in the refrigerator) had germinated. After eight hours of incubation, germination had reached 35% in both cases.

In the lactose formulated conidia and stored for six months under the same above conditions, within two hours of incubation, germination had reached, respectively, 49% and 70%, and within four hours, germination had reached 93% and 95%, respectively. Maximum germination of 99% and 98% were observed after 24 hours of incubation. The results showed that the fungus remained viable for a relatively long period under room temperature but germination was slower with longer time of storage.

Formulation in kaolin and in lactose: both formulations allowed for uniform and mostly individual conidial suspension. When formulated or pure conida were suspended in water, the spores tended to settle rather quickly, requiring continued stearing of the suspension to keep homogeneity.

Compatibility with herbicides and insecticides: The compatibility tests showed that only the herbicide fomesafen (49.6% germination) and the adjuvant energic (0% germination), affected the germination of *Helminthosporium* sp. The control had 96.8% germination.

Greenhouse studies

Incubation time: The fungus was capable of colonizing the leaf tissue withn 6 hours of incubation under saturated moisture. Degree of leaf infection increased with increased time of exposure to high relative humidity and with 12 hours incubation it caused complete defoliation of 32-day old plants in 72 hours. A wilting symptom of the leaves was visible within 12 hours following inoculation.

Phytotoxicity of liquid filtrate: Potted plants showed wilting of the leaves within 12 hours when sprayed with full concentration and 50% dilution of filtered liquid medium collected after 8 days under culture with *Helminthosporium* sp.

Host range: Of all the crops tested, only grain sorghum showed minute (<1mm diameter), hypersensitive, necrotic spots.

Variability in *E. heterophylla* populations: Reaction of milk weed populations to inoculations with *Helminthosporium* sp. varied from 100% susceptibility to 100% immunity. Most of the fields surveyed showed all susceptible plants but many had mixtures of susceptible and resistant plants. The populations with mixed reactions or totally resistant plants were found mostly in west-central region of state of Parana. Among six samples collected from outside of the state of Parana (Sao Domingos, state of Santa Catarina; Lavras, state of Minas Gerais; Sao Grabriel D'Oeste and Chapadao do Sul, state of Mato Grosso do Sul; and Campo Verde, state of Mato Grosso), the samples from Sao Domingos (3) and Chapadao do Sul (Yorinori,1991, unpublished) had mixtures of resistant plants. The other sites had all susceptible plants.

Field studies

Time of inoculation: The time of inoculation (8:00, 12:00 and 16:00), did not affect the degree of infection. Percent defoliation determined six days after inoculation varied from 50 to 58% among the treatments. The infection levels observed 27 days after the inoculation remained almost the same as on the 6th day, indicating that further disease development was hampered by the unfavorable weather that followed.

Inoculum concentration: The effect of inoculum concentration (50, 100, 200 and 300 thousand conidia/ml), determined 8 days after the inoculation, showed significant and increasing percent defoliation with the increase in spore concentration. No defoliation was observed in the control plots. By the 36th day after inoculation, there was no difference in defoliation between the treatments with 50 (56%) and 100 thousand (58%) conidia/ml but they differed from 200 and 300 thousand conidia/ml (both had 73% defoliation). The control plots had 43% defoliation, indicating that there was a later infection spores produced on inoculated plots (3).

DISCUSSION

The results obtained in the studies with *Helminthosporium* sp. showed that the fungus has a potential for biological control of *E. heterophylla* in Brazil. Its virulence, the ability to produce massive amounts of spores under artificial conditions, the resistance to application under dry condition, the longevity under room temperature, the ability to germinate and infect with short incubation period, the compatibility with herbicides and insecticides, are all favorable assets for a biological control agent.

Except for a few unreplicated field tests in the rainy savanna region, all the studies were restricted to the state of Parana where the climatic conditions are, most often, unfavorable for disease development. With further expansion of soybeans and other annual crops (rice, dry beans, sorghum and corn), covering several million hectares in the rainy savanna region, it is most likely that milk weed will continue to be an important problem. It will also be a better target for biological control due to the climatic condition more favorable to the pathogen.

The occurrence of resistant biotypes of *E. heterophylla* may be a hindrance where the weed population is composed solely of resistant biotypes but fungal and chemical herbicides may be integrated where populations are mixed. The possibilities of mixing biological and chemical agents has been shown by the compatibility studies (4). Studies with fungal/herbicide combinations have demonstrated that both fungal and herbicide activities can be enhanced in mixed applications (Gazziero, 1991, unpublished). The mixtures with herbicides, also active against the resistant biotypes of *E. heterophylla*, would prevent the selective survival of the resistant biotpypes.

Previous studies with COLLEGO [*Colletotrichum gloeosporioides* (Penz.) Sacc. f. sp. *aeschynomene*] with acifluorfen resulted in efficient control of northern joint vetch [*Aeschynomene virginica* (L.) B.S.P.] (*susceptible to COLLEGO*) and hemp sesbania [*Sesbania exaltata* (Raf.) Rydb. ex A.W. Hill] in rice and soybean fields (6,7).

Besides the potential for biological control, *Helminthosporium* sp. also offers the possibility for the development of a phytotoxic compound produced as a secondary metabolite which is released in the culture substrate. The fractionation and characterization of this toxic compound is being investigated in the Chemistry Department at the State University of Londrina.

Some of the points that stil need to be improved and further developed to make the use of *Helminthosporium* sp. a routine practice, is the development of an industrial process for massive spore production and perfection of the formulation. It is also essencial to improve the steering mechanism in the spray tank to keep the spore suspension homogeneous; or develop an alternate inert suspension medium of the same density as the spores.

In addition to *E. heterophylla*, two other weed species, *Desmodium purpureum* and *Cassia tora* (sickle pod), are serious weeds in the savanna region and potential targets for biological control (8).

Since the two commercial products, COLLEGO and DeVine (*Phytophthora palmivora*) were developed, a large number of target weeds and potential pathogens have been investigated (9,10,11). Some of the pathogens are already in advanced stages of development (10).

A substantial amount of information on biological control of weeds with plant pathogens has been summarized in the two books dealing specifically with pathogens (10,11). Besides, several hundreds of research papers and review articles, published in many journals, attest to the continued interest of the scientists in this relatively new field of research. The International Symposium on Biological Control of Weeds, organized every three years, has been an excellent forum of debate of the recent advances and difficulties encountered by scientists around the world. The Symposia have shown the wide range of possibilities of biological control, and the restrictions imposed by the regulatory laws have been part of the most serious debates (12). An important trend has also been recently noted with the investment in biological control by the traditional chemical companies.

REFERENCES

1. Lorenzi, H. Plantas Daninhas no Brasil ("Weed Plants in Brazil")(2a. ed.). Editora Plantarum Ltda., Nova Odessa, SP, 1991. 440 p.

2. Gazziero, D.L.P., Almeida, F. S. and Rodrigues, B. N. Controle de plantas daninhas ("Weed control"). In: Recomendacoes Tecnicas para a Cultura da Soja no Parana 1991/92, EMRAPA-CNPSo/OCEPAR, Cascavel, PR, 1991, Documento 47/Boletim Tecnico, 29. pp.88-95 (In Portuguese).

3. Yorinori, J.T. and Gazziero, D.L.P. Control of milk weed (*Euphorbia heterophylla*) with *Helminthosporium* sp. In: Proceedings, International Symposium on Biological Control of Weeds, 7, Rome, 1988, ed. E.S. Delfosse, CSIRO Publications, Victoria, Australia, 1990. pp. 571-576.

4. Yorinori, J.T. Biological control of milk weed (*Euphorbia heterophylla*) with pathogenic fungi. In: Proceedings, *International Symposium on* Biological Control of Weeds, 6, Vancouver, 1984, ed. E.S. Delfosse, Agriculture *Canada, Vancouver, 1985. pp. 677-681.*

5. Viegas, G.P. Indice de Fungos da America do Sul ("Index of Fungi from South America"). Instituto Agronomico, Campinas, SP, 1961. 921 p.

6. Klerk, R.A., Smith, R.J., Jr. and TeBeest, D.O. 1985. Integration of a microbial herbicide into weed and pest control programs on rice (*Oryza sativa*). Weed Sci., 1985, 33, 95-99.

7. Khodayari, K., Smith, R.J., Jr, Walker, J.T. and TeBeest, D.O. Applicators for a weed pathogen plus acifluorfen in soybeans. Weed Tech., 1987, 1, 37-40.

8. Yorinori, J.T. and Gazziero, D.L.P. Prospects of the biological control of weeds in Brazil. IN: Resumos, Simposio de Controle Biologico, 2, Brasilia, 1990, EMBRAPA-CENARGEN, Brasilia, 13, 1990. pp. 72-73.

9. Charudatan, R. The mycoherbicide approach with plant pathogens. In: Microbial Control of Weeds, ed. D.O. TeBeest, Chapman and Hall, New York, 1991. pp. 24-57.

10. Charudatan, R. and Walker, H.L. (eds.). Biological Control of Weeds With Plant Pathogens, Wiley, New York, 1982.

11. TeBeest, D.O. Microbial Control of Weeds, Chapman and Hall, New York, 1991. 284 p.

12. Delfosse, E.S. (ed.). Proceedings, International Simposium on Biological Control of Weeds, 7, Rome, 1988, CSIRO Publications, Victoria, Australia, 1990.

DEVELOPING AN INTEGRATED WEED CONTROL PROGRAM TO PREVENT OR MANAGE HERBICIDE RESISTANT WEEDS IN SOYBEANS

DALE L. SHANER
American Cyanamid Company
P.O. Box 400, Princeton, NJ 08543-0400

ABSTRACT

Herbicide resistant weeds have not been a problem in soybean production. However, concern over the potential for the selection of resistant weeds has increased recently due to changes in soybean production practices and the introduction of several new classes of herbicides that share the same mode of action. Herbicide resistant weeds have not developed in soybeans because of the use of integrated weed management practices which include tank mixing of herbicides with different modes of action, crop rotation and the use of other weed control practices such as cultivation. To prevent resistance from becoming a problem these practices need to continue by educating growers on factors which can lead to the selection of resistant weed populations and maintaining all weed control options for the grower.

INTRODUCTION

In recent years concern has increased over the development of herbicide resistant weed populations. Although herbicide resistance has not been a major concern in soybean production, it does require some attention in order to prevent it from becoming a problem.   There are several reasons why herbicide resistant weed populations have not developed in soybeans including the availability of a wide range of herbicides with different modes of action for weed control (Table 1), the tank mixing of herbicides to control a broad spectrum of weeds, the use of other weed control practices such as tillage, and the rotation of soybeans with other crops with the concomitant rotation of herbicides.

TABLE 1

Mode of Action of Soybean Selective Herbicides

Herbicide Class	Chemical	Mode of Action	Use Pattern	Plants Controlled	
				Monocots	Dicots[b]
Aryloxyphenoxy-propionates	Fluazifop Haloxyfop Quizalofop	AcetylCoA Carboxylase Inhibition	Post[a]	+	
Benzoic Acid	Fenoxaprop Naptalam	Unknown	Pre/Post		+
Biological	Colletotrichum gloeosporoiodes	Disease	Post		+
Bipyridilliums	Paraquat	Photosynthesis Inhibition	Burndown	+	+
Chloracetanilides	Alachlor Metolachlor	Unknown	Pre	+	±
Cyclohexanediones	Sethoxydim Clethodim	AcetylCoA Carboxylase Inhibition	Post	+	
Diazinones	Bentazon	Photosynthesis Inhibition	Post		+
Dinitroanilines	Pendimethalin Trifluralin Ethafluralin	Microtubule Formation Inhibition	PPI/Pre	+	±
Diphenyl Ether	Acifluorfen Fomesafen Lactofen Fluoroglycofen	Protoporphyrinogen Oxidase Inhibition	Post		+

TABLE 1 (cont.)

Herbicide Class	Chemical	Mode of Action	Use Pattern	Plants Controlled	
				Monocots	Dicots
	Glyphosate	5-enolpyruvyl-shikimate-3-phosphate synthase Inhibition	Burndown	+	+
Imidazolinones	Imazaquin Imazethapyr	Acetohydroxyacid Synthase Inhibition	PPI/Pre/Post	+	+
Isoxazolidinone	Clomazone	Pigment Synthesis Inhibition	PPI	+	+
Phenylureas	Linuron	Photosynthesis Inhibition	PPI/Pre/Post		+
Phenoxy Acids	2,4-DB	Unknown	Post		+
Pyridazinone	Norflurazon	Carotenoid Synthesis Inhibition	PPI/Pre	+	+
Sulfonylureas	Chlorimuron Thifensulfuron	Acetohydroxyacid Synthase Inhibition	PPI/Pre/Post		+
Thiocarbamate	Vernolate	Lipid Synthesis Inhibition	PPI	+	
Triazinones	Metribuzin	Photosynthesis Inhibition	Pre		+

aPPI=Preplant Incorporated Application; Pre=preemergence application; Post=postemergence application; Burndown=burndown of emerged weeds before planting crop.
b(+) means primary activity of herbicide class. (±) means class has activity on some species.

However, some of these factors are changing, and these changes could increase the risk of selecting for herbicide resistant biotypes. These changes include the loss of older herbicides, the advent of new cropping practices with decreased levels of cultivation, and the development of new classes of herbicides which have the same mechanism of action. It is important, therefore, that strategies be designed to prevent or delay the selection for herbicide resistant weed populations.

CHANGES IN SOYBEAN PRODUCTION THAT MAY FAVOR HERBICIDE RESISTANCE

Before designing strategies to prevent the development of resistant weed populations, it is necessary to understand the conditions where resistance has developed. All herbicide resistant weed populations have developed where the same herbicide or herbicides with the same mode of action have been used for weed control over an extended period of time [1]. Herbicide resistant weed populations have not developed where either mixtures of herbicides and other weed control practices, crop rotation with concomitant herbicide rotation, or a combination of crop rotations and weed control measures have been the common practice [1]. These practices have been common in most soybean production areas.

However, changes are taking place in soybean culture which may favor the development of herbicide resistant weed populations. One effective means of controlling weeds besides using herbicides has been cultivation. Cultivation for seed bed preparation can eliminate early seedling populations, while cultivation within the crop can control weeds later in the season.

In spite of its usefulness, cultivation in soybean production is decreasing due to three factors. 1). Excessive cultivation can result in soil erosion, compaction, and can effect water resources. In addition, mechanical cultivation requires much more energy compared to chemical weed control and is therefore more expensive to the farmer [2]. Concern over these negative aspects of cultivation has led to minimum and no-till practices. 2) The spread of narrow-row or solid-seeded soybeans has also reduced cultivation options. 3) As the area farmed by a single grower increases, the grower would prefer not to cultivate.

None of these practices would be possible without the use of highly effective herbicides. The down side of these practices is that they reduce the options available to the grower for weed control and increase the selection pressure of herbicides on the weed populations.

In addition, new classes of herbicides being developed for soybeans share the same mode of action (Table 1). These herbicides are highly effective and may control most of a farmer's weed problems by themselves. However, using either a single class of herbicides or those with the same mode of action over an extended period of time could lead to the selection of either resistant weed biotypes or a shift in the weed species from susceptible to tolerant ones. In either case, the farmer will lose effective weed control.

INFORMATION NEEDED TO DEVELOP HERBICIDE RESISTANCE STRATEGIES

One difficulty in designing an effective strategy to combat the development of resistance is quantifying the amount of selection pressure a herbicide is exerting on a weed population. This pressure varies with the use pattern of the herbicide, the area treated, time, and weed species.

For example, although there are triazine-resistant biotypes in more than 100 species, these herbicides continue to be used because they are still effective [3]. This fact indicates that triazines do not exert the same selective pressure on all weed populations even if these populations grow in the same field as resistant biotypes.

Several models have been developed [1,4] which attempt to predict the development of resistance within a population. Although these models are useful, they are only as accurate as the data that go in them. These data include the frequency of the resistant trait within the population, the fitness of the resistant biotype, the reproductive capacity of the species, the seed life of the species in the soil, and the selective pressure or effective kill of the herbicide [5]. In many cases, this information is lacking. Thus the usefulness of these predictive models is limited.

The absence of this fundamental information on basic weed biology and herbicide selective pressure makes it more difficult to develop an effective anti-resistance strategy because there is no way to tell if a particular strategy works or not until after resistance has arisen.

STRATEGIES FOR PREVENTING/MANAGING HERBICIDE RESISTANT WEEDS
AN INTEGRATED APPROACH

There are effective strategies that will prevent or delay the selection for herbicide resistant weed populations. These strategies include:

1. Rotating crops and herbicides with different modes of action.

2. Using the lowest possible use rate of a herbicide for economic weed control. Avoid using high rates of herbicides for "cosmetic" control when such control has no economic value.

3. Combining herbicides with different modes of action but overlapping weed control spectra.

4. Using the minimum number of applications of one particular herbicide per season.

5. Using other weed control methods, such as tillage, as appropriate.

6. Following good agronomic practices such as cleaning equipment and using certified crop seed to prevent the spread of resistant-weed seeds.

The effectiveness of the above strategies is clearly demonstrated in the spread of triazine resistant weeds in Ontario [6]. Stephenson et al. reported that triazine resistant biotypes are a widespread and serious problem in eastern Ontario but are a minor problem in southwestern Ontario, even though corn has been grown in southwestern Ontario for a longer period of time than in the eastern part of the province. Stephenson et al. suggested that the differences between these two parts of Ontario lay in differences in the agronomic practices in the two areas.

In 1988 triazines were applied to 60% of the maize in southwestern Ontario whereas more than 80% of the eastern area received a triazine application. In the southwestern part of the province the triazine was applied with a non-triazine herbicide over 90% of the time which often included a post emergent application. In eastern Ontario only 40% of the of the triazine treatment went on with another non-triazine, and post emergent applications were rare. Furthermore, over 95% of the area in eastern Ontario received manure applications from cattle fed corn silage, while less than 1% of the area in the southwestern part was treated with manure. This practice in the eastern Ontario allowed the rapid spread of resistant weed seed. Farmers cultivated over 40% of the time in southwestern Ontario, while cultivation was rare in the eastern area [6].

This case study in Ontario illustrates the effectiveness of combining different weed control practices to prevent and manage the development of herbicide resistant weed populations. If that is true, why, then, don't all growers follow these same practices? The answer to this question appears to be due to other factors beside weed control which influence a grower's practices.

These other factors include economic and regulatory restraints placed on the producer. Crop rotation, although desirable for many agronomic reasons, may not provide the producer enough monetary return to stay in business, or the crop may be used to support another aspect of the farm, such as dairy production. Also, herbicides are a more cost effective way to control weeds than other practices. Effective herbicides allow more land to be farmed by an individual and free up time for other endeavors. Government programs aimed at reducing soil erosion may force a farmer to depend more heavily on herbicides to control weeds resulting in a concomitant reduction in cultivation.

IMPLEMENTATION OF HERBICIDE RESISTANCE STRATEGIES IN SOYBEANS

An effective herbicide resistance strategy depends on an integrated approach to weed control aimed at reducing selective pressures on weed populations. This integrated approach includes the effective use of herbicides with varying modes of action along with good agronomic practices of crop rotation and timely cultivations. It may also include new methods of cropping such as ridge tilling, herbicide banding, as well as cultural and biological weed control methods. In the past, weed control in soybeans has integrated cultural, mechanical and chemical practices. The key to preventing herbicide resistant biotypes from developing in soybeans is to continue these practices.

One of the first priorities in this process is educating the farmer on the conditions which could lead to the development of herbicide resistant populations. This education has to be done in the context of the current situation in which herbicide resistance is not a problem. The present weed control practices have not selected for resistant weeds and these practices must not be abandoned.

One reason farmers have not had a problem with resistant weeds in soybeans is the availability of herbicides with different modes of action. Knowing the mode of action of herbicides has not been a prerequisite for developing an effective weed control program, but with the proliferation of new herbicides and herbicide premixes, the grower and dealers must become more conversant how herbicides kill plants. To meet this need most agrochemical companies, including Cyanamid, BASF, DuPont, and Sandoz, among others, along with university personnel are working to educate the sales force, dealers and farmers on the mode of action of the herbicides being used and how these herbicides fit into a good weed control system without increasing the potential for selecting for resistant weeds. These companies have also taken definitive positions to actively work to prevent resistance from becoming a problem.

In addition to education of the grower, there is also a need to increase the level of basic research on the genetics and ecology of important weed species in order to better predict the potential for a particular herbicide to select for resistant populations. This research is also necessary in order to develop effective weed control programs that integrate the use of herbicides with mechanical, cultural and biological practices.

Herbicide resistant weeds should not become a major problem in soybeans as long as the integrated weed control practices which have been used in the past continue. The availability of herbicides with different modes of action, crop rotation, and timely tillage all work toward preventing the selection of resistant weed populations. What should be guarded against is developing a dependence on only one means of weed control. The way to combat this dependence is to educate the growers on how to effectively integrate different weed control methods into their farming practices.

REFERENCES

1.   Gressel, J., Why get resistance? It can be prevented or delayed. In Herbicide Resistance in Weeds and Crops, eds. J.C. Caseley, G.W. Cummins, R.K. Atkin, Butterworth-Heinemann Ltd. Oxford, 1991, pp. 1-26.

2.   Hayes, W.A., Minimum Tillage Farming. No-Till Farming Inc. Brookfield, Wisconsin, 1982, pp. 84-93.

3.   LeBaron, H.M., Distribution and seriousness of herbicide resistant weed infestations worldwide. In Herbicide Resistance in Weeds and Crops. eds. J.C. Caseley, G.W. Cummins, R.K. Atkin, Butterworth-Heinemann Ltd. Oxford, 1991, pp. 27-44.

4.   Maxwell, B.D., Roush, M.L. and Radosevich, S.R., Predicting the evolution and dynamics of herbicide resistance in Weed Populations. Weed Tech., 1990, 4, 2-13.

5.   Rubin, B., Herbicide resistance in weeds and crops, progress and prospects. In Herbicide Resistance in Weeds and Crops, eds. J.C. Caseley, G.W. Cummins, R.K. Atkin, Butterworth-Heinemann Ltd. Oxford, 1991, pp. 387-415.

6.   Stephenson, G.R., Dykstra, M.D., McLaren, R.D. and Hamill, A.S., Agronomic practices influencing triazine-resistant weed distribution in Ontario. Weed Tech., 1990, 4, 199-207.

BENTAZON AS A FOUNDATION HERBICIDE FOR WEED MANAGEMENT
SYSTEMS IN CONVENTIONALLY AND REDUCED TILLED SOYBEANS

P. MUNGER, M. LANDES, H. WALTER
BASF Aktiengesellschaft, Agricultural Research Station,
Limburgerhof, Germany

ABSTRACT

Postemergence herbicides are applied extensively in conventional,
reduced, and no-till soybean (Glycine max) production systems in
the United States. Bentazon is an effective tool in each tillage
system either as a component of a total postemergence weed
control program or as a sequential treatment following
applications of a soil-applied herbicide. Bentazon's importance
in U.S. soybean production can be attributed to both the
excellent soybean selectivity of the herbicide and it's
effectiveness in controlling economically important broadleaf
weeds (i.e. Abutilon theophrasti, Xanthium spp., Chenopodium
album, Polygonum spp. and others). Results of field trials showed
that thifensulfuron and imazethypr can be tankmixed with bentazon
to broaden the weed spectrum to include Amaranthus spp. and other
weeds not usually controlled by treatments of bentazon alone.
These tankmixes have not caused the minor leaf necrosis
associated with combinations of bentazon and diphenyl ether
herbicides. Results of field trials showed that total
postemergence grass and broadleaf weed control is obtainable when
sethoxydim is tankmixed with thifensulfuron.
Results of additive studies conducted in the laboratory
suggest that ethoxylated linear alcohols and ammonium may enhance
the cuticular penetration and plant uptake of bentazon.

INTRODUCTION

Soybean production in conservation tillage systems continues to
increase due to economic as well as environmental considerations
(1). These factors and the efforts of soybean producers to comply
with soil management provisions of the 1985 and 1990 Farm Bills
could result in as much as 33 % of the U.S. soybean acreage
included in conservation tillage programs by the year 2000 (2).
Increases in reduced tillage (RT) systems have also been noted in
other important soybean producing countries (3).

The intensive tillage associated with conventional tillage (CT) systems allows the soybean producer to utilize both mechanical and chemical methods of weed control. Herbicides, however, become increasingly important in reduced and no tillage (NT) production systems (4).

Incorporation of herbicides is difficult in reduced tillage systems due to the presence of surface plant residue. The use of preemergence herbicides is also limited in conservation tillage systems where heavy surface residue is present at application. Hence, as the intensity of tillage decreases often the soybean producer must rely solely on postemergence herbicides to control weeds.

Bentazon applied alone or tankmixed with other postemergence herbicides is effective in controlling several economically important weeds in soybeans. Bentazon can be applied subsequent to preplant incorporated herbicides in CT systems or following preemergence treatments in CT, RT, or NT systems to control yellow nutsedge (Cyperus esculentus) or broadleaf weeds that escape soil-applied herbicide treatments.

The postemergence weed spectrum of bentazon can be expanded if tankmixed with other foliarly-applied herbicides. Acifluorfen is commonly applied with bentazon to improve the control of Amaranthus species and other important broadleaf weeds. Other potential tankmix partners are fomesafen and herbicides belonging to the sulfonylurea and imadizolinone classes of chemistry that are effective in controlling broadleaf weeds. The postemergence weed control spectrum can be expanded to include grasses if a gramminicide, such as sethoxydim, is applied in sequence or tankmixed with the broadleaf herbicide combinations.

The performance of bentazon, as well as other postemergence herbicides, can be affected by environmental conditions at application. Efforts to maximize the rate and amount of herbicide uptake across a wide variety of environmental conditions have frequently focused upon spray additives. Field, greenhouse and laboratory research has shown that additives such as urea ammonium nitrate (UAN) and crop oil concentrate (COC) can improve the uptake, effectiveness, and consistency of bentazon; however, the mechanism by which the efficacy of bentazon is improved is poorly understood. Researchers have speculated that these adjuvants facilitate the movement of bentazon across the leaf cuticle and cell membrane, two major barriers to herbicide penetration and uptake (5, 6, 7, 8).

This paper discusses the use of bentazon following soil-applied herbicides or as a component of a total postemergence weed control program in CT, RT, and NT systems. Recent investigations into the effects of additives on the uptake of bentazon are also reviewed.

DISCUSSION

Preplant incorporated or preemergence applications of dinitroaniline or acetanilide herbicides are frequently used in CT, RT, and NT systems. When effective, these herbicides help to alleviate early season, interspecific competition between soybeans and annual grasses and/or small-seeded broadleaf weeds. Subsequent applications of bentazon are effective in controlling weeds such as velvetleaf (Abutilon theophrasti), common ragweed (Ambrosia artemisiifolia), common lambsquarter (Chenopodium album), common sunflower (Helianthus annuus), and common cocklebur (Xanthium pensylvanicum) which are usually not controlled with applications of dinitroaniline or acetanilide herbicides.

Since soil-applied herbicides are less effective in the presence of heavy surface residue, postemergence herbicides become increasingly important as tillage intensity decreases. Frequently, a total postemergence weed control program is the only option available to the RT and NT soybean producer.

Bentazon Tankmixes

Bentazon is effective in controlling many of the economically important broadleaf weeds that infest soybeans. The diphenyl-ethers, acifluorfen and fomesafen can be combined with bentazon to enlarge the broadleaf weed spectrum to include pigweed (Amaranthus spp.), black nightshade (Solanum nigrum), and other weeds not sufficiently controlled by bentazon. Tankmixes of bentazon plus acifluorfen or fomesfan can cause minor necrosis of soybean foliage. Combinations of bentazon with thifensulfuron or imazethapyr have provided a similar spectrum of weed control without the typical leaf "burn" associated with the diphenyl-ethers (Tables 1 and 2).

Tankmixes of bentazon and thifensulfuron at rates of 0,56 and 0,005 kg a.i./ha, respectively, provided excellent postemergence control of velvetleaf, pigweed, common lambsquarter, common sunflower, common cocklebur, and venice mallow (Hibiscus trionum) in field trials (Table 1). The tankmix was more effective in controlling velvetleaf, common ragweed, common lambsquarter, yellow nutsedge, venice mallow, and common cocklebur than thifensulfuron applied alone. Conversely, thifensulfuron improved the control of pigweed and velvetleaf relative to that provided by bentazon.

Increasing the rate of thifensulfuron in the tankmix from 0,0025 to 0,005 kg a.i./ha improved the control tall waterhemp (Amaranthus tuberculatus). The tankmix of the two herbicides caused 10 % soybean injury or less seven days after treatment (DAT). No phytotoxicity was noted 21 DAT.

TABLE 1

Bentazon and thifensulfuron alone and in combination for weed control
in soybeans (42 DAT)

Variable	Benta-zon	Thifen-sulfuron	Bentazon + Thifensulfuron		
kg a.i./ha	1,12	0,005	0,56 + 0,0025	0,56 + 0,005	1,12 + 0,0025
			%		
Crop Injury (7 DAT)	5	11	5	10	7
Abutilon theophrasti	85	84	89	98	99
Amaranthus reteroflexus	41	93	97	99	99
Amaranthus tuberculatus	30	94	88	98	84
Ambrosia artemisifolia	88	33	82	82	92
Chenopodium album	95	89	95	97	99
Cyperus esculentus	99	0	95	96	98
Helianthus annuus	99	98	90	98	98
Hibiscus trionum	98	78	98	98	98
Xanthium spp.	99	79	97	98	99

Weed Growth Stage - 1 to 4 leaves
All treatments with COC (2,3 1/ha) and UAN (9,3 1/ha)

The tankmix of bentazon and imazethapyr at 0,56 and 0,07 kg
a.i./ha, respectively, provided good to excellent control of the
weeds tested (Table 2). Rates of imazethapyr less than 0,07 kg
a.i./ha in the tankmix were not sufficient to provide acceptable
smooth pigweed (Amaranthus hybridus) control. The 0,84 kg a.i./ha
rate of bentazon in the tankmix was necessary for $\geq$ 90 % control
of common lambsquarter, yellow nutsedge, and ladysthumb smartweed
(Polygonum persicaria).

TABLE 2
Bentazon and imazethapyr alone and in combination for
weed control in soybeans (42 DAT)

Variable	Benta- zon	Imaze- thapyr	Bentazon + Imazethapyr		
kg a.i./ha	1,12	0,07	0,56 + 0,07	0,84 + 0,035	1,12 + 0,017
			%		
Crop Injury (7 DAT)	3	1	3	2	2
Abutilon theophrasti	97	95	97	98	98
Amaranthus hybridus	24	79	83	61	54
Amaranthus retroflexus	46	98	98	84	68
Chenopodium album	97	77	87	90	96
Cyperus esculentus	92	42	90	94	96
Polygonum persicaria	100	81	92	97	94
Xanthium pensylvanicum	100	95	100	100	98
Xanthium strumarium	100	99	100	98	100

Weed Growth Stage - 1 to 4 leaves
All treatments with COC (2,3 l/ha) and UAN (9,3 l/ha)

Postemergence broadleaf weed and grass control

Split application or combination treatments of bentazon plus
sethoxydim or bentazon tankmixes with acifluorfen and sethoxydim
have provided consistent, reliable postemergence control of
economically important broadleaf weeds and grasses in soybeans.
Studies have also shown that a three-way tankmix of bentazon,
thifensulfuron, and sethoxydim effectively controls a similar
weed spectrum (Table 3).

**Bentazon plus thifensulfuron plus sethoxydim for
broad spectrum weed control in soybeans (42 DAT)**

Variable kg a.i./ha	Bentazon+ Thifen- sulfuron +X-77 0,84+ 0,005+ 0,5 %	Sethoxy- dim+COC 0,20+ 1,25 %	Bentazon+ Thifensul- furon+Seth- oxydim +X-77 0,84+ 0,0025+ 0,2+0,5 %	Bentazon+ Thifensul- furon+Seth- oxydim+COC 0,84+ 0,0025+ 0,2+0,5 %
			%	
Crop Injury (7 DAT)	5	0	9	8
Abutilon theophrasti	98	0	98	98
Amaranthus hybridus	68	0	77	75
Amaranthus retroflexus	99	0	99	99
Ambrosia artemisiifolia	81	0	80	74
Chenopodium album	98	0	99	99
Echinochloa crus-galli	0	100	99	99
Helianthus annuus	99	0	99	99
Panicum dichotomiflorum	0	99	99	99
Setaria faberi	0	97	95	96
Sinapis arvensis	99	0	99	99
Setaria lutescens	0	100	99	99
Xanthium pensylvanicum	95	0	93	92
Zea mays	0	99	99	99

Weed Growth Stage - 4 to 6 leaves
All treatments applied with UAN (5,6 l/ha)

Sethoxydim applied at 0,2 kg a.i./ha alone and tankmixed with
bentazon and thifensulfuron provided excellent control of
barnyardgrass (Echinochloa crus-galli), fall panicum (Panicum
dichotomiflorum), giant foxtail (Setaria faberi), yellow foxtail
(Setaria lutescens), and volunteer corn (Zea mays) in field
trials. Neither bentazon nor thifensulfuron antagonized the
foliar activity of sethoxydim. Similarly, effects of the tankmix
on broadleaf weed control were also not observed. Visible crop
injury by all treatments 7 DAT was less than 7 %.

Bentazon Uptake

The epicuticular wax component of the leaf cuticle and the plasmalemma are considered to be two major obstacles to the penetration and uptake of postemergence herbicides. In theory, the ideal adjuvant or combination of adjuvants would facilitate penetration through both barriers. As indicated above, additives (UAN, COC, etc.) are known to improve the efficacy of bentazon and bentazon combinations. The reasons for the enhanced activity, however, are not completely understood.

Preliminary data from laboratory studies using isolated cuticles of Chenopodium album suggest that penetration of bentazon through epicuticular wax can be enhanced with ethoxylated linear alcohols (ELO) (9). In these tests, ELO surfactants enhanced the penetration of bentazon greater than ethoxylated derivatives of octylphenol.

Bentazon is a weak acid (pka 3,3) and moves across membranes via diffusion. Once it traverses the plasmalemma, the bentazon molecule accummulates against a concentration gradient in cells by way of an energy-dependent, ion trapping mechanism (10). Liebl (11) found that bentazon uptake increased when ammonium was applied to cultured soybean cells and the surrounding medium was acidified. Hence, it appeared that an ammonium source could factilitate the movement of bentazon across the plasmalemma.

The results of the research with isolated cuticles of Chenopodium album and the work conducted with cultured soybean cells suggest that a combination of ELO and ammonium source additives would promote increased penetration and plant uptake of bentazon.

REFERENCES

1. Buhler, D.D., Philbrook, B.D., and Oplinger, E.S., J. Prod. Agri., 1990, 3, 302 - 308.

2. WEFA and CTIC. U.S. conservation tillage study. 1991. 41 pp.

3. Kiessling, U., Proc. British Crop Protection Conf.- Weeds, 1982. 697 - 702.

4. Lewis, W.M., Jr. Weed control in limited tillage systems, 1985, pp. 41 - 50.

5. Leece, D.R., Aust. J. Pl. Physiol., 1976, 3, 83.

6. Bukovac, M.J., Petracek, P.D., Fader, R.G., and Morse, R.D., J. Exp. Bot., 1971, 22, 598.

7. Kirkwood, R.C., Dalziel, J., Matlib, A., and Somerville, L., Pest Sci., 1972, 3, 307.

8. Richard, E.P., and Slife, F.W., Weed Sci., **27**, 426 – 433.

9   Liebl, R.A. Personal Communication.

10. Sterling, T.M., Balke, N.E., and Silverman, D.S., Plant Physiol. 1990. **92**, 1121 – 1127.

11. Liebl, R.A., Zehr, U.B., Tyker, R.H., Weed Sci. (in press)

INTEGRATING HERBICIDE-RESISTANT CROPS INTO DISCOVERY RESEARCH

JON S. CLAUS
Agricultural Products,
E. I. du Pont de Nemours & Co., Inc.
Wilmington, Delaware, USA

ABSTRACT

Technology is now available to create herbicide-tolerant soybeans. This technology, when combined with environmentally- and toxicologically-sound herbicides and when used wisely in an integrated weed management program, can contribute positively to soybean production. Sulfonylurea-tolerant soybeans will be commercialized in 1993, and by 2000, will be available in most soybean-growing countries. Sulfonylurea-tolerant soybeans will provide crop safety during periods of reduced metabolism of chlorimuron ethyl and thifensulfuron methyl caused by environmental stresses. They will also allow the selection of new soybean-safe herbicides with the optimum biology, toxicology, and environmental properties from the tens of thousands of sulfonylurea herbicides that have been synthesized.

INTRODUCTION

The important role that herbicides play in soybean production is well known. Benefits resulting from the use of herbicides include increased yields, reduced labor costs, improved soybean quality, reduced soil erosion and moisture loss, and less soil compaction. Because of this high benefit to soybean farmers, herbicides are used on most soybean land.

Soybean-selective herbicides have evolved over the last 30 years in many ways. By the mid 1970s, the market was dominated by the pre-emergence and ppi grass and broadleaf herbicides such as trifluralin, alachlor, linuron, and metribuzin. The grass products have good to excellent crop safety and control many species. However, over time, weed shifts have led to some hard-to-control species such as Sorghum. bicolor, S. halepense, and Eriochloa villosa. This weed species shift led to the introduction of several selective postemergence grass products, in the mid to late 1970s and early 1980s such as sethoxydim, fluazifop-butyl, and quizalofop-ethyl. For broadleaf weed control, both linuron and metribuzin are root-uptake products requiring fairly high rainfall for activation. Heavy dependence on these products selected those weed species most likely to escape, such as Xanthium strumarium, Euphorbia heterophylla, Abutylon theophrasti, Cassia obtusifolia, and Ipomeoa spp. Again, the change in weed spectrum led to the discovery of new, selective postemergence broadleaf herbicides, such as bentazon, aciflorfen and chlorimuron-ethyl in the 1970s-80s.

From the first tests in discovery research in 1982, chlorimuron-ethyl was a unique product for the soybean market. It was the first sulfonylurea for commercial use on a broadleaf crop. Importantly, it required 50 to 100 times less chemical to control broadleaf weeds than existing soybean products and was far more effective on legume weeds such as Cassia obtusifolia and Desmodium tortuosum than commercial standards. Chlorimuron-ethyl also controlled more broadleaf weeds with less size restrictions under more environmental conditions than other postemergence products. Because of very favorable biology, toxicology, and environmental attributes, Du Pont commercialized this product in record time -- just four years after first synthesized!

Sulfonylureas are metabolized quickly by tolerant crops (1) under normal growing conditions. However, environmental stress conditions caused by a number of factors such as excess moisture, temperature, insects, soil composition, and nutrient imbalances, can delay metabolism and produce transient phytotoxicity lasting 7 to 14 days. During the same time frame that chlorimuron-ethyl was being developed, Du Pont initiated work on a number of herbicide-resistant crops. Creating herbicide-tolerant crops is, for the most part, not technology limited. In short, the technology works. In soybeans, whole seed mutagenesis was used, to select soybeans that were resistant to levels of chlorsulfuron that would kill normal soybeans. This work led to the selection of W20 ("STS" soybean) as reported by Sebastian et al 1989.

W20 soybeans have a high degree of resistance to both postemergence and pre-emergence applications to a number of sulfonylurea herbicides. "Resistance was monogenic, semi-dominant, and not allelic to any of the previously identified recessive genes hs1, hs2, or hs3 that confer tolerance to SU herbicides. Biochemical tests indicated that the mechanism of resistance is reduced sensitivity of acetolactate synthase to SU inhibition" (2).

The discovery of W20 soybeans gave Du Pont a number of new soybean options. The first is to markedly increase the margin of crop safety to chlorimuron-ethyl under a wide range of environmental conditions. Now, not only are STS soybeans able to metabolize chlorimuron-ethyl but, the ALS enzyme (the site of action of sulfonylurea) is not affected in the resistant line. Under conditions of stress when metabolism is slowed, chlorimuron-ethyl at 10X rates will not cause soybean injury. The second option is with thifensulfuron-methyl.

In 1990, Du Pont registered thifensulfuron-methyl on soybeans. It is extremely active on four important weeds in soybeans. The most important weed is <u>Chenopodium album</u>, which is poorly controlled by chlorimuron-ethyl and other commercial postemergence herbicides in soybeans. However, thifensulfuron- methyl has a narrow safety margin on soybeans and under any stress can cause temporary injury. When W20 is commercialized in 1993, it will greatly increase soybean safety to thifensulfuron-methyl. Thifensulfuron-methyl, because of weed-control spectrum and rapid soil degradation, is the ideal postemergence product in the Northern USA, Canada, and Italy. The third option in soybeans comes from all the SU's Du Pont has synthesized and discarded for lack of suitable soybean selectivity. We have synthesized and screened tens of thousands of SU's in the last 15 years. Most have not shown any soybean selectivity. W20 opened up the opportunity to screen SU's not for selectivity but for the optimum biology, toxicology, and environmental properties.

We have learned a couple important things from this effort. Chlorimuron-ethyl is an outstanding soybean herbicide, and we believe it will continue to be an important herbicide for soybean growers. Thifensulfuron-methyl, because of its short soil residual and activity on key weeds such as <u>Chenopodium</u> and <u>Amaranthus</u>, provides unique and real value to farmers. STS soybeans will enhance and extend the performance and utility of both these excellent herbicides. Additionally, we are now investigating several promising pre-emergence SU's with moderate to short soil residual properties that could be used on W20 soybeans.

CONCLUSION

Selective herbicides will continue to play a major role in weed control. I expect herbicide-resistant crops to also play an important role in discovery research in the future. Our challenge is to continue to strive for finding the "ideal" herbicide. Our criteria for the "ideal" herbicide are:

- cost-effective, flexible, reliable, and convenient to use
- safe to the crop, environment, user and consumer
- low use rate to minimize the amount introduced into the environment
- leaves no harmful residues
- persistence in the crop and soil tailored to desired effects
- high specificity to target organisms
- no off-target effects

- easily integrated with "best management practices"
- does not lead to pest resistance

Herbicide-resistant crops are one more tool that will allow chemists and biologists in discovery research to meet this challenge.

REFERENCES

1. Beyer, E. M., M. J. Duffy, J. V. Hay, and D. D. Schlueter. 1988. Sulfonylurea Herbicides. P117-189. In P.C. Kearney and D. D. Kaufman (et al) **Herbicides: Chemistry, Degradation, and Mode of Action.** Vol. 3 Marcel Dekker, Inc., NY.

2. Sebastian, S. A.; G. M. Fader, J. F. Ulrich, D. R. Forney, and R. S. Chaleff. 1989 **Semi-dominant Soybean Mutation for Resistance to Sulfonylurea Herbicides.** Crop Science, 29: 1403-1408.

WEED CONTROL FOR SOYBEAN IN THE NINETIES

ELLERY L. KNAKE
Department of Agronomy
University of Illinois
Urbana, IL 61801

ABSTRACT

The use of herbicides for soybeans has increased from about 5% of the acreage treated in 1960 to nearly all of the U.S. acreage being treated three decades later. Major grass weeds of soybeans include the foxtails, barnyardgrass, shattercane and johnsongrass. Major broadleaf weeds include pigweed, common lambsquarters, smartweed, velvetleaf, cocklebur, annual morningglories, jimsonweed and nightshade. Soybean yield reductions are proportional to the amount of weeds present. The weeds that start growing at the same time as the crop are the major problem and should be controlled during the first month after planting. The dinitroaniline herbicides remain low cost and popular. The acetanilides also have significant market share. However, the postemergence herbicides for control of grass weeds could increase in use. Chloramben was popular initially but has been phased out. Metribuzin use decreased as bentazon was developed. Use of the imidazolinones and sulfonylureas has increased significantly. The reduced tillage acreage has increased partly due to "conservation compliance". Development of crops with greater herbicide tolerance has been accompanied by concern about herbicide resistant weeds. However the potential for this problem can be reduced with herbicide combinations and herbicide rotations. Judicious use of herbicides and new use techniques can contribute significantly to resource conservation.

INTRODUCTION

With about 9 million acres of soybeans and relatively high yields, Illinois produces nearly one-fifth of the U.S. soybeans. Weed control practices in Illinois are quite representative of especially the midwestern states of the United States.

With the past as prologue to the future, a brief review would indicate that in 1960 only about five percent of the soybean acreage was treated with soil-applied herbicides, primarily with band applications at a cost of two to five dollars per acre. Relatively few postemergence herbicides were available for soybeans. However, by 1990, nearly all soybean acreage was treated with herbicides. Combinations and sequential applications were common (1,2).

Major Weeds i• Soybeans
In much of the midwest, the most prevalent weed is giant foxtail *Setaria faberi* Herrm. In southern areas, Johnsongrass *Sorghum halepense* (L.) Pers. is a major problem. Other grass weeds include yellow foxtail *Setaria glauca* (L.) Beauv., green foxtail *Setaria viridis* (L.) Beauv., barnyardgrass *Echinochloa crus-galli* (L.) Beauv., shattercane *Sorghum bicolor* (L.) Moench and volunteer corn as a weed. Nutsedge can also be a significant problem with yellow nutsedge *Cyperus esculentus* L. predominant in the north and purple nutsedge *Cyperus rotundus* L. in the south.

Broadleaf weeds include pigweed *Amaranthus* spp., common lambsquarters *Chenopodium album* L., pennsylvania smartweed *Polygonum pennsylvanicum* L., velvetleaf *Abutilon theophrasti* Medik., common cocklebur *Xanthium strumarium* L., ivyleaf morningglory *Ipomea hederacea* (L.) Jacq. tall morningglory *Ipomoea purpurea* (L.) Roth and jimsonweed *Datura stramonium* L. In recent years eastern black nightshade *Solanum ptycanthum* Dun. has increased (3).

Effect of Weeds on Soybeans
Many competition or interference studies have been conducted with soybeans. These generally indicate that for the energy consumed by weeds there is a proportional decrease in soybean yield (4). Weeds can compete for moisture, nutrients, and light. Of these factors, light appears to be one of the major factors. Light can affect pod set and as weed intensity increases a decrease in number of pods has been affected more

than size of beans or number of beans per pod. In addition, the allelopathic effect of weeds may affect soybean growth and production(4).

The quality of soybeans can be affected by weeds such as black nightshade with berries about the same size as soybeans which contain a mucilaginous substance. Seed of jimsonweed reportedly has a toxic or hallucinogenic effect and seed of morningglories a purgative effect.

Competition studies with weed growth beginning at various times indicate that those weeds which begin growing early are the most significant and soybeans can soon have the competitive advantage over weeds that begin growth later. Time of removal studies indicate that modest weed growth during the first three or four weeks has little effect on yields if the weeds are then adequately controlled. This suggests that good selective postemergence treatments are quite practical.

Changing Cultural Practices
Soybean culture has changed during the last few decades with a trend toward narrower row spacing. The current trend toward drilled soybeans is expected to continue concomitantly with the interest in less tillage.

Interest in using less tillage to prepare a seedbed for soybeans has increased significantly in recent years and is expected to increase further. This is partly due to "conservation compliance" which requires adoption of soil conserving practices if farmers wish to receive government program benefits. It is also due to availability of no-till drills and the desire of farmers to conserve resources and lower cost of production (5).

Although herbicides are used on nearly all of the soybean acreage in the midwest, rotary hoeing and row cultivation are still quite popular.

History of Herbicide Use
In 1960, herbicides were used on about five percent of the soybeans, with application primarily in 12 to 14 inch bands over the row. Cost was about two to five dollars per acre. By 1990, herbicides were broadcast on nearly all of the acreage with multiple applications common.

With the past as prologue to the future, the evolution of herbicide use is of interest (1). With its irritating properties, CDAA soon gave way to alachlor and then metolachlor was introduced.

Trifluralin was introduced in the mid 60's and it soon gained a major market share. This was accompanied by increased interest in incorporation of herbicides.

For broadleaf weed control naptalam and chloro IPC were soon phased out as chloramben gained a major market share. However, when metribuzin was introduced, it was rapidly accepted as a lower cost alternative for broadcasting even though it generally needed help for control of grass weeds. Growers soon became disenchanted as they noted effect on soybeans from metribuzin. Bentazon postemergence offered relatively good soybean tolerance and was rapidly accepted for control of broadleaf weeds and nutsedge. Acifluorfen found modest use, especially with bentazon to broaden the spectrum of control. With cost a major concern, chloramben use declined and it was phased out.

The discovery of the imidazolinones as farmers were seeking improved crop tolerance led to the rapid introduction of imazaquin. However, especially in the northern areas, residual carryover problems led to replacement of imazaquin with imazethapyr which was aggressively marketed. It was accepted quite well because of relatively good crop tolerance, broad spectrum weed control, flexibility in time and method of application and less concern about carryover to corn.

With the discovery of the sulfonylureas, chlorimuron found a place for soybeans used alone or combined with thifensulfuron postemergence. In combination with metribuzin, chlorimuron was used at a higher rate preemergence than it was postemergence. Giving both burndown and residual, this combination provided very good weed control with a variety of tillage systems. However concern about carryover on high pH soils soon emerged.

In the mid 1980's clomazone was also introduced. However, its potential was soon somewhat limited due to concern about movement outside the target area and carryover.

Trends for Herbicide Use in the Nineties

With the stage set for a further reduction in tillage, farmers in the 90's would like a weed control program for soybeans that provides good soybean tolerance, broad spectrum season-long control, no carryover problem, convenience and modest cost. In addition, there is some concern about the possibility of weed resistance.

For control of grass weeds, many options already exist. Some market share of trifluralin has been lost particularly to pendimethalin and a little to ethalfluralin. However, trifluralin still dominates as an effective low cost herbicide for control of grass weeds. Control of pigweed and lambsquarters is also an important advantage for the dinitroanilines. The convenience of recently introduced dry formulations of trifluralin may help to maintain some market share but the downward trend could continue. Pendimethalin has gained some market share, partly due to aggressive marketing, prepackaged combinations with imidazolinones and adaptability to no-till surface applications.

Unless pressured by such issues as water quality, alachlor could continue to provide considerable flexibility in timing and method of application. A prepackaged combination of alachlor plus trifluralin is conceived by some as a marketing tactic to provide a more competitive "low cost alachlor" in the soybean market. Formulation innovations such as microencapsulation and water dispersible granules may help to prolong the market life of alachlor. With longer residual activity than with alachlor, good flexibility and also having good soybean tolerance, metolachlor can be expected to continue well into the 90's.

The postemergence herbicides for control of grass weeds include sethoxydim, fluazifop-P, quizalofop, fenoxaprop, and clethodim. Although quite effective and having the most activity for residual control, there is little optimism for haloxyfop being registered in the U.S. However, it is reportedly used in some other countries. Although these postemergence herbicides for control of grass weeds are generally quite effective with good soybean tolerance, they have not gained the market share they may deserve. Availability of other low cost options has likely been a major factor.

The recent announcement of a significant price reduction for sethoxydim could have a significant impact to encourage increased use of postemergence herbicides for control of grass weeds, especially if other firms follow suite by reducing price.

With relatively good performance, imazethapyr may be in a position to gain further market share. And in some southern areas, imazaquin could continue. However, the possibility of other imidazolinones with even more favorable characteristics should not be precluded.

With concern about soybean tolerance, metribuzin lost significant market share during the 80's as other options became available. Somewhat ironically, there was some shift to metribuzin plus chlorimuron. This combination provided both good burndown and residual control for no-till. However, significant concern arose about carryover of the chlorimuron on high pH soils (6). Possibilities for helping to alleviate this problem included stressing careful management, replacing part or all of the chlorimuron with a shorter residual sulfonylurea or perhaps combining a reduced rate of chlorimuron plus metribuzin with another herbicide such as clomazone. Introduction of sulfonylurea tolerant corn may also provide another avenue to alleviate carryover concern (7).

Very low rates of chlorimuron and thifensulfuron used postemergence has been accepted in some areas. The potential for sulfonylurea tolerant soybeans is being explored for improving soybean tolerance.

Bentazon lost significant market share in the early 90's. However, characteristics of this compound providing good crop tolerance and relatively good control with no carryover concern might allow an optimist to envision a possible modest comeback with appropriate marketing strategy. Addition of acifluorfen to bentazon for broadening the spectrum of control has generally been accepted as a viable option.

Use of lactofen has been somewhat limited by the perception of effect on soybeans. Limited supply of fomesafen restricted its introduction and observations suggested the need to carefully manage residual. The prepackaged combination of fluazifop-P and fomesafen was likewise restricted by availability but performance suggested potential and availability is scheduled to increase.

Very low rates of 2,4-DB added to some postemergence herbicides offered improved control of such weeds as cocklebur, annual morningglory and giant ragweed in a very cost effective manner.

Although 2,4-DB labeling allowed an alternative to 2,4-D for no-till soybeans, cost of the recommended rates limited acceptance. The use of 2,4-D for no-till soybeans remained quite controversial with some question of legality even though included on some labels other than 2,4-D. Low rates of glyphosate prior to planting offered an alternative to 2,4-D for some weeds. Research has suggested that some treatments such as metribuzin plus chlorimuron, where appropriate, might preclude the need for 2,4-D and make it academic.

As tillage is reduced, some weeds such as hemp dogbane *Apocynum cannabinum* L. often increase. Specialized equipment such as a sponge applicator has been quite cost effective by taking advantage of the height differential with the weed growing taller than the soybeans.

In geographic areas where the cropping system includes both wheat and soybeans, double cropping with soybeans planted immediately after wheat harvest is a common and successful practice. The current repertoire of herbicides allows selection of compounds for both burndown and residual.

Biotechnology Developments
The introduction of sulfonylurea tolerant soybeans holds some promise for improved crop tolerance. With nicosulfuron providing good control of annual grass weeds and primisulfuron better on annual broadleaf weeds, a combination of the two for sulfonylurea tolerant soybeans presents an interesting possibility. Such a combination could offer low rate efficiency, broad spectrum control, appropriate residual, and perhaps certain environmental advantages (8,9).

The possibility of glyphosate tolerant soybeans presents the potential for cost effective weed control with an environmentally desirable herbicide even if two or possibly three applications were needed (7).

The development of corn with imidazolinone and sulfonylurea tolerance could help to alleviate residual concerns. However, continuous use of herbicides with the same or similar mode of action has raised concern about possible development of herbicide resistant weeds. This may not be an insurmountable problem with judicious management that includes herbicide combinations and herbicide rotations.

SUMMARY

As we progress through the nineties, concern about water quality and the environment will influence weed control tactics. Some herbicides associated with such concerns may fall by the wayside. We will not likely stop using herbicides. But we will see some shift in emphasis toward those herbicides perceived to be more environmentally friendly. Biotechnology innovations could usher in new opportunities. We will also see greater precision in selection of herbicides, rates, time and method of application. The farmer of the nineties can expect even more cost effective techniques that provide broad spectrum weed control, good crop tolerance, appropriate residual and convenience. The imagination and creativity of those in discovery, development and marketing will make it happen. We will also take seriously our obligation to develop weed control programs that will help us conserve those resources which the Author of Nature has given us because good planets are hard to find.

REFERENCES

1.   Pike, D.R., McGlamery, M.D. and Knake, E.L., A case study of herbicide use. Weed Technol., 1991. 5:639-646.

2.   Pike, D.R., Glover, K.D., Knake, E.L. and Kuhlman, D.E., Pesticide use in Illinois, results of a 1990 survey of major crops. University of Illinois, College of Agriculture, CES, DP-01-1, 38 pp.

3.   Cline, M.N., Colwell, C.E., Jacobsen, B.J., Knake, E.L., MacMonegle, C.W., Owen, M.D.K., and Pike, D.R., Illinois Pest Profiles, Illinois NHS and University of Illinois, College of Agriculture, 1983, 59 pp.

4.   Knake, E.L., Giant foxtail, University of Illinois, College of Agriculture Bulletin 803, 1990, 22 pp.

5.   Knake, E.L., Weed control systems for lo-till and no-till. University of Illinois, College of Agriculture, CES Circ. 1306, 1990, 12 pp.

6.   Curran, W.S., Knake, E.L. and Liebl, R.A., Corn (Zea mays) injury following use of clomazone, chlorimuron, imazaquin, and imazethapyr. Weed Technol., 1991, 5:539-544.

7.   Duke, S.O., Christy, A.L., Hess, F.D. and Holt, J.S. Herbicide-resistant crops. Comments from CAST, Council for Agricultural Science and Technology, 1991, No. 1991-1, 24 pp.

8.   Knake, E.L., Walsh, J.D. and Pike, D.R., Opportunities and cautions with the sulfonylurea herbicides. North Central Weed Science Society Abstr. 1991, 46:201.

9.   Knake, E.L., Heisner, R.W., Paul, L.E. and Walsh, J.D., Multi-species evaluation of postemergence herbicides. North Central Weed Science Society Res. Rept. 1991, 48:110-116.

INDEX OF CONTRIBUTORS

Printed in the United States
By Bookmasters